Structural Genomics and High Throughput Structural Biology

Structural Genomics and High Throughput Structural Biology

edited by

Michael Sundström

Martin Norin

Aled Edwards

Taylor & Francis
Taylor & Francis Group

Boca Raton   London   New York   Singapore

A CRC title, part of the Taylor & Francis imprint, a member of the
Taylor & Francis Group, the academic division of T&F Informa plc.

Published in 2006 by
CRC Press
Taylor & Francis Group
6000 Broken Sound Parkway NW, Suite 300
Boca Raton, FL 33487-2742

International Standard Book Number-10: 0-8247-5335-6 (Hardcover)
International Standard Book Number-13: 978-0-8247-5335-1 (Hardcover)
Library of Congress Card Number 2005044002

Library of Congress Cataloging-in-Publication Data

Structural genomics and high throughput structural biology / edited by Michael Sundstrom, Martin Norin, Aled Edwards.
    p. cm.
    Includes bibliographical references and index.
  ISBN 0-8247-5335-6 (alk. paper)
    1. Proteins—structure. 2. Proteomics. I. Sundstrom, Michael. II. Norin, Martin. III. Edwards, Aled.

QP551.S815 2005
572'.633—dc22   2005044002

Contributors

Ruben Abagyan
The Scripps Research Institute
La Jolla, California, U.S.A.

Paul D. Adams
Berkeley Structural Genomics Center
Lawrence Berkeley National Laboratory
Berkeley, California, U.S.A.

Cheryl Arrowsmith
Ontario Cancer Institute and Structural Genomics Consortium
University of Toronto
Toronto, Ontario, Canada

John-Marc Chandonia
Berkeley Structural Genomics Center
Lawrence Berkeley National Laboratory
Berkeley, California, U.S.A.

Naomi E. Chayen
Biological Structure and Function Section
Division of Biomedical Sciences
Imperial College London
London, U.K.

Roland L. Dunbrack, Jr.
Institute for Cancer Research
Fox Chase Cancer Center
Philadelphia, Pennsylvania, U.S.A.

Aled Edwards
Structural Genomics Consortium
Banting and Best Department of Medical Research
University of Toronto
Toronto, Ontario, Canada

Henrik Hansson
Department of Cell and Molecular Biology
Uppsala University
Uppsala, Sweden

Sung-Hou Kim
Berkeley Structural Genomics Center
Lawrence Berkeley National Laboratory
and
Department of Chemistry
University of California
Berkeley, California, U.S.A.

Gerard J. Kleywegt
Department of Cell and Molecular Biology
Uppsala University
Uppsala, Sweden

Andrzej Kolinski
Buffalo Center of Excellence in Bioinformatics
University at Buffalo
Buffalo, New York, U.S.A.
and
Faculty of Chemistry
University of Warsaw
Warsaw, Poland

Roman A. Laskowski
European Bioinformatics Institute
Cambridge, U.K.

Weontae Lee
Department of Biochemistry and Protein Network Research Center
College of Science
Yonsei University
Seoul, Korea

Martin Norin
Biovitrum
Department of Chemistry
Stockholm, Sweden

Bernhard Rupp
Department of Biochemisty and Biophysics
Texas A & M University
College Station, Texas, U.S.A.
and
Macromolecular Crystallography and Structural Genomics Group
Lawrence Livermore National Laboratory
University of California
Livermore, California, U.S.A.

J. Michael Sauder
Structural GenomiX, Inc.
San Diego, California, U.S.A.

Dong Hae Shin
Berkeley Structural Genomics Center
Lawrence Berkeley National Laboratory
Berkeley, California, U.S.A.

Jeffrey Skolnick
Buffalo Center of Excellence in Bioinformatics
University at Buffalo
Buffalo, New York, U.S.A.

Michael Sundström
Structural Genomics Consortium
University of Oxford
Oxford, U.K.

Guoli Wang
Institute for Cancer Research
Fox Chase Cancer Center
Philadelphia, Pennsylvania, U.S.A.

Weiru Wang
Plexxicon Inc.
Berkeley, California, U.S.A.

Adelinda Yee
Ontario Cancer Institute and Structural Genomics Consortium
University of Toronto
Toronto, Ontario, Canada

Yang Zhang
Buffalo Center of Excellence in Bioinformatics
University at Buffalo
Buffalo, New York, U.S.A.

Preface

Structural genomics is a technology- and methodology-driven field with the simple aim to determine many protein structures at a lower cost per structure than by using traditional approaches and to populate protein fold space to as high a degree of completeness as possible. The field emerged in the mid to late 1990s from distinct pilot projects primarily in Canada, Japan, and the U.S., and its inception was spurred by the abundance of genomic information and the methodological/technological advances made in protein expression, computational methods, biomolecular NMR and X-ray crystallography. The various pilot projects had shared technological goals, but the field struggled to find a common scientific objective in large part because the various pilot projects were launched with different objectives in mind and to meet the goals of different funding agencies.

Almost ten years later, structural genomics still has, at its core, technology development but has also "crystallized" around two coherent scientific themes — discovery and applied structural genomics. The technological development effort (Chapter 2, Chapter 3, Chapter 4, and Chapter 5) is viewed with great enthusiasm by the general scientific community because of the interest in adapting structural genomics-derived technology for their own scientific projects. Since the late 1990s perhaps the most significant improvements have been made in the analysis of X-ray diffraction data; the integrated packages that identify heavy atom sites, phase X-ray diffraction data and build the protein chain are truly remarkable.

The scientific objectives of discovery structural genomics are akin to those of traditional structural biology, which seek to achieve greater understanding of well-studied biological systems. In fact, discovery structural genomics is targeted to study genes/proteins of unknown structure in order to discover new information about folding, specific structural features, or function. In this guise, discovery structural genomics gives structural biologists a chance, for the first time, to generate hypotheses about new proteins, rather than to work on proteins preordained to be interesting.

The relative merits of discovery structural genomics and structural biology can be, and have been, debated. In some respects, the debate is slightly clouded by the fact that there is no commonly accepted definition of structural genomics or a consensus of its objectives. In this way, pundits have the opportunity to both define structural genomics and the reasons for its existence or demise. Taking our definition of discovery structural genomics, with tens of thousands of proteins of unknown structure and function to be analyzed, it is not unreasonable to suggest that in the long haul, discovery structural genomics may have a greater impact on our understanding of biology than will traditional structural biology. The parallel strategy adopted by structural genomics is already generating more structural information per unit time per person than other approaches. In addition, discovery structural genomics is less subject to the mercurial trends of biomedical research, which tend to focus researchers into narrowly defined areas largely defined by funding possibilities. With these different views in mind, we have devoted five chapters (Chapter 1, Chapter 6, Chapter 7, Chapter 8, and Chapter 9) to review the premises and the promises of structural genomics from the discovery view and how the generated structural data is used in computational approaches.

Applied structural genomics (Chapter 12) describes the focused application of the structural genomics technologies to proteins of known function, but of interest to the biotechnology and pharmaceutical industry. This effort often targets families of proteins of known relevance to drug discovery (e.g., protein kinases) and also applies the structural genomics suite of technologies to generate structures of a given protein in complex with a number of different ligands such as chemical inhibitors.

Finally, with any field devoted to generating experimental information comes the problem of data visualization and analysis. The computational problems raised by structural genomics are not unique, and it could be argued that they have yet to emerge. Structural genomics efforts are contributing only 10% of the new structures to the PDB, hardly an unmanageable deluge. However, in the future, we can expect that this rate will increase and that an increasing number of structures will be deposited without companion biological experimentation. The analysis and interpretation of these new structures must be driven computationally; we have devoted two chapters (Chapter 10 and Chapter 11) to these impending problems.

Finally, it is always difficult to compile an up-to-date book in a technology-driven field. What we have attempted to do is to provide a snapshot of the present and some view to the future of the field. The contributors to this book are each at the forefront of their fields, and it will be fun to read this book again in five years to see how well they each prognosticated.

Aled Edwards

Martin Norin

Michael Sundström

The Editors

Michael Sundström, Ph.D. is the chief scientist for the Structural Genomics Consortium (SGC) at the University of Oxford; he has held this position since July 2003. The SGC determines the three-dimensional structures of human proteins of medical relevance, and places them in the public domain without restriction. Currently, the Oxford site has more than 60 staff members and efforts are focused on oxidoreductases, phosphorylation dependent signaling, and integral membrane proteins.

Dr. Sundström earned his Ph.D. at the Uppsala Biomedical Center in 1992. After postdoctoral training at the Karolinska Institute, he accepted a position at Kabi-Pharmacia in Stockholm as a scientist in protein crystallography and was later appointed head of the protein crystallography group. In 1998, he moved to the Pharmacia oncology research site in Milan, Italy as director of structural chemistry and informatics and in addition served as chairman for the oncology research review committee. In 2001, he was appointed research director of Actar AB (a start-up drug discovery company), and later joined Biovitrum AB.

SELECTED PUBLICATIONS

Norin, M. and Sundström, M. Structural proteomics: developments in structure-to-function predictions. *Trends Biotechnol.* 20(2):79–84 (2002).

Norin, M. and Sundström, M. Protein models in drug discovery. *Curr. Opin. Drug Discov. Devel.* 4(3):284–90 (2001).

Dalvit, C. et al. Identification of Compounds with binding affinity to proteins via magnetization transfer from bulk water. *J. BioMolecular NMR.* 68:65–68 (2000).

Sundström, M. et al. The Crystal Structure of Staphylococcal Enterotoxin D Type D reveals Zn2+ Mediated Homodimersation. *EMBO J.* 15(24):6832–6840 (1996).

Sundström, M. et al. The Crystal Structure of an Antagonist Mutant of Human Growth Hormone in Complex with its Receptor at 2.9 Å Resolution — a Comparison with the Native Hormone Receptor Complex at 2.5 Å. *J. Biol. Chem.* 271(15): 32197–32203 (1996).

Martin Norin, Ph.D. is the director and head of the chemistry department at Biovitrum AB, a biotech company in Stockholm, Sweden that focuses on the development of innovative drugs for the treatment of metabolic diseases, inflammation as well as niche products.

Dr. Norin earned his Ph.D. in biochemistry at Royal Institute of Technology (KTH), Stockholm, in 1994 and thereafter accepted a position as computational chemist/molecular modeler at Kabi-Pharmacia. He gradually assumed larger responsibilities within Pharmacia such as section manager for computational chemistry, project leader, and eventually director and head of the structural chemistry department within Metabolic Diseases/Pharmacia, and subsequently at Biovitrum (a Pharmacia spin-off company). Since 2005, he has been in charge of the chemistry department at Biovitrum, responsible for the work of 75 scientists in medicinal, analytical, and computational chemistry, as well as chemInformatics.

SELECTED PUBLICATIONS

Lewis, M.D. et al. A novel dysfunctional growth hormone variant (Ile179Met) exhibits a decreased ability to activate the extracellular signal-regulated kinase pathway. *J. Clin. Endocrinol. Metab.* 89(3):1068–75 (2004).

Millar, D.S. et al. Novel mutations of the growth hormone 1 (GH1) gene disclosed by modulation of the clinical selection criteria for individuals with short stature. *Hum. Mutat.* 21(4):424–40 (2003).

Norin, M. and Sundström M. Structural proteomics: developments in structure-to-function predictions. *Trends Biotechnol.* 20(2):79–84 (2002).

Norin M. and Sundström M. Protein models in drug discovery. *Curr. Opin. Drug Discov. Devel.* 4(3):284–90 (2001).

Forsberg G. et al. Identification of framework residues in a secreted recombinant antibody fragment that control production level and localization in Escherichia coli. *J. Biol. Chem.* 272(19):12430–6 (1997).

Aled Edwards, Ph.D. is chief executive of the Structural Genomics Consortium, an Anglo-Canadian collaboration created for the purpose of substantially increasing the number of protein structures of relevance to human health available in the public domain. He is Banbury Professor of Medical Research in the Banting and Best Department of Medical Research at the University of Toronto.

Dr. Edwards received his Ph.D. in Biochemistry from McGill University and did his postdoctoral training at Stanford University. He has co-authored over 120 scientific publications and has helped place over 200 unique protein structures into the public domain. Dr. Edwards is the co-founder and chief scientific advisor of Affinium Pharmaceuticals, a structure-based drug discovery company in Toronto. He is also a co-founder of Jaguar Biosciences, a company focusing on diagnosis and treatment of hepatitis C.

SELECTED PUBLICATIONS

Saridakis, V. et al. The structural basis for methylmalonic aciduria: the crystal structure of archeael ATP:cobalamin adenosyltransferase. *J. Biol. Chem.* 279:23646–53 (2004).

Kimber, M.S. et al. Data mining crystallization databases: Knowledge-based approaches to optimize protein crystal screens. *Proteins* 51:562–568 (2003).

Yee, A. et al. An NMR Approach to Structural Proteomics. *Proc. Nat. Acad. Sci. U.S.A.* 99:825–1830 (2002).

Cramer, P. et al. Architecture of RNA Polymerase II and Implications for the Transcription Mechanism. *Science* 288:640–649 (2000).

Mer, G. et al. The solution structure of the C-terminal domain of the RPA32 subunit of human replication protein A: Interaction with XPA. *Cell* 103:449–456 (2000).

Christendat, D. et al. Structural Proteomics of an Archeon: A global survey of protein expression, solubility and structure. *Nat. Struct. Biol.* 7:903–909 (2000).

Bochkarev, A. et al. Crystal structure of the single-stranded DNA binding domain of replication protein A bound to DNA. *Nature* 385:176–181 (1997).

Table of Contents

1 Overview of Structural Genomics: Landscape, Premises, and Current Direction

Sung-Hou Kim, Dong Hae Shin, Weiru Wang, Paul D. Adams, and John-Marc Chandonia

CONTENTS

1.1 THE LANDSCAPE OF SEQUENCE GENOMICS

Worldwide genome sequencing projects have produced an explosive growth of sequence information from many living species. Complete sequence information for more than 200 genomes is accessible through public domain sources, and about several times more genomes are currently being sequenced worldwide (http://maine.ebi.ac.uk:8000/services/cogent/stats.html;http://www.ncbi.nih.gov/Genomes/, http://www.tigr.org/tdb/). Although the currently known genome sizes range, in order of magnitude, from 10^6 to 10^{11} DNA base pairs, the number of genes is estimated to range only from 10^3 to 10^5 per organism. Taking the estimated 13.6 million species of living organisms on Earth into account, there are 10^{10} to 10^{12} different proteins encoded by organisms from three

1

domains of life (eukaryotic, prokaryotic, and archaea). Thus, the protein universe is vast and diverse (Table 1.1).

Analyses of human genome yield an estimate of about 20,000 genes [Venter et al. 2001, Lander et al. 2001, and subsequent estimates] in *Homo sapiens*, which is only about 40 times greater than the smallest sequenced genome, *Mycoplasma genitalium*, encoding ~470 genes. In higher organisms the number of proteins, however, may be an order of magnitude more than the number of genes due to alternative splicing, posttranslational modification and processing, etc. When this multiplicity is taken into account, the protein universe is even more vast and diverse.

The functions of most of these proteins have not been characterized experimentally. Each protein functions at both molecular and cellular levels. There is a strong indication that each protein performs one or more functions at both levels. The molecular function or functions of a protein can be defined as one or more chemical and/or physical functions that the protein performs, i.e., as an enzyme, a molecular switch, or a structural unit, etc. A cellular function can be accomplished by a combination of the molecular functions of many proteins, sometimes in a particular time-sequence or as a network. Each protein may participate in one or more cellular functions such as biosynthetic pathways, signaling pathways, or more complex molecular machinery, etc. Currently, the most powerful methods of inferring the functions (molecular and/or cellular) of a new protein sequence are to assign the previously characterized functions of other proteins that have high sequence similarity to the new sequence. This is based on the commonly accepted assumption that sequence similarity implies likely homology, and thus structural and functional similarity. More elaborate methods such as Pfam [Bateman et al. 1999; Bateman et al. 2004; Sonnhammer et al., 1997], SMART [Ponting et al., 1999], ProMap [Yona and Linial, 2000], and others further cluster the sequence space into families of domains of similar sequences, and suggest that the members of a given sequence family share their functions. However, the sequence based functional prediction at genomic scale yields annotations for less than 50% of the genes, of which perhaps as much as 20% may be in error. Furthermore, almost all annotations are transitive annotations from a very small number of experimentally verified annotations. Thus, they are subject to validation by experiment and/or collaborating inferences. The remaining 50% of predicted genes are considered to encode proteins with unknown (more correctly, un-inferable) functions, or hypothetical proteins. Three-dimensional

---

TABLE 1.1
Estimates of Number of Proteins and Folds

	Estimated Number
Genome size (base pairs)	$10^6 - 10^{11}$
Range of genes in an organism	$0.5 \propto 10^3 - 20 \propto 10^3$
Living organisms on Earth	$1.3 \propto 10^7$
Size of the protein universe on Earth	$\sim 0.7 \propto 10^{10} - 2.6 \propto 10^{11}$
Protein fold motifs	$\sim 10^4$
Protein fold motifs known	$\sim 10^3$

(3-D) structures of proteins may provide the most important additional information for both.

1.2 PREMISES OF STRUCTURAL GENOMICS

One of the grand challenges of understanding the rapidly growing sequence genomics information has led to the emergence of a new discipline, functional genomics, to determine the functions of genes on a genome-wide scale. Structure genomics is a sub-discipline that promises to experimentally reveal the three-dimensional structures and provide structure-based inferences of molecular functions for representatives of gene products in all sequence families, especially proteins of unknown structure and functions. Combining structural and sequence information about a protein is currently the most reliable method of inferring its molecular functions, which, in turn, can help to infer the molecular functions of other members of the protein sequence family to which it belongs. Furthermore, structural genomics can help to identify remote evolutionary relationships among sequence families that are unrecognizable by sequence comparison alone. Thus, structural genomics is aimed at obtaining a systems view of the protein structural world at an organism-scale as well as global level to understand the relationship between the structure and molecular functions of all components in cells.

How many protein structures and their molecular functions need to be determined to get a global view of the protein world in terms of folding patterns and molecular functions?

1.2.1 THE PROTEIN FOLD SPACE

1.2.1.1 Concept

As mentioned earlier, the protein universe is vast and perhaps contains as many as trillions of different proteins. When grouped by sequence similarity, some sequence families have a large number of members, but most have a very small number of members, thus revealing an extreme value distribution of protein sequence family sizes. The distribution of family sizes for Pfam [Bateman et al. 2004] families is shown in Figure 1.1. Two lines of simplification are considered:

1. In such a distribution, the majority of proteins belong to a relatively small number of large families.
2. Structures in the protein structure database (Protein Data Bank; PDB) [Abola et al., 1997 and Berman et al., 2000] reveals that a large number of proteins consist of one or more structural domains, architectural motifs, also called folds.

Each structural domain performs one or more chemical reactions and/or physical interactions either by itself or in conjunction with other structural domains in the same protein. It is generally accepted that the majority of naturally occurring protein domains adopt primarily one or a limited number of stable folds. While

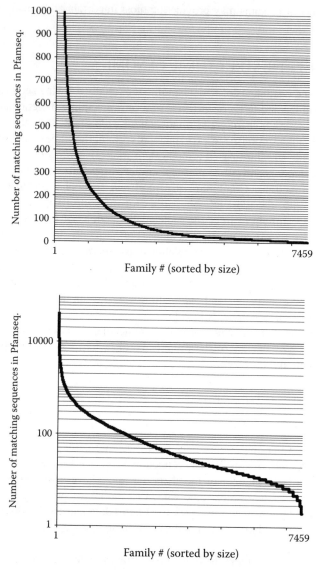

FIGURE 1.1   Protein family size distribution. The number of protein sequences from Pfam-seq 14.0 classified in each family in Pfam 14.0 is shown. Families are sorted by size. The median family size is 33. (a) nonlogarithmic plot of family sizes, in which families larger than 1000 members are off the scale. (b) logarithmic plot of the same data.

the precise number remains uncertain, current estimates suggest that fewer than 10,000 distinct domain folds may account for most of the proteins in nature [Vitkup et al., 2001].

Examination of the protein structure database (Protein Data Bank; PDB) [Abola et al., 1997 and Berman et al., 2000] revealed the following important facts: although over 26,000 protein structures (of which about 4,000 structures are nonredundant based on the CE algorithm [Shindyalov and Bourne,1998]) have been experimentally determined, they belong to only about 1,000 fold families [Hubbard, 1997; Murzin et al., 1995; Brenner et al., 1997, Holm and Sander, 1998; Orengo et al, 1999]. A fold of a peptide chain was considered redundant if it differed from any other folds according to the following criteria: DALI Z-score > = 4.0 or RMSD < 3.0Å for the number of aligned residue positions > = 70% of the length of this chain.

Furthermore, high sequence similarity implies similarity in three-dimensional structure and, in general, in molecular function. However, the opposite is not necessarily true. Similarity in three-dimensional structure has been observed in the absence of detectable sequence similarity. Such proteins are termed *remote homologs* because homology between them cannot be reliably inferred in the absence of the structure. Remote homologs often share the same or related molecular functions, although sometimes they have apparently unrelated molecular functions.

The above observations have led to a concept of protein fold space [Holm and Sander, 1993; Michie et al., 1996; Hou et al., 2003]. Conceptually the protein fold space can be depicted as in Figure 1.2a: the majority of the protein sequences in the protein universe can be grouped into a finite set of sequence families. Each "island" represents a protein sequence family with its members having sequence similarity. Two sequence similarity islands are joined, sometimes, into a larger family because they are remote homologs of each other and have similar structures. Thus, the objective of structural genomics is to obtain the three-dimensional structures of one or more representatives from each sequence family and infer their molecular functions. Because of convergent and divergent evolution, each fold family may be associated with one or more of a finite number of molecular functions. Once the molecular functions of each family (a dictionary of fold vs. molecular function) are assigned, a given cellular function can then be represented as a linear and/or branched network of molecular functions (Figure 1.2b). This concept is similar to an electronic circuit diagram consisting of interconnected electronic components, where various circuits generate signal outputs (corresponding to cellular functions) and each component within a given circuit has its own function (corresponding to molecular functions).

1.2.1.2 Representation of the Protein Fold Space

Pioneering efforts to represent protein fold space has been made based on several different assumptions [Holm and Sander, 1993; Michie et al., 1996]. A recent unbiased (not derived from human-curated fold classifications such as SCOP or CATH) computational analysis of protein folds [Hou et al., 2003] reveals that protein fold space can be represented approximately in three-dimensional space. In this representation (Figure 1.3), the 500 most common protein folds are segregated into four elongated regions approximately corresponding to the four protein fold classes (α, β, $\alpha+\beta$, and α/β), and each region appears to be distributed roughly according to the evolutionary timeline. A major objective of structural genomics, then, is to

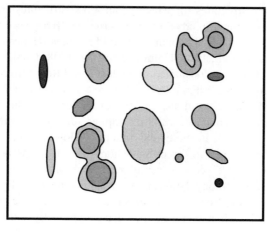

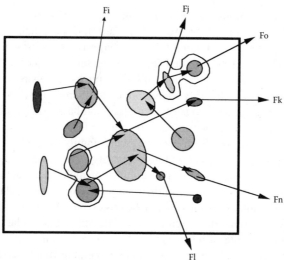

FIGURE 1.2 (a) Conceptual protein fold space: Each island schematically represents a family of protein domains with sequence similarity. Each member of the family has one of a relatively small number of molecular (biochemical and biophysical) functions associated with the family. Some of the sequence families are joined (lower left and upper right) to indicate that they have the same three-dimensional fold, although there are no significant sequence similarities between the joined families. A current estimate is that there are between a few thousands and 10,000 fold families in the protein fold world. A small, evolutionarily minimized organism is expected to have a smaller set of fold families, a fold basis set, compared with a higher organism. Asterisks represent those proteins whose 3-D structures have been determined experimentally. (b) Each cellular (genetic and physiological) function is represented by a collection of molecular functions, shown as a set of connected and branched arrows. A hypothetical biosynthetic pathway (e.g., Fl) for an amino acid may be represented by a network of ordered enzymes synthesizing the intermediates of the amino acid. A signal transduction pathway (e.g., Fn) may be represented by a set of connected arrows representing the proteins in the pathway in the order in which the signal is transmitted. Some pathways may be cross-talking to each other, as indicated by a bifurcated arrow (e.g., Fj and Fo).

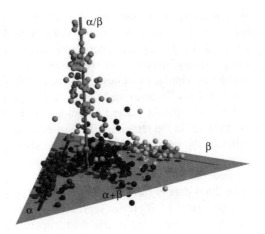

FIGURE 1.3 A 3-D map of the protein fold space. About 500 most common protein folds are represented. Each sphere represents a protein fold family. Although the map was constructed without fold class assignment of each fold, protein folds that belong to the α, β, and α/β SCOP classes segregate approximately into three elongated clusters mostly around three separate axes, α(red), β(yellow), and α/β(cyan). Most protein folds of the α+β class (blue) fall between the α and β axes and on or near the plane defined by the α and β axes. The α and β axes approximately intersect at one point, the origin, indicated by the green ball. This protein fold map appears to have an embedded evolutionary clock in it. (From Hou, J., Sims, G.E., Zhang, C., and Kim, S.-H., *Proc. Natl. Acad. Sci. U.S.A.*, 100, 5, 2003.)

fill the less populated areas of the fold space as represented in this map, and to relate each fold family to its molecular functions.

1.2.2 Fold and Functional Space Coverage

1.2.2.1 Family and Fold Space Coverage

Recent studies estimate that the proteins encoded by about 40 to 50% of the open reading frames (ORFs) of known genomic sequences have homologs in the current Protein Data Bank (PDB) [Chandonia and Brenner, 2004]. Of the largest 5000 Pfam [Bateman et al., 2004] families, at least 2108 have one or more representatives with an experimentally determined three dimensional structure. Structural representatives for Pfam families of previously unknown structure are being solved at a rate of about 20 per month, even as the total number of structures solved continues to increase exponentially. Solving the structure of one representative from the 5000 largest Pfam families would allow accurate fold assignment for approximately 68% of all prokaryotic proteins (covering 59% of residues) and 61% of eukaryotic proteins (40% of residues) [Chandonia and Brenner, 2004]. Assessing the fold representation depends very heavily on protein fold classification methods, for which there are several [Murzin et al., 1995; Pascarella and Argos, 1992; Orengo et al., 1994; Gibrat, 1996; Holm and Sander, 1996; Schmidt et al., 1997].

In an analysis of PDB entries deposited through 1997, fewer than 25% of newly solved protein families actually represented novel folds according to the SCOP classification [Murzin et al.,1995]; approximately 50% were identified as remote homologs of proteins of known structure on the basis of the structure, and the remaining families had a fold similar to a known structure (although the homology could not be clearly established) [Brenner and Levitt 2000]. More than 85% of the entries deposited by then were closely related homologs of at least one protein of known structure. Only proteins in the remaining 14% of entries had, on the basis of lack of sequence similarity, the potential to be fundamentally new and structurally unrelated. A very minor portion, 3.4%, of the entries represented new folds by SCOP [Murzin et al.,1995] standards.

An update based on 2004 data is shown in Figure 1.4. In recent years (2000 to 2004), less than 2% of deposited PDB entries represented a new fold according to SCOP standards. Only 6.3% represented a new fold, superfamily, or family; the remaining 92.7% of deposited entries were studies of proteins from families in which the structure of at least one member had previously been determined experimentally. In contrast, structures determined by nine National Institutes of Health (NIH)-funded structural genomics centers and the nine international centers (enumerated below) that currently report their results to TargetDB [Chen et al. 2004] averaged 6.2% new fold and 27.4% new fold, superfamily or family.

At the current rate of more than 5000 PDB depositions per year, the conventional structural biology approach is unlikely to achieve comprehensive coverage of protein fold space in the near future. A systematic approach biased to discover new structures

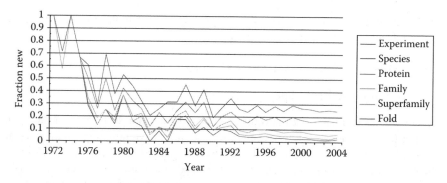

FIGURE 1.4 Revision of the Brenner and Levitt analysis with present data. Each line shows the cumulative fraction of domains from PDB entries deposited in each year from 1972 to 2004 that was classified in SCOP 1.67 as the first experimentally determined structure in a given category at each level in SCOP. For example, the superfamily line (second from bottom) indicates the fraction of domains that represented a new superfamily within a previously represented fold. Partial data (through 15 May) were used for 2004. Obsolete PDB entries were classified in the same way as the domains of their superseding entries. Only proteins from the main (first seven) classes of SCOP were considered, and only a single representative of several domains in a given PDB entry with identical classifications (e.g., homodimers) was used. Each PDB entry was considered a separate experiment.

and folds is currently taken by most structural genomic centers in order to achieve the objectives of structural genomics on a reasonable timescale and cost. Furthermore, a well-thought-out strategy is needed to cover fold space with coarse granularity to understand fold distribution and evolution, as well as with fine granularity to be able to accurately model 3-D structures of all members of a sequence family using comparative modeling. Although all members of each sequence family are expected to adopt roughly the same fold, the evolutionary diversity within a family is often too large to allow accurate modeling of all sequences from a single solved structure. Current state-of-the-art comparative modeling methods are able to produce models of medium accuracy (about 90% of the main chain modeled to within 1.5 Å RMS error) when sequence identity (ID) between the model and the template is at least 30%. Below 30% ID, however, alignment errors increase rapidly and become the major source of modeling error [Baker and Sali, 2001].

1.2.2.2 Functional Space Coverage

In addition to increasing coverage of fold space, the high throughput structural genomics approach can also be applied to cover functional space of medical and biological interest in a time- and cost-efficient way. In cases where a protein has validated medical and therapeutic value — e.g., proteins strongly implicated in human diseases such as cancer and metabolic disorders or essential proteins found only in pathogenic microbes — structural information is needed for specific proteins, or a member of a particular subfamily. This level of coverage will increase our understanding of these proteins' biochemical functions, enabling, for example, drug design [Burley and Bonanno, 2002] and prediction of functionally important epitopes [Laskoswski et al., 2003]. Estimates indicate that this will require an order of magnitude more targets than for coarse-grain coverage, so this fine-coverage option should be used sparingly.

1.3 INTERNATIONAL EFFORTS

Because of the vast scope of the structural genomics tasks, there is a strong international trend to coordinate the efforts by various groups in the field. At the Second International Structural Genomics Meeting held at the Airlie Conference Center, Airlie, Virginia, U.S.A., a general consensus was reached on the Agreed Principles and Procedures (http://www.nigms.nih.gov/news/meetings/airlie.html#agree), which stated that it is time that the structural biology community proposes the large scale mapping of protein structure space. "This structural genomics initiative aims at the discovery, analysis, and dissemination of three-dimensional structures of protein, RNA and other biological macromolecules representing the entire range of structural diversity found in nature. A broad collection of structures will provide valuable biological information beyond that which can be obtained from individual structures."

Recent technological advances in X-ray crystallography and NMR spectroscopy allow structure determination of selected targets at an astonishing speed compared to the past [Adams et al., 2003; McPherson, 2004; Segelke et al., 2004]. Nevertheless,

the size and complexity of genomes still preclude the possibility of structural determination of every gene product experimentally. Therefore the specific goal of structure genomics projects is broadly focused on:

- Determining by experimental methods a representative set of macromolecular structures, including medically important human proteins and proteins from important pathogens and model organisms.
- Providing molecular models of proteins with sequence similarity to known
  protein structures, in order to significantly extend the coverage of structure
  space.
- Deriving functional information from these structures by experimental
  and computational methods.

The currently available methods are inadequate to facilitate the primary goals of
structural genomics. Further method development is a critical factor that will enable
the success of structural genomics in the long run. The international agreement has
included the following aspects as goals of structural genomics: (a) methods of
selecting representatives of protein families based on enhancement of structure space
coverage, or functional significance; (b) high-throughput methods for production of
target proteins suitable for structure determination; (c) methods for high throughput
data collection; (d) methods for automated determination, validation, and analysis
of 3-D structures; (e) methods for homology-based modeling, related methods and
validation of modeled structures; (f) informatics systems to optimize and support
the process of structure determination; (g) bioinformatics methods for assessing
biological function based on structure and other linked biological information
sources; (h) methods for more challenging problems of production and structure
determination such as those involving membrane proteins and multimolecular complexes.

The Department of Energy in the United States, RIKEN in Japan, and National
Research Council (NRC) in Canada took a pioneering role supporting a few small
pilot projects in 1996. Full scale pilot projects were supported by RIKEN starting
in 1997, by the U.S. Department of Energy in 1998, and by the National Institute
of General Medical Science (NIGMS) of NIH in 2000, followed by many countries
in Europe and other continents (http://www.isgo.org). The International Structural
Genomics Organization (ISGO) was created with the aim of fostering international
cooperation and communication in structural genomics. Table 1.2 illustrates the
publicly funded structural genomics initiatives around the world based on a survey
of the centers and publicly available data on each.

As the worldwide structural genomics project is still in its pilot stage, most
resources are currently spent on technology development [Dieckman et al., 2002;
Karain et al., 2002; Pedelacq et al., 2002; Busso et al., 2003; DiDonato et al., 2004;
Nguyen et al., 2004]. However, a considerable number of structures is already
coming out of the pipelines at a much faster rate than before. According to TargetDB
[Chen et al., 2004], as of November 2004, the worldwide structure genomics effort
has collectively produced 1662 protein structures, where 1242 structures were deter-

TABLE 1.2
Structural Genomics (SG) Centers

SG Center	Objective	Website
NIH Centers		
Berkeley Structural Genomics Center (BSGC)	Structural complement of minimal organisms *Mycoplasma genitalium* and *Mycoplasma pneumoniae*	http://www.strgen.org
Joint Center for Structural Genomics (JCSG)	Structural genomics of *T. maritima* and *C. elegans*	http://www.jcsg.org
Midwest Center for Structural Genomics (MCSG)	Novel protein folds from all kingdoms. Current targets are chosen from large sequence families of unknown structure	http://www.mcsg.anl.gov
New York Structural Genomics Research Consortium (NYSGRC)	Novel structural data from all kingdoms of life with emphasis on medically relevant proteins. Current targets are enzymes with unknown structure	http://www.nysgrc.org
Northeast Structural Genomics Consortium (NESG)	Novel folds of eukaryotic proteins including *S. cerevisiae, C. elegans, D. melanogaster, Homo sapiens,* or tractable prokaryotic homologs	http://www.nesg.org
Southeast Collaboratory for Structural Genomics (SECSG)	Structural proteomes of *P. furiosis, H. sapiens, and C. elegans.*	http://www.secsg.org
TB Structural Genomics Consortium	Structures of *M. tuberculosis* proteome, with emphasis on functionally important proteins	http://www.doe-mbi.ucla.edu/TB
Center for Eukaryotic Structural Genomics (CESG)	Novel eukaryotic proteins, with *A. thaliana* as a model genome	http://www.uwstructuralgenomics.org
Structural Genomics of Pathogenic Protozoa Consortium (SGPP)	Structural genomics of protozoan pathogens	http://depts.washington.edu/sgpp
International Centers (* indicates the center currently reports results to TargetDB)		
Montreal–Kingston Bacterial Structural Genomics Initiative (*BSGI)	Novel representatives for protein families	http://sgen.bri.nrc.ca/bsgi

TABLE 1.2
Structural Genomics (SG) Centers (continued)

SG Center	Objective	Website
	NIH Centers	
Structure 2 Function Project (*S2F)	Functional characterization of *Haemophilus influenzae* proteins	http://s2f.umbi.umd.edu
Structural Proteomics in Europe (*SPINE)	Structures of a set of human proteins implicated in disease states	http://www.oppf.ox.ac.uk
Protein Structure Factory (*PSF)	Structure of human proteins	http://www.rzpd.de/psf
RIKEN Structural Genomics/Proteomics Initiative (*RIKEN)	Structural genomics of *Thermus thermophilus* HB8 and an archaeal hyperthermophile, *Pyrococcus horikoshii* OT3	http://www.rsgi.riken.go.jp
Oxford Protein Production Facility (*OPPF)	Targets of biomedical interest: Human proteins, cancer and immune cell proteomes, and Herpes viruses	http://www.oppf.ox.ac.uk
Yeast Structural Genomics (*YSG)	Proteins from yeast	http://genomics.eu.org
Israel Structural Proteomics Center (*ISPC)	Proteins related to human health and disease	http://www.weizmann.ac.il/~wspc
Mycobacterium Tuberculosis Structural Proteomics Project (*XMTB)	Proteins from *Mycobacterium tuberculosis* and related mycobacteria	http://xmtb.org
Structure 2 Function Project (*S2F)	Proteins from *Haemophilus influenzae*	http://s2f.carb.nist.gov
Bacterial Targets at IGS-CNRS, France (*BIGS)	Proteins from the bacteria *Rickettsia*, as well as proteins with unique species-specific sequences (ORFans) from *Escherichia coli*	http://igs-server.cnrs-mrs.fr/Str_gen
Marseilles Structural Genomics Program (*MSGP)	Structural genomics of bacterial, viral, and human ORFs of known and unknown function	http://afmb.cnrs-mrs.fr/stgen
The Korean Structural Proteomics Research Organization	Structural genomics of *Mycobacterium tuberculosis* and *Helicobacter pylori* leading to drug discovery	http://kspro.org
National Centers of Competence in Research (NCCR)	Structures of membrane proteins, and analysis of intermolecular interactions in supramolecular assemblies	http://www.structuralbiology.unizh.ch
North West Structural Genomics Centre, U.K. (NWSGC)	Host/pathogen interactions, leading to the development of more effective drugs for human health and pathogen biology	http://www.nwsgc.ac.uk

Ernest Laue Group	Structural analysis of protein-protein interactions that are important in cellular control	http://www.bio.cam.ac.uk/~edl1
Ontario Centre for Structural Proteomics	Novel folds from thermophilic archaea, *Methanobacterium thermoautotrophicum*, *Helicobacter pylori*, and other bacterial, eukaryotic, and archeal proteins	http://www.uhnres.utoronto.ca/proteomics
Montreal Network for Pharmaco-Proteomics and Structural Genomics	Function and structure of genes and proteins that can be used in developing new drugs	http://www.genomequebec.com/asp/dirProgrammes/projetsDetails.asp?id=c1p14&l=e

mined using X-ray crystallography and 420 by NMR. 938 structures were produced by the nine NIH-funded centers, and 724 by other international centers.

1.4 METRICS AND LESSONS LEARNED FROM PSI PILOT PHASE

A wealth of information has been generated during the past four years by structural genomics researchers. The efforts in the U.S. funded by the NIH Protein Structure Initiative (PSI) have been well documented on the NIH website (http://nigms.nih.gov/psi). In 2004, the PSI is currently nearing the end of its five-year pilot phase. We describe below the metrics and lessons learned from the experience of two distinctly different approaches taken by PSI centers during the pilot phase (four years for seven centers and three years for two centers). In one of these approaches, the objective has been to obtain dense coverage of protein structure space from a minimal organism containing about 450 genes. In the other approach, the goal was to obtain as many new structures as possible from sparse coverage of the protein structure space of one or more organisms with several thousands or more genes each.

- The steps required to proceed from cloning a gene encoding a protein to determining its 3-D structure can be divided into two distinct categories. First, those steps where the underlying science and technologies are well understood, such as in cloning, X-ray diffraction data collection, structural solution by computational crystallography, and NMR data collection and structure solution. These steps are *mostly automated* and amenable for high throughput operation. Second, those steps where the underlying science is not known or only partially known, and the outcome of the processes is unpredictable. These steps require *multi-variable screenings*, and are not automated at present with single exception of crystallization screening.
- The *single-path (low-hanging fruit) approach*, whereby, for a large number of diverse genes, one single optimized path is taken from cloning to structure determination, has been capable of producing ~20 to 150 structures per center (Figure 1.5a), but has less than a *5% success rate* (Figure 1.5b) on average in discovering structures of *unique full-length proteins*, the proteins without sequence homology to those of known structures in the Protein Data Bank (PDB).
- The *multi-path approach*, where feedback loops and multi-factor screenings are employed for one or more critical steps in the path for *challenging proteins* that failed by a single-path approach, has a *24% or higher success rate* for discovering structures for full-length proteins of *unknown function or structure* (Figure 1.5b, BSGC results). A similar success rate, of about 15%, can be achieved if targets are prescreened to select only those with *known function* even with single-path approach (Figure 1.5b, NYSGRC results).

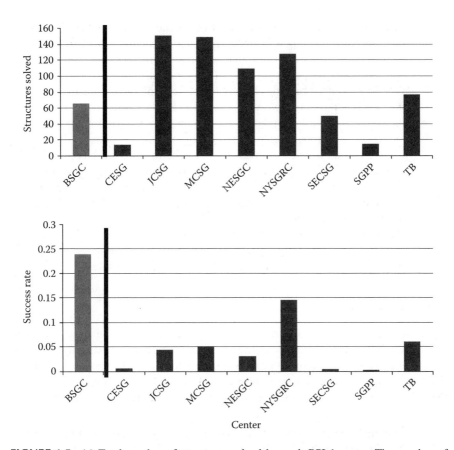

FIGURE 1.5   (a) Total number of structures solved by each PSI-1 center. The number of unique structures solved varies, depending on the number of targets selected, selection criteria of each center, and success rate. BSGC (light-colored bar) selected about 500 full-length unique proteins of the minimal organisms *M. genitalium* and *M. pneumonia*, and other PSI centers (dark-colored bars) selected several thousand targets each from multiple large organisms or one very large organism (e.g., *Arabidopsis*). (b) Success rate on the ordinate axis measured by the number of structures determined divided by the number of clones made by the 9 PSI-1 centers on the abscissa (http://www.nigms.nih.gov/psi). The *single-path approach*, taken by all PSI-1 centers (dark-colored bars) has about 5% success rate on average for all target proteins, but the *multi-path approach* taken by BSGC (in orange) shows about *24%* success rate for target proteins. A similar success rate of ~15% by the NYSGRC was achieved if targets were prescreened to select for only those with *known function*, even with single-path approach. Abbreviations: Berkeley Structural Genomics Center (BSGC); Center for Eukaryotic Structural Genomics (CESG); Joint Center for Structural Genomics (JCSG); Midwest Center for Structural Genomics (MCSG); Northeast Structural Genomics Center (NESGC); New York Structural Genomics Research Center (NYSGRC); South East Center for Structural Genomics (SECSG); Structural Genomic for Pathogenic Protozoa (SGPP); TB consortium (TB).

- Approximately half of the structures of unique proteins revealed a *new fold*, and the remaining half are *remote homologs* (similar structure without sequence similarity) of proteins of known structure (BSGC results).
- Approximately half the structures of *hypothetical proteins* (proteins that have no sequence homologs among the proteins of known function) infer one or a few possible *molecular functions* (BSGC results).
- On average, ~1/3 of a proteome are predicted to be *membrane proteins*, and there are no robust heterologous over-expression systems available at present, and no metrics are available for gene-to-structure success rate.
- On average, ~1/3 of a proteome is present as components of *stable molecular complexes* in the cell (Gavin et al., 2002), and many of these component proteins may be insoluble or may not properly fold in the absence of their complex partners.
- The *protein fold space* can be mapped in three-dimensional space based on pair-wise structural similarities (Hou et al., 2003), thus providing a platform for representing all PSI and other protein structures.

1.5 PROSPECTS

By analogy to the tremendous impact that sequence genomics databases have had on all disciplines, ranging from basic life sciences to drug design and curing diseases, structural genomics is capable of providing information complementary to sequence genomics data as well as unique information not available by other methods. From the collective results and experiences of the PSI pilot centers, it is clear that:

- It is safe to assume that, to a first approximation, the majority of structural genomics target proteins are *challenging proteins* that are not suited to a single-path approach and can only be solved with a reasonable success rate using a *multi-path approach*.
- A significant portion of the procedures required — from cloning to structure determination — are already *automated* or *parallelized* to the extent that it is feasible to attempt, with some additional automation of multi-variable screening steps, near-complete coverage of the protein fold space on a global scale. This would provide coarse-grain resolution coverage for most of the large protein families, and fine-grain resolution coverage — allowing fairly accurate comparative models to be constructed — of subfamilies and small families of biological and medical importance.

ACKNOWLEDGMENT

Some of the concept development and works cited were supported by the grant from NSFDBI-0114707 and NIH Protein Structure Initiative Grant GM62412.

REFERENCES

Abola E.E., Prilusky J., and Manning N.O., Protein Data Bank archives of three-dimensional macromolecular structures, *Methods Enzymol.*, 277, 556–571, 1997.

Adams, M.W., Dailey, H.A., DeLucas, L.J., Luo, M., Prestegard, J.H., Rose, J.P., and Wang, B.C., The Southeast Collaboratory for Structural Genomics: a high-throughput gene to structure factory, *Acc. Chem. Res.*, 36, 191–198, 2003.

Baker, D. and Sali, A., Protein structure prediction and structural genomics, *Science*, 294, 93–96, 2001.

Bateman A., Birney E., Durbin R., Eddy S.R., Finn R.D., and Sonnhammer E.L., Pfam 3.1, 1313 multiple alignments and profile HMMs match the majority of proteins, *Nucleic Acids Res.*, 27, 1, 260–262, 1999.

Bateman, A., Coin, L., Durbin, R., Finn, R.D., Hollich, V., Griffiths-Jones, S., Khanna, A., Marshall, M., Moxon, S., Sonnhammer, E.L., Studholme, D.J., Yeats, C. and Eddy, S.R., The Pfam protein families database, *Nucleic Acids Res*, 32 Database issue, D138–141, 2004.

Berman, H.M., Westbrook, J., Feng, Z., Gilliland, G., Bhat, T.N., Weissig, H., Shindyalov, I.N., and Bourne, P.E., Protein Data Bank, *Nucleic Acids Res.*, 28, 235–242, 2000.

Boyd, D., Schierle, C., and Beckwith, J., How many membrane proteins are there?, *Protein Sci.*, 7, 201–205, 1998.

Brenner, S.E., Chothia, C., and Hubbard, T.J., Population statistics of protein structures: lessons from structural classifications, *Curr. Opin. Struct. Biol.*, 7, 3, 369–376, 1997.

Brenner, S.E. and Levitt, M., Expectations from structural genomics, *Protein Sci.*, 9, 1, 197–200, 2000.

Burley, S.K. and Bonanno, J.B., Structural genomics of proteins from conserved biochemical pathways and processes, *Curr. Opin. Struct. Biol.*, 12, 383–391, 2002.

Busso, D., Kim, R. and Kim, S.H., Expression of soluble recombinant proteins in a cell-free system using a 96-well format, *J. Biochem. Biophys. Methods*, 55, 233–240, 2003.

Chandonia, J.-M. and Brenner, S.E., Implications of structural genomics target selection strategies: Pfam5000, whole genome, and random approaches, *Proteins*, 58, 166–79, 2005.

Chen, L., Oughtred, R., Berman, H.M., and Westbrook, J., TargetDB: A target registration database for structural genomics projects, *Bioinformatics*, 20, 16, 2860–2862, 2004.

DiDonato, M., Deacon, A.M., Klock, H.E., McMullan, D., and Lesley, S.A., A scaleable and integrated crystallization pipeline applied to mining the Thermotoga maritima proteome, *J. Struct. Funct. Genomics*, 5, 133–146, 2004.

Dieckman, L., Gu, M., Stols, L., Donnelly, M.I., and Collart, F.R., High throughput methods for gene cloning and expression, *Protein Expr. Purif.*, 25, 1, 1–7, 2002.

Gibrat, J.F., Madej, T., and Bryant, S.H., Surprising similarities in structure comparison, *Curr. Opin. Struct. Biol.*, 6, 377–385, 1996.

Holm, L. and Sander, C., Protein structure comparison by alignment of distance matrices, *J. Mol. Biol.*, 233, 1, 123–138, 1993.

Holm, L. and Sander, C., Mapping the protein universe, *Science*, 273, 595–602, 1996.

Holm, L. and Sander, C., Touring protein fold space with Dali/FSSP, *Nucleic Acids Res.*, 26, 316–319, 1998.

Hou, J., Sims, G.E., Zhang, C., and Kim, S.-H., A global representation of the protein fold space, *Proc. Natl. Acad. Sci. U.S.A.*, 100, 5, 2386–2390, 2003.

Hubbard, T.J., Murzin, A.G., Brenner, S.E., and Chothia, C., SCOP, a structural classification of proteins database, *Nucleic Acids Res.*, 25, 1, 236–239, 1997.

Karain, W.I., Bourenkov, G.P., Blume, H., and Bartunik, H.D., Automated mounting, centering and screening of crystals for high-throughput protein crystallography, *Acta Crystallogr. D. Biol. Crystallogr.*, 58, 10–1, 1519–1522, 2002.

Lander, E.W., et al., Initial sequencing and analysis of the human genome, *Nature*, 409, 860–921, 2001.

Laskowski, R.A., Watson, J.D., and Thornton, J.M., From protein structure to biochemical function?, *J. Struct. Funct. Genomics*, 4, 167–177, 2003.

McPherson, A., Protein crystallization in the structural genomics era, *J. Struct. Funct. Genomics*, 5, 3–12, 2004.

Michie, A.D., Orengo, C.A., and Thornton, J.M., Analysis of domain structural class using an automated class assignment protocol, *J. Mol. Biol.*, 262, 2, 168–185, 1996.

Murzin, A., Brenner, S.E., Hubbard, T., and Chothia, C., SCOP: a structural classification of proteins for the investigation of sequences and structures, *J. Mol. Biol.*, 247, 536–540, 1995.

Nguyen, H., Martinez, B., Oganesyan, N., and Kim, R., An automated small-scale protein expression and purification screening provides beneficial information for protein production, *J. Struct. Funct. Genomics*, 5, 23–27, 2004.

Orengo, C.A., Jones, D.T., and Thornton, J.M., Protein superfamilies and domain superfolds, *Nature*, 372, 631–634, 1994.

Orengo, C.A., Pearl, F.M., Bray, J.E., Todd, A.E., Martin, A.C., Lo Conte, L., and Thornton, J.M., The CATH database provides insights into protein structure/function relationships, *Nucleic Acids Res.*, 27, 275–279, 1999.

Pascarella, S. and Argos, P., A databank merging related protein structures and sequences, *Protein Eng.*, 5, 121–137, 1992.

Pedelacq, J.D., Piltch, E., Liong, E.C., Berendzen, J., Kim, C.Y., Rho, B.S., Park, M.S., Terwilliger, T.C., and Waldo, G.S., Engineering soluble proteins for structural genomics, *Nat. Biotechnol.*, 20, 9, 927–932, 2002.

Ponting C.P., Schultz J., Milpetz F., and Bork P., SMART: identification and annotation of domains from signalling and extracellular protein sequences, *Nucleic Acids Res.*, 27, 1, 229–232, 1999.

Schmidt, R., Gerstein, M., Altman, R., LPFC: an Internet library of protein family core structures, *Protein Sci.*, 6, 246–248, 1997.

Segelke, B.W., Schafer, J., Coleman, M.A., Lekin, T.P., Toppani, D., Skowronek, K.J., Kantardjieff, K.A., and Rupp, B., Laboratory scale structural genomics, *J. Struct. Funct. Genomics*, 5, 147–157, 2004.

Shindyalov, I.N. and Bourne, P.E., Protein structure alignment by incremental combinatorial extension (CE) of the optimal path, *Protein Eng.*, 11, 9, 739–747, 1998.

Sonnhammer, E.L., Eddy, S.R., and Durbin, R., Pfam: a comprehensive database of protein domain families based on seed alignments, *Proteins*, 28, 3, 405–420, 1997.

Venter, J.C., et al., The sequence of the human genome, *Science*, 291, 1304–1351, 2001.

Vitkup, D., Melamud, E., Moult, J., and Sander, C., Completeness in structural genomics, *Nat. Struct. Biol.*, 8, 6, 559–566, 2001.

Yona, G., Linial, N., and Linial, M., ProtoMap: automatic classification of protein sequences and hierarchy of protein families, Nucleic Acids Res., 28, 1, 49–55, 2000.

2 Purifying Protein for Structural Biology

Aled M. Edwards

CONTENTS

2.1 OVERVIEW

Purification of sufficient numbers and amounts of proteins is commonly thought to be one of the more important steps in high-throughput structure determination. In recent years, many groups have honed methods to produce purified prokaryotic proteins in large scale. As international projects inevitably move to focus on eukaryotic and membrane proteins, there are new requirements to develop cost-effective methods to produce these proteins. This review will describe existing methods for producing bacterial proteins, some new strategies to produce bacterial proteins more economically, and approaches that might be used for eukaryotic and membrane proteins.

2.2 PROTEIN EXPRESSION

Expressing large amounts of soluble, recombinant protein in a cost-effective manner is the most challenging step of structural proteomics. It is commonly believed that if a protein can be purified to homogeneity and concentrated to millimolar concentrations, the probability of generating a protein crystal or a good sample for nuclear magnetic resonance (NMR) spectroscopy is high. The high-throughput purification strategies, which will be discussed, can be implemented only once the expression and solubility issues are tackled.

2.2.1 ESCHERICHIA COLI: THE CONSENSUS EXPRESSION SYSTEM

For reasons of expense and simplicity, *Escherichia coli* (*E. coli*), a gram-negative bacterium, is the workhorse system to express recombinant proteins for structural biology. It is believed that prokaryotic proteins are more easily expressed in *E. coli* because of common folding pathways and mechanisms. As a result, prokaryotic proteins have been used as fodder to develop the automation of large-scale expression and purification.

All large-scale approaches to protein purification also involve recombinant proteins fused to a protein tag (affinity tag). Most fusion proteins are linked to the recombinant protein via a sequence of amino acids that are recognized and cleaved by a specific protease. Although the expression and solubility of a given protein can be affected quite significantly by the choice of the fusion protein, most laboratories use hexahistidine tags as the fusion of choice. Hexahistidine tags are less likely to adversely affect crystallization or NMR data collection, and indeed sometimes are required for crystallization [1]. A large-scale, well-controlled study will be required to determine the relative advantages of leaving or removing the fusion protein prior to crystallization or NMR studies. At the present state of knowledge, the most prudent approach is to try both forms of the protein.

2.2.1.1 Insufficient Protein Expression and Solubility: The Bane of Structural Biologists

Despite the considerable effort in developing *E. coli* expression plasmids and cell lines, most full-length eukaryotic proteins are not produced in sufficient quantities to suit structural biologists. The proteins are not expressed or are insoluble because:

- They employ a set of codons not commonly used in *E. coli*.
- They are not good substrates for the bacterial folding machinery, in which case there is an aggregation of folding intermediates. Proteins that comprise multiple domains fall into this class.
- An intrinsic property of the protein causes aggregation. For example, very hydrophobic patches on the surface may interact at high concentrations.
- They require posttranslational modifications, such as N-linked glycosylation, that do not occur in *E. coli*.
- They require an obligate co-factor or additional protein for proper folding.

Over the past years, much effort has been devoted to tackle each of these problems and is discussed below.

2.2.1.1.1 Codon Usage

For proteins whose codon composition varies from that used in *E. coli*, the cells can be supplemented with genes encoding tRNAs that recognize the low abundance codons [2]. There are now engineered bacterial strains such as the Codon Plus system from Stratagene that are augmented with genes for rare tRNAs [3]. The strategy has proven remarkably effective in many instances.

2.2.1.1.2 Insolubility

There are several strategies to deal with proteins that do not fold well in *E. coli* and that are produced in an insoluble form. One method is to induce them to fold properly by the co-expression of chaperones in the bacterial cell [4,5]. However, more often than not, this process has been ineffective and the proteins ultimately must be expressed in a different system.

A second strategy is to purify the protein in denatured form and then renature the purified protein. This approach has been used sporadically for decades, but it is not in routine use largely because the effectiveness of the method has not been quantified. Recently, Maxwell and colleagues [6] tested a generic denaturation/refolding protein purification procedure to assess the number of structural samples that could be generated using such a strategy. Seventy small proteins that were expressed in the insoluble fraction in *E. coli* were purified from the cell pellet in guanidine. They were then refolded using dialysis against a single buffer. Surprisingly, 31 could be renatured and had detectable secondary structure, as assessed using circular dichroism. Seven of these previously insoluble proteins provided good HSQC NMR spectra; the others could not be sufficiently concentrated for NMR studies. Twenty-five small proteins that were expressed in the soluble fraction were also purified from the cell pellet in guanidine, in order to assess how many normally soluble proteins could be purified using a denaturation/refolding protocol. Twenty-two of these proteins could be purified in soluble and folded form using the denaturing strategy, for a recovery rate of 88%. Interestingly, the denaturing approach yielded 20 samples that provided good HSQC NMR spectra, as opposed to 17 that were generated when the proteins were purified from the soluble fraction. Certainly, for smaller proteins, the denaturation/refolding protocol is an efficient way to generate structural samples for high-throughput studies of small proteins.

If a protein is insoluble when expressed in *E. coli*, another strategy is to attempt to study the equivalent protein from another organism. Orthologues are thought to have largely similar three-dimensional structure and differ predominantly in surface properties. Changes in the surface properties of proteins are known to dramatically affect the expression, solubility, and propensity to crystallize. Indeed, many crystallographers have successfully used directed mutagenesis of surface residues to induce crystallization of a given protein. To quantify the value of pursuing the analysis of orthologues, Savchenko and colleagues [7] expressed 68 pairs of homologous proteins from *E. coli* and *Thermotoga maritima*. A sample suitable for structural studies was obtained for 62 of the 68 pairs of homologues under standardized growth and purification procedures. Fourteen samples (8 *E. coli* and 6 *T. maritima*) generated NMR spectra of a quality suitable for structure determination and 30 samples (14 *E. coli* and 16 *T. maritima*) formed crystals. Only three samples (1 *E. coli* and 2 *T. maritima*) both crystallized and had excellent NMR properties. The conclusions from this work were: (1) the inclusion of even a single orthologue of a target protein increased the number of structural samples almost twofold; (2) there was no clear advantage to the use of thermophilic proteins to generate structural samples; and (3) the use of both NMR and crystallographic approaches almost doubled the number of structural samples.

The orthologue study showed that natural sequence variation can influence expression and solubility in *E. coli*. It is also possible to generate soluble sequence variants of a protein using genetic selection. In one example of this strategy, Waldo and colleagues [8] fused a variety of coding sequences N-terminal to the coding region of the green fluorescent protein (GFP), which is known to fold poorly in *E. coli*. Insoluble proteins were found to inhibit the folding of GFP; proteins that were soluble did not affect the folding of GFP. The solubility of the fusion protein could therefore be monitored by measuring the fluorescence of the bacterial colonies after transformation. A similar strategy used chloramphenicol acetyl transferase as the fusion protein and antibiotic resistance as the read-out[9]. Although elegant in design, it is unclear how successful these methods will be in practice. Despite the fact that these methods have now been in practice for several years, there are relatively few structures in the literature. It is not clear whether this is because there are unexpected difficulties in the methods or whether they have not yet been put into widespread use.

Cell-free expression in bacterial or wheat germ lysates may also enable the production of proteins unable to be expressed in *E. coli*. Bacterial cell-free expression was developed as an alternative to bacterial expression decades ago, but such methods were unable to rival intracellular expression in producing sufficient amounts of recombinant protein. There is now a resurgence of interest in the use of cell-free methods because the problem with inefficient expression has been overcome by incorporating continuous flow methods [10]. This would be a significant advance because cell-free expression may allow for the soluble expression of those proteins that are unable to fold properly in bacteria or in eukaryotic cells. However, the widespread use of cell-free protein expression will only come after the method has been compared directly with bacterial expression for a large number of proteins, and this has not yet been done.

Proteins that comprise multiple domains are difficult substrates for the bacterial folding machinery because folding intermediates of the different domains have a propensity to aggregate. A general and powerful strategy that simplifies the analysis of multidomain proteins is to produce and study them as individual domains. Experience has shown that a single domain of a protein can often be expressed in bacteria, whereas the intact, multi-domain protein cannot. Over the past decade, this approach has yielded structural information for dozens of eukaryotic proteins.

Experimental delineation of putative domain boundaries (limited proteolysis coupled with mass spectrometry) [11] is highly effective. This method is successful because unstructured portions of proteins are relatively more susceptible to proteolysis than are protein domains. As a result, the domains are stable intermediates during the proteolysis of a protein and can subsequently be identified. To identify the borders of these intermediates, they can be isolated for mass spectrometry and their masses determined. Then, the corresponding portion of the protein can be identified using computational methods and the stable fragment sub-cloned for recombinant expression.

Sequence-based approaches can also be used to determine putative domain boundaries, particularly with the advent of genomic information for sometimes dozens of homologues. In this method, a multiple sequence alignment is used to identify highly conserved portions of proteins. The boundaries of the areas of sequence conservation are thought to indicate the boundaries of the functional domains, which can also correspond to structural domains. Secondary structure predictions are also used to guide the selection of the exact N- and C-termini of the conserved protein fragment. The exceptions to the correspondence of functional and structural domains are when functionally conserved regions of proteins do not adopt a stable tertiary structure (such as a protein-interaction motif that folds only when bound to its partner).

2.2.1.1.3 Posttranslational Modifications

In *E. coli*, it is difficult to produce proteins that require nonprokaryotic posttranslational modifications for proper folding or function. One exception is the expression of some protein kinases, which are often auto-phosphorylated. It is possible to generate homogeneously dephosphorylated kinases by co-expressing protein phosphatases.

2.2.1.1.4 Protein Co-Expression

Finally, a protein that requires another protein for expression and folding can be co-expressed in the same bacterial cell from the same plasmid [12]. The proteins are encoded either in a single RNA or by a set of tandem genes. The co-expression strategy is highly effective, but it requires considerable knowledge of the protein target.

2.3 EXPRESSION OF EUKARYOTIC PROTEINS

Most of the current structural proteomics efforts have focused on full-length prokaryotic proteins (Bacteria and Archaea). Eukaryotic proteins present the structural biologist with more challenges, which include:

- Protein heterogeneity due to posttranslational modifications. Eukaryotic proteins are often modified and in most cases, it is not simple to reconstitute the modifications in a recombinant protein.
- Increased frequency of protein interactions. It is thought that a larger percentage of eukaryotic proteins participate in protein-protein interactions. As a result, some proteins are unable to be produced in the absence of other proteins, and some are unfolded in the absence of their interaction partner.
- Multiple domains. Eukaryotic proteins are generally larger than bacterial proteins. The increase in size results not because eukaryotic proteins are inherently larger but rather because eukaryotic proteins comprise, on average, a larger number of domains than do prokaryotic proteins.

Although these factors present challenges for the structural biologist, they are not unique to eukaryotic proteins, but only more common. For example, there are bacterial proteins that are modified (H-NS), that require another protein to fold and be expressed (the β subunit of RNA polymerase) and that comprise multiple domains (GlmU).

As a result, it may be best not to classify a protein by its biological source, but rather by its characteristics for structural biology. The following classifications might be more appropriate:

- Unmodified intracellular single-domain proteins. These proteins, whether derived from prokaryotes or eukaryotes, are relatively straightforward targets for structural biology.
- Unmodified intracellular multidomain proteins. These proteins are also quite amenable to structural biology, once the individual domains have been identified using limited proteolysis/mass spectrometry or bioinformatics.
- Protein complexes. Again, these represent a highly tractable set of targets and do not require new technology, provided the interacting proteins are known with confidence.
- Proteins that require posttranslational modification to adopt a homogeneous structure. These targets are the most difficult for structural efforts in which automated approaches are to be employed because each target is a unique case and requires significant knowledge of the biological system.

The real challenges in structural biology are to prepare sufficient quantities of purified, concentrated proteins for structural efforts. The ability to do this depends on the properties of the individual protein rather than its biological source. Perhaps the most significant practical difference between prokaryotic and eukaryotic proteins is that it is more tedious to derive expression clones for eukaryotic proteins. The presence of introns within eukaryotic genes means that cDNA, rather than genomic DNA, must be the source of the expression constructs. Accruing a library of expression vectors for eukaryotic proteins is costly both in time and in reagents.

Importantly, acquiring a construct to express most eukaryotic proteins is only the beginning of the molecular biology phase of the project. Over the past decade, structural biologists have gained an appreciation that the first expression construct for any given protein domain rarely generates a protein that will crystallize or behave well in NMR experiments. The basis for this belief is that the probability of generating a suitable structural sample is thought to increase if the unstructured parts of the protein are eliminated. As a result, it is common to screen dozens of variants of a given protein before identifying one that it suitable for structural determination. These include sequence variants or variants with different N- and C-termini. As one example, several fragments of the DNA-binding domain from the EBNA1 protein from Epstein-Barr virus form crystals, but only the fragment lacking the unstructured, acidic C-terminal domain formed well-ordered crystals suitable for structure determination [13]. Thus, the focus of cloning efforts in structural proteomics will not be the generation of cDNA clones but to develop cloning systems that can rapidly generate multiple constructs for any given protein.

2.4 PROTEIN PURIFICATION

Once adequate amounts of protein are expressed, there are two general strategies to purify prokaryotic proteins for structure determination.

The first and most commonly used approach is simply a parallelization of common laboratory practices. While not technologically impressive, many groups have shown that this approach is highly effective. By combining benchtop chromatography with judicious use of FPLC or HPLC, one scientist is readily able to purify up to 20 affinity-tagged proteins per week, starting from cell pellets. A team of three scientists can therefore purify thousands of proteins per year without need for advanced technological development or expensive automation; it is difficult to imagine a more cost-effective strategy. However, although the purification process is highly efficient, it is important to note that this approach requires a proportional investment in cell growth and general laboratory maintenance, and also requires teams of highly skilled scientists.

The second approach to purifying proteins for structure determination is to use largely automated methods. This method has been employed mostly in the biotechnology community because of the significant capital investment that was required to produce the custom automation. At least three automated, large-scale purification systems currently exist. Syrrx (San Diego, CA; www.syrrx.com) employs a system that was developed at the Genomics Institute of the Novartis Research Foundation (http://web.gnf.org). It combines robotic centrifugation and sonication with a parallel column chromatography system capable of purifying 96 to 192 proteins per day [14]. In this system, other steps including desalting and concentration are carried out off-line before the samples are ready for crystallization. Affinium Pharmaceuticals (Toronto, ON; www.afnm.com) developed the ProteoMax™ system to process cell extracts to purified, concentrated protein samples. This system clarifies the cell lysate, performs the column chromatography, desalts, and concentrates; in optimal cases, the purified protein is ready for structural studies. Structural Genomix (SGX; San Diego, CA; www.stromix.com) has developed a modular automated platform in which, wherever possible, commercially available robotic systems were incorporated into the technology platform. Highly parallel large-scale purification is carried out using automated serial column chromatography (Ni-ion affinity, ion exchange, and gel filtration).

2.5 PERSPECTIVE

Structural proteomics aims to provide structural information for every protein in every organism. The public efforts in the U.S. are focused on completing the universe of protein folds to provide sufficient information to create accurate models for every protein. Although numbers vary, it is estimated that the effort will require between 5,000 and 10,000 structures before this goal is accomplished. In the first five years of the Protein Structure Initiative, the aggregate tally of new structures is about 1000, though the rate with which structures are being deposited is increasing. It would be realistic to set a completion date of 2015 for the project.

Another aim of structural proteomics is to provide more accurate information about proteins for drug discovery purposes. In this area, the current modeling methods are not sufficiently useful for computational chemists, and high-resolution structural information is preferred. The Structural Genomics Consortium (SGC) is a public/private effort whose aim is to put structural information for proteins of therapeutic interest into the public domain, regardless of their structural similarity to other proteins. For drug discovery, it is the differences among proteins of similar structure that underscore selectivity, and for biology, it is the differences among similar proteins that underscore biological complexity. The SGC has a three-year aim to create the infrastructure to generate 200 high-resolution structures per year, and in its first year has already placed the structures of 50 human proteins into the public domain.

Considerable resources have been placed in selected structural proteomics centers around the world. Is the investment justified? It is difficult to say. The sole purpose of creating centers of activity is to achieve economic efficiency, and this is the most important metric by which the centers should be judged. Technological advances are critical, but only if they significantly reduce the cost per structure. To date, the numbers of structures emerging from the various international efforts has not justified the investment. However, the centers are in a building phase and should not be expected to be producing at maximal rates. Over the next three years, it will be important to judge the centers with economics in mind. It may be revealed that the existing methods of structure determination are in fact the most cost effective, that no new technological improvements are necessary and that the field should simply get on with the job.

REFERENCES

1. Kimber, M.S. et al., Data mining crystallization databases: knowledge-based approaches to optimize protein crystal screens, *Proteins,* 51, 562–8, 2003.
2. Kane, J.F., Effects of rare codon clusters on high-level expression of heterologous proteins in Escherichia coli, *Curr. Opin. Biotechnol.,* 6, 494–500, 1995.
3. Carstens, C.P. and Waesche, A., Codon bias-adjusted BL21 derivatives for protein expression, *Strategies,* 12, 49–51,1999.
4. Luo, Z.H. and Hua, Z.C., Increased solubility of glutathione S-transferase-P16 (GST-p16) fusion protein by co-expression of chaperones groes and groel in Escherichia coli, *Biochem. Mol. Biol. Int.,* 46, 471–7, 1998.
5. Hayhurst, A. and Harris, W.J., Escherichia coli skp chaperone coexpression improves solubility and phage display of single-chain antibody fragments, *Protein Expr. Purif.,* 15, 336–43, 1999.
6. Maxwell, K.L., Bona, D., Liu, C., Arrowsmith, C.H., and Edwards, A.M., Refolding out of guanidine hydrochloride is an effective approach for high-throughput structural studies of small proteins, *Protein Sci.,* 12, 2073–80, 2003.
7. Savchenko, A. et al., Strategies for structural proteomics of prokaryotes: Quantifying the advantages of studying orthologous proteins and of using both NMR and X-ray crystallography approaches, *Proteins,* 50, 392–9, 2003.
8. Waldo, G.S., Standish, B.M., Berendzen, J., and Terwilliger, T.C., Rapid protein-folding assay using green fluorescent protein, *Nat. Biotechnol.,* 17, 691–5, 1999.
9. Maxwell, K.L., Mittermaier, A.K., Forman-Kay, J.D., and Davidson, A.R., A simple in vivo assay for increased protein solubility, *Protein Sci.,* 8, 1908–11, 1999.
10. Spirin, A.S., Baranov, V.I., Ryabova, L.A., Ovodov, S.Y., and Alakhov, Y.B., A continuous cell-free translation system capable of producing polypeptides in high yield, *Science,* 242, 1162–4, 1988.

11. Koth, C.M. et al., Elongin from Saccharomyces cerevisiae, *J. Biol. Chem.* 275, 11174–80, 2000.

12. Bochkarev, A., Bochkareva, E., Frappier, L., and Edwards, A.M., The crystal structure of the complex of replication protein A subunits RPA32 and RPA14 reveals a mechanism for single-stranded DNA binding, *Embo. J.*, 18, 4498–504, 1999.

13. Bochkarev, A., Bochkareva, E., Frappier, L., and Edwards, A.M., The 2.2 A structure of a permanganate-sensitive DNA site bound by the Epstein-Barr virus origin binding protein, EBNA1, *J. Mol. Biol.*, 284, 1273–8, 1998.

14. Lesley, S.A., High-throughput proteomics: protein expression and purification in the postgenomic world, *Protein Expr. Purif.*, 22, 159–64, 2001.

3 Protein Crystallization: Automation, Robotization, and Miniaturization

Naomi E. Chayen

CONTENTS

ABSTRACT

Protein crystallization has gained a new strategic and commercial relevance in structural proteomics, which aims to determine the structures of thousands of proteins. X-ray crystallography plays a major role in this endeavour, but crystallography is totally dependent on the availability of suitable crystals. Automation and high throughput is essential for dealing with the overwhelming numbers of proteins that are handled. There have been major advances in the automation of protein preparation and of the X-ray analysis once diffraction quality crystals are available. However, these advances have not yet been matched by equally good methods for the crystallization process itself. Producing high quality crystals has always been the bottleneck to structure determination, and with the advent of structural proteomics this problem is becoming increasingly acute. The task of overcoming the bottleneck to produce suitable crystals is tackled by two approaches. The first approach relies on empirical techniques that are based mainly on trial and error while the second approach involves attempts to understand the fundamental principles of the crystallization process and to apply this knowledge to design methodology for obtaining high-quality crystals. This chapter highlights the advances in experimental methods that have been made to both approaches, focusing on recent progress in relation to miniaturization and automation for structural proteomics.

3.1 BACKGROUND

Completion of the Human Genome Project offers the potential of identifying a host of genetic disorders and the hope to design therapies for treating them. The genes encode proteins that are the targets for drugs. The function of these proteins is determined by their three-dimensional structure, thus a detailed understanding of protein structure is essential to develop therapeutic treatments and to engineer proteins with improved properties. Producing high quality crystals has always been the bottleneck to structure determination, and it is still not understood why some proteins crystallize with ease while others stubbornly refuse to produce suitable crystals [1].

Due to the important role of structural genomics/proteomics, the pressure to produce crystals is becoming greater than ever. As a result, the science of crystallization is gathering a new momentum and becoming a rapidly developing field. This is evidenced by the overwhelming growth in attendance at crystallization conferences, a high demand for practical courses in crystal growth, and the increasing numbers of commercial companies selling crystallization kits and tools [2].

3.1.1 THE SEARCH FOR CRYSTALLIZATION CONDITIONS (SCREENING)

Searching for crystallization conditions of a new protein has been compared with looking for a needle in a haystack. There have never been any set rules or recipes as to where to start off when attempting to crystallize a new protein, hence efficient methods were required to point the experimenter in the right direction. There is generally no indication that one is close to crystallization conditions until a crystalline precipitant or the first crystals are obtained. A major aid to get started is by using multi-factorial screens, in other words by exposing the protein to be crystallized to numerous different crystallization agents. Once a lead is obtained as to which conditions may be suitable for crystal growth, the conditions can generally be fine-tuned by making variations to the parameters (precipitant, pH, temperature, etc.) involved.

The idea of screening was introduced in the late seventies [3] but did not become popular because the numerous experiments that were necessary for screening required repetitive pipetting, which was laborious, time consuming, and boring [4]. Screening procedures took on a new dimension in the 1990s with a surge in development of automatic means of dispensing crystallization trials [e.g., 5–8]. However, most of the robots were not widely used, and for several years thereafter progress in automation of crystallization did not advance much further. With the appearance of structural proteomics, automation has become vital to progress, and in the last four years automation has become increasingly popular, attracting large investment worldwide. As a result, sophisticated apparatus for conducting automatic high throughput screening of crystallization trials using sub-microlitre quantities of sample have been developed [e.g., 9–14].

The past four years have seen some of the greatest achievements in the field of protein crystallization. The ability to dispense trials consisting of nanolitre volumes in a high throughput mode has cut the time of setting up experiments from weeks to minutes, a scenario that was unimaginable a few years ago [1]. Even more incredible is the revelation that diffracting crystals can be produced from protein samples in volumes as small as 5 to 20 nanolitre [10,13].

3.2 AUTOMATION AND MINIATURISATION OF SCREENING PROCEDURES

Since many proteins of interest are often available in limited supply, there was demand for techniques that would rapidly obtain as much information as possible on a protein while using minimum amounts of material.

3.2.1 THE MICROBATCH METHOD

The first semi-high-throughput automated system to dispense crystallization trials of less than 1 µl was designed in 1990 to deliver batch trials under oil [8]. The method was named *microbatch* to define a micro-scale batch experiment. Microbatch trials consisting typically of 0.2 to 2 µl drops of a mixture of protein and crystallizing agents are generated by computer controlled micro-dispensers (IMPAX and ORYX, Douglas Instruments) and are dispensed and incubated under paraffin oil to prevent evaporation of these small drops. These systems have two modes of action: one is used to screen numerous potential crystallization conditions and the other is used for fine-tuning the most promising screening conditions by changing concentrations and pH in small steps [8,15]. Batch/microbatch is mechanically the simplest method of crystallization, and therefore it lends itself easily to automation and high throughput. Indeed, the microbatch method has recently been adapted by several laboratories and Genomics Consortia for high-throughput screening experiments utilising a variety of apparatus. Luft et al. [9] were the first to adapt the microbatch method to high-throughput (HTP) by using a large bank of syringes (387), which dispensed 0.4 μl volumes into 1536-well micro-assay plates. This enabled them to dispense 40,000 trials per day — a major breakthrough in 1999. The RIKEN Genomic/Proteomics High Throughput Factory at Spring-8 and several other companies and academic laboratories in the U.S. and Europe use IMPAX and ORYX for their crystallization by operating several machines simultaneously. More recently another apparatus using ink jet technology has been designed for delivering microbatch trials of 15 nanolitre volumes. It has a throughput of 500 experiments per hour [16]. The apparatuses described above are in continuous use and have produced new crystals. Other devices for applying the microbatch technique have also been developed [e.g., 17,18].

3.2.1.1 The Effect of Different Oils

In a batch experiment, the protein to be crystallized and the crystallizing agents are mixed at their final concentrations at the start of the experiment, hence supersaturation is achieved upon mixing. Consequently crystals will only form if the precise conditions have been correctly chosen. This is in contrast to other crystallization methods (based on diffusion) in which the protein solution is undersaturated and conditions are changing from the time of set up until equilibrium is reached. This dynamic nature of the diffusion methods enables a self-screening process to take place (dashed lines on Figure 3.1) in each diffusion trial [19,20].

A modification of the original microbatch method provides a means of simultaneously retaining the benefits of a microbatch experiment while gaining the inherent advantage of the self-screening process of a diffusion trial [21]. This modification is based on the fact that water evaporates through different oils at different rates. Paraffin oil allows only a negligible amount of evaporation through it during the average time required for a crystallization experiment. In contrast, silicone oils allow free diffusion of water. A mixture of paraffin and silicone oils permits partial diffusion, depending on the ratio at which they are mixed [21].

It has been shown that for screening purposes it is preferable to use silicone oil or a mixture of paraffin and silicone oils [21–23]. This enables the evaporation of the crystallization trials, thereby leading to a higher number of hits and faster formation of crystals compared to trials that are set under paraffin oil. Similar effects can be obtained by varying the thickness of the paraffin oil covering the trials [20]. In the case of optimisation, where the conditions need to be known and stable, the trials must be covered by paraffin oil.

Microbatch can be used for almost all the known precipitants, buffers, and additives including detergents. The oils are compatible with the common precipitants such as salts, polyethylene glycols (PEG), and jeffamine MPD, as well as glycerol and ethanol. Microbatch, though, cannot be used for crystallization trials containing small volatile organic molecules such as dioxane, phenol, or thymol since these molecules dissolve into the oil [19,20].

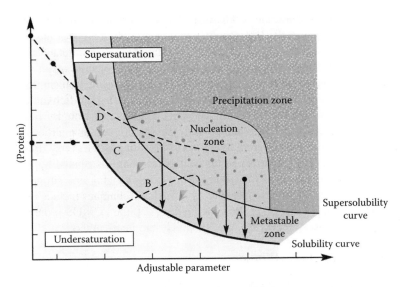

FIGURE 3.1 Schematic illustration of a protein crystallization phase diagram. The adjustable parameter can be precipitant or additive concentration, pH, temperature etc. The four major crystallization methods are represented, showing their different routes of reaching the nucleation and metastable zones, assuming the adjustable parameter is precipitant concentration. The black circles represent the starting conditions. Two alternative starting points are shown for free interface diffusion and dialysis because the undersaturated protein solution can contain either protein alone, or protein mixed with a low concentration of the precipitating agents. (A) Batch (B) Vapour diffusion (C) Dialysis (D) Free interface diffusion. The solubility is defined as the concentration of protein in the solute that is in equilibrium with crystals. The supersolubility curve is defined as the line separating conditions where spontaneous nucleation (or phase separation, precipitation) occurs from those where the crystallization solution remains clear if left undisturbed. (From Chayen, N.E., *Curr. Opionions Struct Biol.*, 14, 2004. With permission.)

3.2.2 Diffusion Techniques

3.2.2.1 Vapour Diffusion

Although microbatch is the simplest method of crystallization, it is a relatively new technique. Thus many experimenters still prefer to use vapour diffusion which has proved to be very successful over more than 40 years. Hence there have also been considerable developments in scaling down and automation of the popular vapour diffusion methods, namely hanging and sitting drops. Syrrx, who has set up millions of crystallization trials, uses a system for dispensing 20 nl to 1 microlitre sitting drops in 96 well plates with a throughput of thousands of experiments per hour. Structural Genomix has their own proprietary apparatus for generating vapour diffusion trials. A system developed at Lawrence Berkley National laboratory [24] dispenses 480 hanging drops of 20 to 100 nl per hour. The Berlin Protein Structure Factory dispenses 250 nanolitre hanging drops in 96 well plates using solenoid inkjet dispensing technology [12]. The Oxford Protein Production Facility employs the commercial Genomic Solutions' CARTESIAN range of protein crystallization workstations that deliver mostly 100 nanoliter volumes of proteins and mother liquor reagents, also in 96 well plates [11]. The workstations use proprietary technology that combines high-speed actuating capability of a micro-solenoid inkjet valve with the high-resolution displacement of a syringe pump. By synchronising the valve, the pump, and x-y positioning, one can dispense hundreds of different conditions of crystallization within minutes [25]. The ORYX system of Douglas Instruments performs vapour diffusion trials (mainly sitting drops) as well as microbatch, while the Tecan and the Gilson Cyberlab robots focus on sitting and hanging drop set-ups of 100 nanolitre — 1 microlitre, and the Mosquito robot (distributed by Molecular Dimensions Ltd.) dispenses 20 to 100 nl drops using hanging and sitting drops as well as microbatch. Other systems applying diffusion techniques include the Robins Hydra Plus One microdispenser [26], the CyBio [27], the PCMOS protein crystallization microfluidic systems [28], the Rhombix, and many others.

3.2.2.2 Free Interface Diffusion

The free interface diffusion method (FID) in which protein and precipitant solutions are carefully superimposed and left to slowly mix diffusively, has previously been underused due to handling difficulties and the requirement for large quantities of sample. In the past two years this method has experienced a revolutionary revival with the design of new ways of conducting FID trials. Experiments can now be performed in an X-ray capillary using 1 µl of sample [29,30]. In some cases, the crystals need not be removed from the capillary at any stage, since cryo protection, soaking with heavy atoms, flash-cooling, and data collection can in principle all be performed in the capillary [31]. A crystallization cassette has also been designed for holding many capillaries that can be put into contact with various precipitant solutions for the use of this method within a high-throughput environment [31].

A breakthrough in the miniaturisation and automation of the FID method has been made with the development by Hansen et al. [10] who designed a microfluidic chip for rapid screening of protein crystallization conditions. The chip is comprised of a multi-layer silicon elastomer and has 480 valves operated by pressure. The valves are formed at the intersection of two channels separated by a thin membrane. When pressure is applied to the top channel it collapses the membrane into the lower channel. When the pressure is removed, the membrane springs open. 144 trials are performed simultaneously using 3 µl of protein sample, which gets distributed throughout the chip (by the action of the valves). Each trial contains 5 to 20 nanoliters of protein sample. The authors report that this device has produced diffracting crystals in volumes containing 5 nanolitre of protein solution, including new crystals that were not detected by other crystallization methods. Crystals can be directly extracted from the chip and placed into the X-ray beam [10]. Figure 3.2 shows an example of a protein crystallized in Fluidigm's next generation 4 × 96 screening chip that consumes only 4.8 µl for 384 experiments. The figure demonstrates the classic crystallization pattern of free interface diffusion where precipitate, crystals of various sizes, and clear solution are simultaneously present along the gradient. It is interesting to note that until recently, the free interface diffusion method has been

FIGURE 3.2 Crystals of an AAA ATPase illustrating the classic crystallization pattern of free interface diffusion. Crystals are shown in a TOPAZ 4.96 screening chip as imaged on the AutoInspeX Workstation, courtesy of Fluidigm Corporation, South San Francisco, CA. The area of the protein channel is $700 \times 100 \times 10$ microns. Protein was run at a concentration of 20 mg/ml and arrayed against a TOPAZ OptiMix 1 reagent screen. (Protein is courtesy of J. Davies and W. Weis, Stanford University, Palo Alto, CA.)

used more in microgravity than on Earth, because of difficulties in setting up trials due to gravity on earth. The new experimental designs, and especially the chip, overcome these difficulties while at the same time enabling trials to be performed automatically using minute quantities of sample.

New commercial crystallization robots applying all methods of crystallization are continuously appearing in the marketplace.

It is interesting to note that crystals form faster in nano drops compared with those formed in microlitre drops. Moreover, often the nanolitre scale screens have identified crystallization conditions that were not apparent in larger drops [10,13,24].

3.3 IMAGING AND MONITORING OF CRYSTALLIZATION TRIALS

Dispensing trials with such amazing speed and ease has raised the issue of observation and monitoring of the vast numbers of trials. Major effort is currently being invested in designing image processing equipment for automated follow-up and analysis of the results [e.g., 31–35]. The ideal system is yet to be designed, but progress towards this is being made in great strides. Full automation of the visualisation and monitoring of trials is expected to be the next major breakthrough in the field of crystallization.

3.4 GENOMICS/PROTEOMICS PROJECTS: CURRENT STATE OF THE ART

In spite of the impressive advances described above, the crystallization problem has not been solved. High throughput has not led to high output.

Table 3.1 shows the sum of results from 17 structural genomics/proteomics projects worldwide as of 1 June 2004, demonstrating the success rate of getting from cloned protein to structure determination (the numbers for each individual project vary as they are dealing with different genomes and have started at different times). The results demonstrate that out of 28,855 clones and 17,178 expressed proteins, 5432 proteins were soluble and 5998 were purified. Of these, 1086 good crystals were produced leading to the determination of 721 structures at that time. It transpires that over half of the trials that were set up from pure soluble proteins

TABLE 3.1
Summary of All Major Structural Genomics Projects 1 June 2004

Cloned	28,855
Expressed	17,178
Soluble	5,432
Purified	5,998
Diffracting crystals	1,086
Structure defined	721

Source: From http://targetdb.pdb.org

had produced crystals, but only about half of those crystals were diffracting ones. It is evident that the largest failure rate in the Structural Genomics process is in the step between making purified protein and obtaining *diffraction quality* crystals. It was initially expected that if proteins could be made soluble and highly pure, there would be no problem in getting them to crystallize. However, this is not the case. The results are indicating that even when proteins can be expressed and rendered highly pure and soluble, this still does not guarantee a yield of useful crystals. Some proteins (referred to as low-hanging fruit) do crystallize during the initial screening stage, but many trials are yielding micro crystals or low-ordered ones. The problem is how to convert those into useful analysable crystals. It is evident that screening procedures are not sufficient to produce the expected output, and further effort is required to find crystallization strategies that will deliver the desired results [36].

The conversion of crystals into useful ones requires the design of optimisation techniques that go beyond the usual fine tuning of the initial conditions. In order to achieve this, one must move away from the approach that relies on trial and error, and make attempts to control the crystallization process. Such techniques require intellectual input and do not readily lend themselves to automation. To date most optimisation trials (other than merely changing the conditions around those found by screening) were done manually using selected individual proteins. In order to become useful to the structural proteomics effort, there is an urgent need to miniaturise and automate these methods and adapt them to cope with the vast numbers of leads resulting from the screening procedures [37].

The next section describes a number of ways to achieve this.

3.5 AUTOMATION OF OPTIMISATION EXPERIMENTS BASED ON THE FUNDAMENTAL PRINCIPLES OF CRYSTALLIZATION

3.5.1 UTILISATION OF CRYSTALLIZATION PHASE DIAGRAMS

Crystallization is a phase transition phenomenon. Crystals grow from an aqueous protein solution when the solution is brought into supersaturation [38,39]. The solution is brought to a supersaturated state by varying factors such as the concentrations of precipitant, protein, additives, and/or pH, temperature etc [40–42].

A phase diagram for crystallization shows which state (liquid, crystalline, or amorphous solid [precipitate]) is stable under a variety of crystallization parameters.

It provides a method for quantifying the influence of the parameters such as the concentrations of protein, precipitant(s), additive(s), pH, and temperature on the production of crystals. Hence phase diagrams form the basis for the design of crystal growth conditions [4,38–42].

Crystallization proceeds in two phases, nucleation and growth. A prerequisite for crystallization is the formation of nuclei around which the crystal will grow. Nucleation requires a higher degree of supersaturation than for growth [40–44]. Once the nucleus has formed, growth follows spontaneously.

Figure 3.1 illustrates a typical crystallization phase diagram consisting of four zones representing different degrees of supersaturation: a zone of high supersaturation where the protein will precipitate; a zone of moderate supersaturation where spontaneous nucleation will take place; the metastable zone (just below the nucleation zone) of lower supersaturation where crystals are stable and may grow but no further nucleation will take place (the conditions in this region contain the best conditions for growth of well-ordered crystals); a zone of undersaturation where the protein is fully dissolved and will never crystallize [1].

In an ideal experiment, once nuclei have formed, the concentration of protein in the solute will drop, thereby leading the system into the metastable zone where growth should occur, without the formation of further nuclei [38–44]. Unfortunately the ideal experiment does not often occur, and there is a need to manipulate the system in order to direct it in the way that the experimenter would like it to proceed. The aim is to control crystal growth by separating the phases of nucleation and growth, i.e., to start the process at conditions that induce nucleation and thereafter transfer the nuclei to metastable conditions.

Although phase diagrams offer the basis for growing protein crystals in a rational way, they are not often employed because they are time consuming to produce and require large amounts of protein sample. Accurate quantitative phase diagrams require numerous solubility measurements, (solubility being defined as the concentration of protein in the solute which is in equilibrium with crystals) and reaching this state of equilibrium can take several months [38] because macromolecules diffuse slowly.

The area of conditions called the metastable zone lies between the solubility and supersolubility curves on the crystallization phase diagram. The supersolubility curve is defined as the line separating conditions where spontaneous nucleation (or phase separation, precipitation) occurs from those where the crystallization solution remains clear if left undisturbed.

The supersolubility curve is less well defined than the solubility curve, but experimentally, it is found to a reasonable approximation, much more easily. It has been shown that for practical purposes, it is sufficient to obtain the supersolubility curve. To construct it, one must set up many crystallization trials, varying at least two parameters (e.g., protein concentration vs. precipitant concentration, protein concentration vs. pH, etc.) and plot their outcomes on a two-dimensional parameter grid. The supersolubility curve can be obtained rapidly using robots and can aid in separation of nucleation and growth using seeding and other means. A diagram containing the supersolubility curve (but not the solubility curve) is called a working phase diagram [43,45].

An example of a working phase diagram that was generated for the purpose of seeding was constructed applying the automated microbatch technique to dispense 6 protein concentration values vs. 24 different precipitant concentrations [46]. Conditions for microseeding were predicted from that phase diagram and subsequent to seeding, crystals of a bacterial enzyme diffracting to 2Å were grown reproducibly. Experiments were performed in 2 μl drops; these days such experiments could easily be performed using smaller volume drops [43,46]. Similar phase diagrams were constructed automatically in order to find the metastable conditions for successfully inserting specially designed external nucleants [47].

3.5.2 DYNAMIC SEPARATION OF NUCLEATION AND GROWTH

Seeding is the most widely used method for separation of nucleation and growth, and it has proved to be very successful [48]. However, seeding manoeuvres are not readily amenable to automation since they often involve handling of fragile crystals. Methods that would bypass such difficulties were sought.

An easier way of conducting trials with a similar outcome to seeding is by dilution. This can be performed either in vapour diffusion or in microbatch; however it is much simpler to automate the process in microbatch. Moreover, the results are more reliable and reproducible in a batch system where the volume and composition of a trial drop are known and remain constant [20].

The microbatch technique was used to establish a working phase diagram for the enzyme carboxypeptidase G_2 [49]. The concentrations of the protein and precipitant were varied, while pH and temperature were kept constant. The conditions for nucleation (i.e., conditions that would produce crystals if the trials were left undisturbed) were found by means of an automated system, and 2 μl drops were set up as microbatch trials under these conditions. At various time intervals after set-up of the experiments, the robot was programmed to automatically insert the dispensing tip into the drops and add buffer or protein solution, thereby diluting the trials. Single diffracting crystals were routinely attained, equivalent to the best, very rarely obtained without employing the dilution procedure [43,49].

Dilution in vapour diffusion has so far been achieved manually with 0.7 – 1μl drops. In the case of hanging drops, the coverslips holding the drops were incubated for some time over reservoir solutions that normally give many small crystals. After a given time (selected by reference to when the first crystals were seen in the initial screens), the coverslips were transferred over reservoirs with lower precipitant concentrations that would normally yield clear drops. The working phase diagrams that were required in order to obtain the information as to which conditions the drops should be transferred, were generated using a robot [45]. This technique has produced significant improvement in crystal order where other techniques had failed [e.g., 45,50]. Figure 3.3 (a) shows an example of a working phase diagram with the transfers indicated by the arrows. The resulting c-phycocyanin crystals (Figure 3.3 (c)) reproducibly diffracted to 1.45 Å, the highest resolution ever obtained for crystals of this protein [51]. The transfer procedure has not yet been automated but can easily be automated by using robots that dispense hanging drop trials.

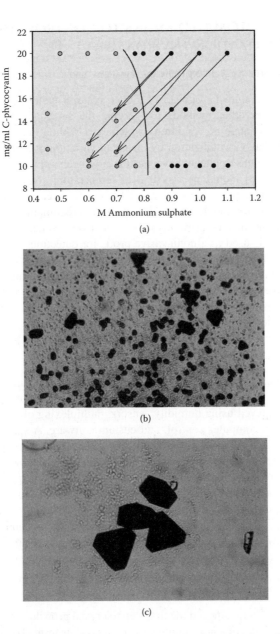

FIGURE 3.3 Crystallization of c-phycocyanin by dynamic separation of nucleation and growth. (a) The working phase diagram for c-phycocyanin. The arrows correspond to transfer from spontaneous nucleation conditions (black circles) to metastable conditions (open circles). (b) Typical crystals grown in spontaneous nucleation conditions. (c) Crystals diffracting to 1.45 Å obtained by transfers shown in (a). Size of largest crystal 0.15 × 0.25 × 0.25 mm. The photographs were taken at the same magnification. (From Chayen, N.E., *Curr. Opinions Struct Biol.*, 14, 2004. With permission.)

3.6 OTHER AUTOMATED MEANS TO CONTROL THE CRYSTALLIZATION ENVIRONMENT

3.6.1 SOLUBILITY AS A FUNCTION OF TEMPERATURE

The use of temperature is a powerful tool to control nucleation and growth. It is advantageous over other methods for varying the supersaturation because using temperature the change of conditions is achieved without adding anything to the crystallization trial. Crystallization by means of variation of temperature and application of temperature gradients have successfully been applied to protein crystallization [e.g., 52–53] including a system of temperature controlled micro reactors [54]. Most of these procedures have not been automated save for a system that contains miniature thermoelectric devices under microcomputer control that change temperature as needed to grow crystals of a given protein. A static laser light scattering probe is used as a noninvasive probe for detection of aggregation events signalling when the temperature needs to be altered. A control/follower configuration has been used to demonstrate that information obtained by monitoring one chamber can be used to effect similar results in the remaining chambers [55]. Since not all proteins are sensitive to temperature change, this method only applies to a limited category of proteins.

3.6.2 CONTROL OF EVAPORATION

Vapour-diffusion systems have been constructed that can simultaneously vary the evaporation rate for up to 40 different experiments by means of an evaporative gas flow. This is achieved using humidity sensors, multiplexing of several chambers to one sensor, microcomputer control, and custom software. A static laser light scattering sensor can be used to detect aggregation events and trigger a change in the evaporation rate for a growth solution. This approach has shown that varying the evaporation rate of the growth solutions results in a smaller population of larger crystals than obtained without dynamic response [55].

In the case of microbatch, evaporation can be controlled by setting up the trials under a very thin layer of paraffin oil, which allows evaporation and hence nucleation to take place by concentration of the drop. Adding more oil to the crystallization plate after some time will seal the trials and arrest evaporation [37].

3.6.3 HETEROGENEOUS NUCLEATION

Nucleation is a prerequisite and the first step to crystal growth, yet excess nucleation causes the production of a large number of small crystals instead of a smaller number of useful ones [40–42]. The ability to control the nucleation stage of crystallization could restrict the number of nuclei formed in a sample, thereby providing only a few nucleation centres that should grow into single, ordered crystals [56].

Heterogeneous nucleation can be caused by a host of factors (Table 3.2) including a solid material in the crystallization medium on which the growing crystal forms. This may happen on the wall of the crystallization vessel [57–59], on a crystalline surface [60], on a foreign particle [e.g., 47,61,62] or on a biological

---

TABLE 3.2
Factors Affecting Nucleation

• Seeding
• Epitaxy
• Charged surfaces
• Magnetic and electric fields
• Mechanical means — e.g., pressure, vibration
• Concentration of precipitant, protein, additives
• Temperature, pH
• Container walls
• Kinetics of the crystallization process

---

contaminant [63]. Not all, but some of those parameters can be controlled if the experiment is designed with care.

A common source of uncontrolled heterogeneous nucleation are solid surfaces such as the walls of the container that holds the crystallization trial [57]. Excess nucleation caused by contact with the container wall has previously been avoided by creating a container-less environment. This was achieved by floating the crystallization drop between the layers of two oils. The bottom layer is a high-density fluorinated silicon fluid (1.2 g cm^{-3}), and the top layer is a standard low-density (0.84 g cm^{-3}) silicon fluid that is placed above the high-density oil layer. The two oils are immiscible and the aqueous crystallization drop floats at the interface, thereby not touching the container wall [58,59]. The application of this technique results in reduction of the numbers of crystals grown between the oils by as much as tenfold with a size increase of up to three times of those grown in contact with the container [58]. Nevertheless, in the context of structural proteomics this procedure has a number of major disadvantages:

- It is not automated.
- A minimum drop size of 2 µl is required.
- It is difficult to harvest the crystals.

A modification of the original method has overcome these drawbacks: Automation, miniaturisation, and high-throughput can now be achieved by replacing the lower layer (previously consisting of the high-density oil) with a much cheaper hydrophobic surface of high-vacuum silicone grease or Vaseline [37]. The grease covers the bottom of the crystallization plate onto which the trial drops (0.3 µl) are automatically dispensed using a robot. The greased surface provides a stable interface to the upper layer, which consists of a low-density oil, usually paraffin (0.84 g/cm^3). This prevents crystals from migrating to the walls, making them much easier to harvest. The crystals are lifted directly out of the drop with a loop or a spatula. Commercial container-less kits are also available from Molecular Dimensions, U.K. and Hampton Research, U.S.A.

Other manual methods to influence nucleation, such as ultra filtration and addition of external nucleants in a controlled manner, have not yet been automated.

3.6.4 CRYSTALLIZATION IN GELS

Crystallization in gels presents an additional means to promote or inhibit nucleation, depending on the type of gel applied [64–66]. It has also been shown that the presence of gels can improve crystal quality by reducing sedimentation and convection, in some way mimicking a microgravity environment [64–66]. Crystallization in gels can be performed in batch, microdialysis, vapour diffusion, and FID [64, 67–69] as well as in the gel acupuncture method where the crystals grow inside an X-ray capillary that is inserted into a gel directly [70] or via an oil layer [71]. The application of gels to the crystallization of proteins has been pursued for more than 15 years, yet the method remains underused. This may be due to the relatively complicated procedures required when applying gels to crystallization trials and also to the large quantities (mostly 10 µl) of sample needed. The last two years have seen major improvements in the use of gels by way of miniaturization and automation. The Granada Crystallization Box [26] enables the growth of crystals in 1 µl volumes of gel inside capillaries, thus combining the advantages of growth in gel with those of the FID method. An even simpler way is to set up gelled trials in microbatch, which, for the first time, has enabled the automatic dispensing of sub-microlitre gelled drops in a high throughput mode [37]. The experiments are performed using a computer controlled dispensing system in which precipitant, buffer, protein, and additives are put into different syringes. By placing a gel solution while it is still in liquid form, into one of the syringes, it is possible to automatically dispense microbatch trials that form the gel/crystallization mixtures in final volumes of 0.3 to 2 µl. This can be achieved during a comparable timescale and with the same ease as conventional automated microbatch trials. Comparison of crystals grown in gels with those grown in standard trials show a significant improvement in crystal order. Harvesting crystals from the gelled drops is performed in the same way as from standard microbatch trials.

3.7   AUTOMATION AND MINIATURISATION OF THE CRYSTALLIZATION OF MEMBRANE PROTEINS

Advances in methodology for miniaturising and automating the crystallization of membrane proteins involve the lipidic cubic phase (*in cubo*) [72,73] and crystallization of membrane proteins under oil [74].

3.7.1 CRYSTALLIZATION *IN CUBO*

Widespread use of the *in cubo* method has been severely hindered by its tediousness and the large amounts of protein required. A means to alleviate these problems has been developed [72,73] for setting up *in cubo* trials using standard microwell plates and ca 200 nl of lipidic cubic phase compared to microlitre quantities used previously. Crystallization set ups are prepared using syringes that are assembled into a semi-automatic dispenser and are dispensed into microwell plates such as Terazaki plates. This overcomes the awkward manual pipetting procedure. The plates are sealed with clear transparent tape and are stored at different temperatures. Diffracting crystals of bacteriorhodopsin were obtained using this procedure.

3.7.2 CRYSTALLIZATION UNDER OIL

The first automation and miniaturisation of the crystallization of membrane proteins was achieved by setting the trials in microbatch under oil. The idea of crystallizing membrane proteins under oil had first been received with scepticism due to doubts regarding the suitability of an oil-based method for crystallizing lypophilic compounds. However, in the last six years an increasing number of proteins with a variety of different detergents have been crystallized in microbatch under oil, in some cases yielding better or different crystal habits to those grown by standard crystallization methods [74]. Examples of these are: a chlorophyll binding protein 43 (CP43) of the photosystem II complex from spinach [75], photosystem I [Nechushtai personal communication] and Light harvesting complex II from Rhodopseudomonas acidophila [McDermott personal communication]. Some crystals, such as F_1c_{10} of *ATP Synthase* [76] and the native outer membrane phospholipase A (OMPLA) [77] could only be obtained in microbatch and not by vapour diffusion methods. In all the above cases both screening for initial conditions, and optimisation thereafter, were performed using a robot, consuming far less protein sample than that required for manual trials. Miniaturisation and automation of membrane protein trials applying vapour diffusion and FID are currently underway.

3.8 KEY ISSUES AND NEXT STEPS

Structural proteomics involves many steps in order to reach from gene to structure. Crystallization is a crucial rate-limiting step in this chain of events. It is becoming increasingly evident that the current high-throughput procedures for crystallizing proteins do not always produce the expected output of high-quality crystals required for structure determination by X-ray crystallography. This is because almost all of the effort to date has been channelled into one aspect, that of automated screening procedures. Screening is of utmost importance. It is necessary in order to find the initial conditions under which a given protein is likely to crystallize, but screening is not enough. It needs to be complemented by an equally important procedure in crystal production, namely crystal optimisation. The enormous number of proteins that need to be dealt with indicate that, like screening, optimisation must be adapted to high throughput. Without this we will soon run out of resources and be left with a backlog of useless microcrystals.

In the rush towards structural genomics/proteomics, optimisation techniques have been somewhat neglected, mainly because it was mistakenly hoped that large-scale screening would alone produce the desired results. In addition, optimisation has relied on particular individual methods that are often difficult to automate and to adapt to high throughput. The last four years have seen enormous progress by way of automating and miniaturising crystallization trials. Both the quantities of sample and the time scale of conducting the experiments have been reduced by several orders of magnitude. The following phase, which is the image-capture and analysis of the crystallization drops, is currently progressing with impressive speed. The next crucial step is to simplify and automate the more complicated optimisation procedures and adapt those to high-throughput experiments. This is not an easy task, which is only just beginning to advance. Several examples have been given in this chapter, but we still have a long way to go. Once well underway, the combination of automated screening with further development of automated crystal optimisation methods promises to increase the output of structural proteomics and to make a significant impact on the scientific standards in this field.

REFERENCES

1. Chayen, N.E., Turning protein crystallization from an art into a science, *Curr. Opinions Struct Biol.*, 14, 577–583, 2004.
2. Chayen, N.E., Crystallization of Proteins, *Z. Kristallog.*, 217, 292–293, 2002.
3. Carter, C.W., Jr. and Carter, C.W., Protein crystallization using incomplete factorial experiments, *J. Biol. Chem.*, 254, 12219–12223, 1979.
4. Chayen, N.E., Boggon, T. J., Cassetta, A., Deacon, A., Gleichmann, T., Habash, J., Harrop, S.J., Helliwell, J.R., Nieh, Y.P., Peterson, M.R., Raftery, J., Snell, E.H., Hadener, A., Niemann, A.C., Siddons, D.P., Stojanoff, V., Thompson, A.W., Ursby, T., and Wulff, M., Trends and challenges in experimental macromolecular crystallography, *Quart. Rev. Biophys.*, 29, 3, 227–278, 1996.
5. Rubin, B., Talatous, J., and Larson, D., Minimal intervention robotic protein crystallization, *J. Cryst. Growth*, 110, 156–163, 1991.
6. Oldfield, T.J., Ceska, T.A. and Brady, R.L., A flexible approach to automated protein crystallization, *J. Appl. Cryst.*, 24, 255–260, 1991.
7. Sadaoui, N., Janin, J., and Lewit-Bently, A., TAOS: an automatic system for protein crystallization, *J. Appl. Cryst.*, 27, 622–626, 1994.
8. Chayen, N., Shaw Stewart, P.D., Maeder, D.L., and Blow, D.M., An automated system for microbatch protein crystallization and screening, *J. Appl. Cryst.*, 23, 297–302, 1990.
9. Luft, J.R., Wolfley, J., Jurisica, I., Glasgow, J., Fortier, S., and DeTitta, G.T., Macromolecular crystallization in a high throughput laboratory: the search phase, *J. Cryst. Growth*, 232, 591–595, 2001.
10. Hansen, C.L. Skordalakes. E., Berger, J.M., and Quak, S.R., A robust and scalable microfluidic metering method that allows protein crystal growth by free interface diffusion, *PNAS 99*, 16531–16536, 2002.
11. Walter, T.S., Brown, D.J., Pickford, M., Owens, R.J., Stuart, D.I., and Harlos, K., A procedure for setting up high-throughput nanolitre crystallization experiments. I. Protocol design and validation, *J. Appl. Cryst.* 36, 308–314, 2003.
12. Mueller; U., Nyarsik, L., Horn, M., Rauth, H., Przewieslik, T., Saenger, W., Lehrach, H., and Eickhoff, H., Development of a technology for automation and miniaturization of protein crystallization, *J. Biotechnol.*, 85, 7–14, 2001.
13. Stevens, R., High throughput protein crystallization, *Curr. Opn. Struct. Biol.*, 10, 558–563, 2000.
14. Rupp, B., Maximum-likelihood crystallization, *J. Struct. Biol.*, 142, 162–169, 2003.
15. Chayen, N., Shaw Stewart, P.D., and Blow, D.M., Microbatch crystallization under oil — a new technique allowing many small-volume crystallization trials, *J. Crystal Growth*, 122, 176–180, 1992.

16. DeLucas, L.J., Bray, T.L., Nagy, L., McCombs, D., Chernov, N., Hamrick, D., Cosenza, L., Belgovskiy, A., Stoops, B., and Chait, A., Efficient protein crystallization, *J. Struct. Biol.*, 142, 188–206, 2003.

17. Jaruez-Martinez, G., Steinmann, P., Roszak, A.W., Isaacs, N.W., and Cooper, J.M., High-throughput screens for postgenomics: studies of protein crystallization using microsystems technology, *Analytical Chemistry*, 74, 3505–3510, 2002.

18. Bodenstaff, E.R., Hoedemaeker, F.J., Kuil, M.E., de Vrind, H.P.M. and Abrahams, J.P., The prospects of protein nanocrystallography, *Acta Cryst.*, D58, 1901–1906, 2002.

19. Chayen, N.E., The role of oil in macromolecular crystallization, *Structure*, 5, 1269–1274, 1997.

20. Chayen, N.E., Comparative Studies of Protein Crystallization by Vapour Diffusion and Microbatch, *Acta Cryst.*, D54, 8–15, 1998.

21. D'Arcy, A., Elmore, C., Stihle, M., and Johnston, J.E., A Novel Approach to Crystallising Proteins Under Oil, *J. Cryst.Growth*, 168, 175–180, 1996.

22. D'Arcy, A., MacSweeney, A., and Habera, A., Modified microbatch and seeding in protein crystallization experiments, *J. Synchrotron Rad.*, 11, 24–26, 2004.

23. D'Arcy, A., MacSweeny, A., Stihle, M., and Haber, A., The advantages of using a modified microbatch method for rapid screening of protein crystallization conditions, *Acta Cryst.*, D59, 396–399, 2003.

24. Santarsiero, B.D., Yegian, D.T., Lee, C.C., Spraggon, G., Gu, J., Scheibe, D., Uber, D.C., Cornell, E.W., Nordmeyer, R.A., Kolbe, W.F., Jin, J., Jones, A.L, Jaklevic, J.M., Schultz, P.G. and Stevens, R.C., An approach to rapid protein crystallization using nanodroplets, *J. Appl. Cryst.*, 35, 278–281, 2002.

25. Brown, J., Walter, T.S., Carter, L., Abrescia N.G.A., Aricescu, A.R., Batuwangala, T.D., Bird, L.E., Brown, N., Chamberlain, P.P., Davis, S.J., Dubinina, E., Endicot, J., Fennelly, J.A., Gilbert, R.J.C., Harkiolaki, M., Hon, W-C., Kimberley, F., Love, C.A., Mancini, E.J., Manso-Sancho, R., Nichols, C.E., Robinson, R.A., Sutton, G.C., Schueller, N., Sleeman, M.C., Stewart-Jones, G.B., Vuong, M., Welburn, J., Zhang, Z., Stammers, D.K., Owens, R.J., Jones, E.Y., Harlos, K., Stuart, D.I., A procedure for setting up high-throughput nano-litre crystallization experiments. II. Crystallization results, *J. Appl. Cryst.*, 36, 315–318, 2003.

26. The Robbins® First Edition Hydra®-Plus-One Microdispenser for rapid setup of protein crystallization plates, *Acta Cryst.*, D58, 1760, 2002.

27. Proven automation technology transfer to a new application, *Acta Cryst.*, D58, 1760, 2002.

28. PCMOS — Protein Crystallization Microfluidic Systems, *Acta Cryst.*, D58, 1763, 2002.

29. García-Ruíz J.M., Gonzalez-Ramirez, L.A., Gavira, J.A., Otálora, F., Granada Crystallization Box: a new device for protein crystallization by counter-diffusion techniques, *Acta Cryst.*, D 58, 1638–1642, 2002.

30. Tanaka, H., Inaka, K., Sugiyama, S., Takahashi, S., Sano, S., Sato, M., and Yoshitomi, S., A simplified counter diffusion method combined with a 1D simulation program for optimizing crystallization conditions, *J. Synchrotron Rad.*, 11, 45–48, 2004.

31. Ng, J.D., Gavira, J.A., García-Ruíz, J.M., Protein crystallization by capillary counterdiffusion for applied crystallographic structure determination, *J. Struct. Biol.*,142, 218–231, 2003.

32. Luft, J.R., Collins, R.J., Fehrman, N.A., Lauricella, A.M., Veatch, C.K., DeTitta, G.T., A deliberate approach to screening for initial crystallization conditions of biological macromolecules, *J. Struct. Biol.*, 232, 170–179, 2003.

33. Elkin, C.D. and Hogle, J.M., *J. Cryst. Growth,* 232, 563–572, 2001.
34. Jurisica, I., Rogers, P., Glasgow, J.I., Collins, R.J., Wolfley, J.R., Luft, J.R., and DeTitta, G.T., Improving objectivity and scaldbility in protein crystallization, *IEEE Intell. Syst.,* 16, 26–34, 2001.
35. Wilson, J., Towards the automated evaluation of crystallization trials, *Acta Cryst., Section D, Bio. Cryst.,* D58, 1907–1914, 2002.
36. Chayen, N.E., Protein crystallization for genomics — throughput versus output, *J. Struct Functl. Genomics,* 4, 115–120, 2003.
37. Chayen, N.E. and Saridakis, E., Protein Crystallization for Genomics: Towards High-Throughput Optimisation Techniques, *Acta Cryst. D* 58, 921–927, 2002.
38. Ataka, M., Protein crystal growth: an approach based on phase diagram determination, *Phase Transitions,* 45, 205–219, 1993.
39. Mikol, V. and Giege, R., The physical chemistry of protein crystallization, in *Crystallization of Nucleic Acids and Proteins. A Practical Approach,* A. Ducruix and R. Giege, eds., IRL Press of Oxford University Press, Oxford, 1992, 219–239.
40. McPherson, A., *Crystallization of Biological Macromolecules,* Spring Harbor Laboratory Press, Cold Spring Harbor, 1999.
41. Ducruix, A. and Giegé, R., eds., *Crystallization of Nucleic Acids and Proteins, A Practical Approach,* 2nd ed., Oxford University Press, Oxford, 1999.
42. Bergfors, T., ed., *Protein Crystallization: Techniques, Strategies, and Tips,* International University Line, La Jolla, CA, 1999.
43. Chayen, N.E., Methods for separating nucleation and growth in protein crystallization, *Prog. Biophys. Molec. Biol.,* 88, 329–337, 2005.
44. Ries-Kautt, M. and Ducruix, A., (1992). Phase diagrams, in *Crystallization of Nucleic Acids and Proteins. A Practical Approach,* A. Ducruix and R. Giege, eds., IRL Press of Oxford University Press, Oxford, 195–218.
45. Saridakis E., Chayen, N.E., Systematic improvement of protein crystals by determining the supersolubility curves of phase diagrams, *Biophys. J.,* 84, 1218–1222, 2003.
46. Korkhin, Y.M., Evdokimov, A. and Shaw Stewart, P.D., Crystallization of a protein by microseeding after establishing its phase diagram, *Application Note I Douglas Instruments,* www.douglas.co.uk, 1995.
47. Chayen, N.E., Saridakis, E., El-Bahar, R., and Nemirovsky, Y., Porous silicon: an effective nucleation-inducing material for protein crystallization, *J. Mol Biol.,* 312, 591–595, 2001.
48. Stura, E.A., Seeding, in *Protein Crystallization: Techniques, Strategies, and Tips,* T.M. Bergfors, ed., International University Line, La Jolla, CA, 141–153, 1999.
49. Saridakis, E.E.G., Shaw Stuart, P.D., Lloyd, L.F., and Blow, D.M., Phase diagram and dilution experiments in the crystallization of carboxypeptidase G2, *Acta Cryst.,* D 50, 293–297, 1994.
50. Saridakis, E. and Chayen, N.E., Improving protein crystal quality by decoupling nucleation and growth in vapour diffusion, *Protein Sci.,* 9, 755–757, 2000.
51. Nield, J., Rizkallah, P., Barber, J., and Chayen, N.E., The 1.45 Å three-dimensional structure of C-Phycocyanin from the Thermophilic Cyanobacterium Synechococcus Elongatus, *J. Struct. Biol.,* 141, 149–155, 2003.
52. DeMattei, R.C. and Feigelson, R.S., The solubility dependence of canavalin on pH and temperature, *J. Cryst. Growth,* 110, 34–40, 1991.
53. Rosenberger, F., Howard, S.B., Sowers, J.W., and Nyce, T.A. Temperature dependence of protein solubility — determination and application to crystallization in X-ray capillaries, *J. Cryst. Growth,* 129, 1–12, 1993.

54. Berg, M., Urban, M., Dillner, U., Mühlig, P., and Mayer, G., Development and characterization of temperature-controlled microreactors for protein crystallization, *Acta Cryst.,* D58, 1643–1648, 2002.
55. Collingsworth, P.D., Bray, T.L., Christopher, G.K., Crystal growth via computer controlled vapour diffusion, *J. Cryst. Growth,* 219, 283 – 289, 2000.
56. Blow, D.M., Chayen, N.E., Lloyd, L.F., and Saridakis, E., Control of nucleation of protein crystals, *Protein Sci.,* 3, 1638–1643, 1994.
57. Yonath, A., Mussing, J., and Wittman, H.G., Parameters for crystal growth of ribosomal subunits, *J. Cell. Biochem.,* 19, 145–155, 1982.
58. Chayen, N.E., A novel technique for containerless protein crystallization, *Protein Eng.,* 9, 927–929, 1996.
59. Lorber, B. and Giege, R., Containerless protein crystallizaiton in floating drops: application to crystal growth monitoring under reduced nucleation conditions, Proc. Sixth Int. Conf. on Crystall. of Biological Macromolecules, *J. Cryst. Growth,* 168, 204–215, 1996.
60. McPherson, A. and Schlichta, P., Heterogeneous and epitaxial nucleation of protein crystals on mineral surfaces, *Science,* 239, 385–387, 1988.
61. Malkin, A.J., Cheung, J., and McPherson, A., Crystallization of satellite tobacco mosaic virus. 1. nucleation phenomena, *J. Cryst. Growth,* 126, 544–554, 1993.
62. Falini, G., Fermani, S., Conforti, G., and Ripamonti, A., Protein crystallization on chemically modified mica surfaces, *Acta Cryst.,* D58, 1649–1652, 2002.
63. Chayen, N.E., Radcliffe, J., and Blow, D.M., Control of nucleation in the crystallization of lysozyme, *Protein Sci.,* 2, 113–118, 1993.
64. Robert, M.-C., Vidal, O., Garcia-Ruiz, J.-M., and Otálora, F., Crystallization in gels and related methods, in *Crystallization of Nucleic Acids and Proteins,* Ducruix, A. and Giegé, R., eds., Oxford University Press, Oxford, 1999, 149–175.
65. Cudney, B., Patel, S., and McPherson, A., Crystallization of macromolecules in silica gels, *Acta Cryst.,* D 50, 479–483, 1994.
66. Biertümpfel, C., Basquin, J., Suck, D., and Sauter, C., (2002) Crystallization of biological macromolecules using agarose gel, *Acta Cryst.,* D58, 1657–1659, 2002.
67. Thiessen, K.J., The use of two novel methods to grow protein crystals by microdialysis and vapour diffusion in an agarose gel, *Acta Cryst.,* D 50, 491–495, 1994.
68. Bernard, Y., Degoy, S., Lefaucheux, F., and Robert, M.C., A gel-mediated feeding technique for protein crystal growth from hanging drops, *Acta Cryst.,* D 50, 504–507, 1994.
69. Sica, F., Demasi, D., Mazzarella, L., Zagari, A. Capasso, S. Pearl, L.H., D'Auria, S., Raia, C.A., and Rossi, M., Elimination of twinning in crystals of *Sulfolobus solfataricus* alcohol dehydrogenase holo-enzyme by growth in agarose gels, *Acta Cryst.,* D 50, 508–511, 1994.
70. Garcia-Ruiz, J.M., and Moreno, A., Investigations on protein crystal growth by the gel acupuncture method, *Acta Cryst.,* D 50, 484–490, 1994.
71. Moreno, A., Saridakis, E., and Chayen, N.E., Combination of oils and gels for enhancing the growth of protein crystals, *J. Appl. Cryst.,* 35, 140–142, 2002.
72. Nollert, P., From test tube to plate: a simple procedure for the rapid preparation of microcrystallization experiments using the cubic phase method, *Acta Cryst.,* D 35 637–640, 2002.
73. Caffrey, M., Membrane protein crystallization, *J. Struct. Biol.,* 108–132, 2003.
74. Chayen, N.E., Crystallization of membrane proteins in oils, in *Methods and Results in Crystallization of Membrane Proteins,* Iwata, S., ed., International University Line, La Jolla, CA, 2003, 131–139.

75. Hankamer, B., Chayen, N.E., De Las Rivas, J., and Barber, J., Photosysten II core complexes and CP-43 crystallization trials, in *First Robert Hill Symposium on Photosynthesis,* Barber, J., ed., Imperial College, London, 1992, 11–12.
76. Stock, D., Leslie, A.G.W., and Walker, J.E., Molecular Architecture of the Rotary Motor in ATP Synthase, *Science,* 286, 1700–1705, 1999.
77. Snijder, H.J., Barends, T.R.M., and Dijkstra, B.W., Chapter 16 in oils, in Methods and Results in Crystallization of Membrane Proteins, Iwata, S., ed., International University Line, La Jolla, CA, 2003, 265–278.

4 NMR Spectroscopy in Structural Genomics

Weontae Lee, Adelinda Yee, and Cheryl H. Arrowsmith

CONTENTS

4.1 INTRODUCTION

NMR spectroscopy is an alternative method to X-ray crystallography for the structure determination of small to medium sized proteins (<25 kDa) in aqueous or micellar solutions. Recent progress in computational and experimental NMR techniques has improved the efficiency of biological NMR research [1–4]. Ever improving strategies for uniform and selective biosynthetic incorporation of stable isotopes (^{15}N, ^{13}C and ^{2}H) into recombinant proteins combined with newer multidimensional (3-D and 4-D) heteronuclear NMR experiments allows one to resolve resonance overlap for larger proteins and complexes than ever before.

The application of multidimensional heteronuclear NMR experiments enables one to extend the protein size up to potentially 35 to 65 kDa for structure determination by NMR spectroscopy [4]. This size regime covers about a third to a quarter of the nonmembrane proteins in many microbial proteomes [5] and many independently folded domains of larger proteins such as signaling modules and nucleic acid binding domains [6].

In the context of structural genomics, the NMR approach has proven to be successful and economical when applied to small proteins or domains [5,7]. A primary advantage of NMR is that protein samples can be assessed for their amenability to NMR analysis in a rapid and parallel fashion. The ^{15}N-HSQC spectrum of a uniformly ^{15}N-labeled protein, which can be acquired within minutes to hours once a protein is purified (and in some cases even without purification), is a reliable indicator of whether a structure can be determined by NMR methods. This diagnostic step, early in the structure determination process, allows one to focus resources on those samples that are highly likely to produce structures even though the subsequent data analysis and structure determination steps often take longer than, for example, those of a high quality MAD diffraction data set. This attractive feature of the NMR method combined with the continuing fast pace of NMR technology development means that NMR will continue to play a useful role in structural genomics.

NMR is a preferred method for small proteins lacking methionine residues required for MAD phasing of crystallographic data. Furthermore, because of the different nature of the sample preparations for NMR and crystallography (the need for the maintenance of a stable monodisperse solution of protein as opposed to the ordered crystallization of a protein), the use of both methods allows one to obtain structural samples from a wider range of proteins than either method alone [8]. Here we review the NMR sample pipeline as implemented in our labs and discuss emerging technologies that promise to shorten the pipeline from gene to structure by NMR.

4.2   THE SAMPLE NMR PIPELINE

Our preference is to express proteins destined for NMR analysis in *E. coli* (BL21 DE3) with an N-terminal His$_6$ tag that can be cleaved (with thrombin or TEV protease) if necessary. Eubacterial and archaeal proteins are expressed directly in 2x M9 media supplemented with ZnCl$_2$, thiamine, biotin, and with ^{15}NHCl as the sole nitrogen source. We typically grow 36 individual 500 ml cultures simultaneously in incubator shakers. The ^{15}N labeled proteins are purified in parallel (18 at a time) using the batch method with a nickel NTA affinity resin. The proteins purified in this manner are typically 85 to 95% pure and are then further concentrated by ultrafiltration/dialysis into an NMR compatible buffer. Proteins are screened at high salt concentrations 450 to 500 mM (NaCl); we find that most proteins are more soluble at high salt. Although it is not unusual for some of the proteins to precipitate either during the purification process or during concentration, we are generally able to isolate enough protein (>3mg) for an NMR sample for roughly half of the eubacterial and archaeal proteins put through this process. For eukaryotic proteins, the losses due to poor expression or solubility in *E. coli* are more substantial, requiring a small scale test expression step before scale-up.

^{15}N-HSQC spectra are acquired at 25°C and classified into the following categories; Good, Promising, Poor or Unfolded (Figure 4.1). Good spectra are those that contain the expected number of peaks (roughly one per nonproline residue) and for

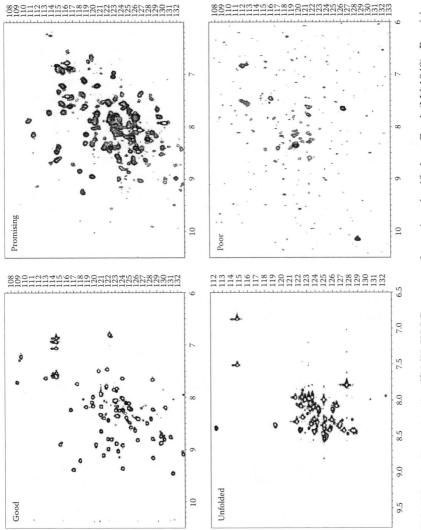

FIGURE 4.1 Representative ¹⁵N-¹H HSQC spectra of proteins classified as Good (Mth1048); Promising (Mth0445); Unfolded (Mth0042); and Poor (Mth0313). All HSQC spectra were acquired at 25°C.

which the peaks are of uniform intensity and well dispersed throughout the spectrum. Proteins with these spectral characteristics are very likely to yield high quality data for the additional 7 to 10 multi-dimensional multi-nuclear NMR spectra that are required for a structure determination. Promising spectra are those that have many but not all of the above features (for example, uniform well-dispersed peaks, but too few or too many peaks). It is sometimes possible to coax such promising samples into good samples by changing the solution conditions (for example, the salt concentration or pH). Poor spectra generally have far too few peaks, and those that are observed are usually poorly dispersed and broad. The Poor category is the single largest category of samples and likely reflects proteins that are highly aggregated and those with conformational heterogeneity. Soluble, but unfolded proteins are readily identified because the peaks are sharp and clustered in a narrow region of the ^{1}H dimension of the HSQC (7.5–8.5 ppm). The unfolded category is the smallest in our experience.

Until recently the His$_6$ tag was only removed by proteolytic cleavage for promising spectra in the hopes that this may improve the sample quality. In our experience many residues in the His$_6$ tag are not observable in the NMR spectrum and thus for most cases it does not interfere with the manual assignment of resonances. It may, however, cause difficulties for some automated assignment procedures and because of this we now recommend removing the tag if it does not adversely affect the spectrum and/or solubility.

We have tested the use of protein refolding methods to increase the number of structural samples for NMR. We found that approximately 10% of our small proteins that are insoluble when expressed in *E.coli* could be recovered in a soluble form that gave good quality NMR spectra using a Guanidinium Chloride extraction followed by refolding into an NMR compatible buffer [9]. This study also showed that for our small protein targets, the ^{15}N-HSQC spectra are very similar for those proteins purified by both native and denaturing protocols, suggesting that the 3-D structure of the refolded protein is the same as those purified by native means.

Proteins that yield a good HSQC spectrum are targeted for structure determination by NMR. For these samples, salt is titrated out in order to achieve a minimum salt concentration that would keep the protein in solution and still give good quality spectra. A new protein sample is prepared with uniform ^{13}C and ^{15}N incorporation and purified via Ni NTA chromatography and usually a subsequent ion exchange column. A full series of triple resonance experiments for sequential assignment of backbone and side chain ^{1}H, ^{15}N, and ^{13}C resonances is then collected [7] along with ^{15}N and ^{13}C edited NOESY spectra for structural restraints [10,11]. Additional data such as residual dipolar couplings [12] and scalar coupling and/or chemical-shift based dihedral angle restraints [13,14] can be readily added to increase the convergence of structure determination programs and improve the accuracy and precision of structures. If the protein is small and reasonably soluble, and the spectra have little or modest levels of spectral overlap, we have found the reduced dimensionality experiments of Szyperski and coworkers to be a useful set of experiments for sequence specific assignments [15]. The additional information encoded by the splitting of cross peaks in these spectra can facilitate the assignment process, and

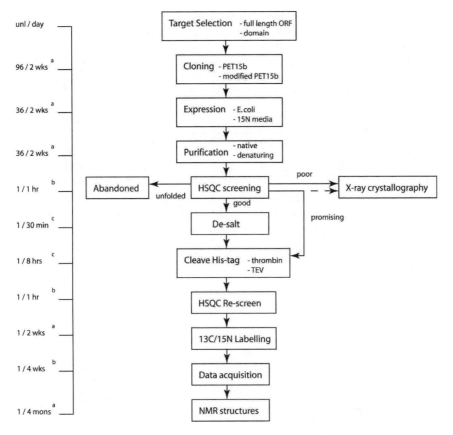

FIGURE 4.2 Schematic diagram of the sample flow in our NMR structural proteomics project. The approximate time requirement for each step is indicated. a(superscript) per person; b(superscript) per instrument; c(superscript) per batch. $^{13}C/^{15}N$ labeled samples are often prepared one sample at a time because the final NMR sample condition differs slightly from sample to sample. Data acquisition time is an average, see Yee et al. (reference 7, supplementary figures) for the details.

the differential splitting of otherwise degenerate peaks also eliminates ambiguities. Figure 4.2 shows a schematic of the protocol we use in NMR structural proteomics with the approximate time required for each step in the protocol.

4.3 RECENT DEVELOPMENT OF NMR TECHNIQUES FOR STRUCTURAL PROTEOMICS

Protein NMR spectroscopy is still a rapidly evolving field with promising developments on many fronts that are likely to improve the speed and quality of NMR data and structure determination. Here we briefly summarize those methods that will have the greatest impact on structural genomics.

4.3.1 ULTRA HIGH FIELD MAGNET

With conventional NMR equipment (500–600 MHz, conventional probes), an average NMR structure determination requires about 5 to 6 weeks of data collection for a well behaved, reasonably soluble protein of up to ~20 kDa [16]. This is due to the intrinsic low sensitivity of NMR and the need to sample to high digital resolution in the indirect dimensions of multidimensional spectra to resolve resonance overlap. To overcome the low sensitivity problem, samples are often prepared at relatively high concentration (about 1 mM), which leads to nonspecific aggregation for many proteins, severely limiting the number of proteins that can be studied by NMR. The introduction of NMR magnets with higher field strength (≥800 MHz) provides both greater sensitivity and higher resolution. For example, the detection sensitivity of a 900 MHz magnet is expected to be four ~ five times greater than at 500 MHz and the resolution is twice that at 500 MHz. The shortened experimental time increases the numbers of data sets that can be acquired per instrument, and the increased resolution will facilitate the data analysis leading to an overall increase in the rate at which structures can be determined, provided a constant supply of good quality samples can be generated.

4.3.2 CRYOGENIC/CHILLED PROBE TECHNOLOGY

The current state-of-the-art NMR probes (the part of the instrument that transmits and detects the radiofrequency signals into and from the sample) have low temperature devices designed to decrease thermal noise of the probe and related electronics, improving the signal-to-noise ratio by a factor of three to four compared to conventional probes. The improved sensitivity can be coupled with any magnetic field strength, allowing the study of many proteins with limited solubility or poor levels of expression that could not have been addressed by NMR in the past. Alternatively, for proteins that can be studied at higher concentrations (~1 mM), a complete data set for structure determination can now be collected in about a week [17]. Thus, one 600 MHz instrument could potentially generate enough data for about 50 structures of small proteins per year. This technology is also especially powerful for rapid screening of a large number of proteins and buffer conditions for the quality of their HSQC, which will be required in order to generate a constant supply of samples for the new generation of more sensitive instruments [7].

4.3.3 TRANSVERSE RELAXATION OPTIMIZATION SPECTROSCOPY (TROSY)

The original triple resonance NMR techniques developed in the early 1990s are not readily applied to structure determination of proteins with molecular weights over ~25 kDa, due to rapid transverse relaxation mechanisms that occur in large proteins. This is mainly caused by the slow tumbling motion of larger proteins in solution, resulting in signal loss, line broadening, and overall reduced sensitivity [18]. TROSY techniques select for a component of the nuclear magnetization that is less sensitive to such adverse relaxation effects during chemical shift evolution, providing much better sensitivity and line width for large proteins [19]. The

TROSY technique has already been successfully applied to larger proteins with sizes of 65 to 100 kDa [20]; however, it has been proposed that TROSY could be readily applicable to proteins with molecular weight of 300 kDa. Although the TROSY effect is most pronounced at ultra-high field strength (~1 GHz) applied to very large proteins, it is also very helpful for the analysis of medium sized proteins in the range of many drug targets (30–50kDa) and at 600–700MHz field strengths, which are more common in many NMR labs. The application of TROSY methods will be very effective in determining the structure of membrane proteins in slowly tumbling micelles [21] or for the study of oligomeric proteins with small subunit sizes [22].

4.4.4 RESIDUAL DIPOLAR COUPLING (RDC)

The traditional NMR method for structure determination of proteins is based on three-bond J coupling and NOE data, providing constraints to generate a three-dimensional fold of the protein. However, resonance assignment and collection of an adequate number of NOEs for structure generation have always been difficult due to overlap of proton resonances, especially for large proteins.

Recently, NMR experiments have been performed for proteins in slightly anisotropic environments, in which the large one-bond internuclear dipolar couplings are no longer averaged to zero. The residual couplings on the order of 10^{-3} measured in these anisotropic environments can provide the average orientation of atomic bonds within a protein relative to the molecular alignment frame [1]. Recent application of a variety of methods for weak magnetic alignment of the protein achieved with phage or bicelle media allows measurement of residual dipolar couplings (^{1}H-^{15}N and ^{1}H-^{13}C) in a wide variety of proteins [23]. These dipolar couplings provided information about the relative domain orientations of the protein by examination of the principal axes of the tensors. In addition, residual dipolar coupling information can serve as a conformational constraint to generate low-resolution structures for the rapid determination of protein folds [24] or for the cross validation of structures determined with other methods.

4.4.5 AUTOMATED DATA ANALYSIS

One of the major bottle-necks for NMR structure determination is sequence specific assignment of resonance frequencies and the subsequent extraction of distance information from NOESY spectra (see Figure 4.2). Therefore, an integrated platform for automated data analysis would be valuable for NMR-based structural genomics research. Recently, programs such as Autoassign [25], GARANT [26], MAPPER [27], and MUNIN [28] have been developed for automated resonance assignment. These programs can successfully assign about 80~90% of the backbone atoms with a set of high quality triple-resonance spectra. However, there still is no robust automated procedure for assignment of side chain resonances and the subsequent transfer of the assignment to the peaks in NOESY spectra. ARIA [29] and NOAH [30] programs are designed to solve part of this problem by performing both structure calculation and NOE assignment simultaneously. Recently, the

Montelione group has developed a program, AutoStructure [31], which is a combined package of Autoassign and DYANA [32]. Continued development of these types of programs will likely improve the speed and quality of NMR structure determination for small proteins. The recent development of G-matrix Fourier transform (GFT) NMR and related methods also hold promise to improve automated data analysis [31,41].

4.4.6 ISOTOPE LABELING TECHNIQUES

4.4.6.1 Isotope Labeling of Protein

The NMR structure determination of large proteins has greatly benefited from the use of stable isotope labeling (^{15}N, ^{13}C and ^{2}H) of the target protein. The overlapped proton resonances are readily separated into three or four dimensions using ^{15}N and/or ^{13}C chemical shifts [33]. However, despite of the utility of ^{15}N, ^{13}C-labeling technique, large proteins (> 25 kDa) normally suffer from both proton-mediated dipolar relaxation and the heteronuclear spin relaxation effect, resulting in poor signal-to-noise ratio. Perdeuteration of the side chains of large proteins, replacing all carbon-bonded protons with deuterium, takes advantage of the 6.7-fold lower gyromagnetic ratio of deuterium relative to that of proton. Nowadays, it is an essential step to increase sensitivity and resolution of NMR spectra of larger proteins, and it also enhances the TROSY effect for the remaining protons, such as the exchangeable amide proteins.

4.4.6.2 Selective Protonation of Methyl Group and Segmental Labeling

The Kay group [34] has developed a protocol for the production of ^{15}N, ^{13}C, and ^{2}H-labeled proteins with selective protonation of methyl groups of alanine, valine, leucine, and isoleucine residues. In these otherwise perdeuterated proteins, the NMR relaxation properties of the protonated methyl groups are favorable, resulting in reasonably well-resolved ^{13}C–^{1}H correlations and narrow line-widths. This approach provides crucial distance information within the hydrophobic core of proteins through the detection of NOEs between methyl groups and between methyl and amide protons.

Another labeling approach is segmental labeling, which can be an efficient method to determine a protein structure with two structural domains by dividing a large target protein into parts of manageable size. The protein sample is produced through the ligation of labeled and unlabeled polypeptides using a peptide-splicing element such as an intein, which codes for a protein that catalyzes its excision out of its host protein and ligates the flanking sequences of the host protein with a peptide bond [35]. Each half of the protein is expressed as fusion proteins with the appropriate intein peptide sequence in the appropriate isotopic media. The purified fusion domains are combined in a denatured form and refolded to generate an active intein that can catalyze the splicing reaction to ligate the two target polypeptides each with differing isotopes incorporated. However, this method is restricted to the study of proteins with structurally independent domains, or those in which there is a known loop in which the ligation sequence can be inserted.

4.4.6.3 Cell Free System

Large-scale production of target proteins normally requires tedious series of sub-cloning, expression, and purification steps, as described above. The newly developed cell-free expression, system offers many advantages over the traditional cell-based methods such as speed, convenience, amenability to automation and the avoidance of problems due to toxicity of the gene product. This method has been pioneered for structural genomics by Yokoyama and colleagues [36], who recently demonstrated yields of 6 mg of protein in 15 ml of *E coli*-derived transcription-translation reaction mixture [37]. This methodology is particularly promising for structural genomics because of the ease of sample preparation and amenability to automation. However, for NMR applications it requires the use of expensive labeled amino acids in the reaction mixture. If the cost savings due to automation and increased productivity of the simpler cell-free expression method are found to outweigh the costs of the more expensive reagents, it may prove to be the method of choice for NMR-based structural proteomics.

4.4 NMR-BASED FUNCTIONAL GENOMICS

As structural genomics projects around the world gather momentum, it seems clear that soon hundreds to thousands of new structures will be generated from these projects, many of which will be products of un-annotated or uncharacterized genes. A final, but important application of NMR in this field will be the biochemical characterization of proteins. Many of the NMR screening methods developed in the pharmaceutical industry [38,39] for monitoring the binding of small molecules to proteins will find important applications in functional genomics by NMR. NMR-based screening of proteins against a targeted or broadly defined functional library of cofactors, substrates, and other biologically relevant small molecules, will reveal clues to help classify a protein's biochemical activities and to identify molecules that can modulate their activities. For proteins that yield good HSQC spectra and for which the amide resonances have been assigned, the HSQC-based screening method of Fesik and colleagues is particularly powerful, in that it reveals not only the existence of an interaction with another molecule, but it also identifies the residues that are involved. Mapping this information onto an existing structure (determined by X-ray or NMR) will go a long way toward understanding the significance of the interaction. This method is also applicable to monitoring of protein–protein or protein–peptide interactions.

For the many proteins generated by structural genomics projects that do not yield good HSQC spectra, including large proteins not normally targeted for NMR analysis, techniques that monitor the NMR properties of the small molecules are the method of choice [39,40]. Because these methods focus on the small molecule, they are more sensitive and require smaller amounts of protein, making it feasible to screen all soluble proteins if desired. Thus, it may be desirable to integrate an

NMR based screening tool into the pipeline of structural genomics projects in order to leverage the value of the large numbers of proteins that are being produced. In addition to generating new functional information, the identification of small molecule ligands will also likely facilitate the crystallization and/or NMR properties of proteins and thereby facilitate the structure genomics pipeline.

REFERENCES

1. Bax, A., Weak alignment offers new NMR opportunities to study protein structure and dynamics, *Protein Sci.*, Jan. 12, 1, 1–16, 2003.
2. Fernandez, C. and Wider, G., TROSY in NMR studies of the structure and function of large biological macromolecules, *Curr. Opin. Struct. Biol.*, Oct. 13, 5, 570–80, 2003.
3. Kay, L.E., Clore, G.M., Bax, A., and Gronenborn, A.M., Four-dimensional heteronuclear triple-resonance NMR spectroscopy of interleukin-1 beta in solution, *Science*, Jul. 27, 249, 4967, 411–4, 1990.
4. Sattler, M. and Fesik, S.W., Resolving Resonance Overlap in the NMR Spectra of Proteins from Differential Lantahnide-Induced Shifts, *J. Am. Chem. Soc.*, 119, 7885–7886, 1999.
5. Christendat, D., Yee, A., Dharamsi, A., Kluger, Y., Savchenko, A., Cort, J.R., Booth, V., Mackereth, C.D., Saridakis, V., Ekiel, I., Kozlov, G., Maxwell, K.L., Wu, N., McIntosh, L.P., Gehring, K., Kennedy, M.A., Davidson, A.R., Pai, E.F., Gerstein, M., Edwards, A.M., Arrowsmith, C.H., Structural proteomics of an archaeon, *Nat. Struct. Biol.*, Oct. 7, 10, 903–9, 2000.
6. Pearl, F.M., Martin, N., Bray, J.E., Buchan, D.W., Harrison, A.P., Lee, D., Reeves, G.A., Shepherd, A.J., Sillitoe, I., Todd, A.E., Thornton, J.M., Orengo, C.A., A rapid classification protocol for the CATH Domain Database to support structural genomics, *Nucleic Acids Res.*, Jan. 1, 29, 1, 223–7, 2001.
7. Yee, A., Chang, X., Pineda-Lucena, A., Wu, B., Semesi, A., Le, B., Ramelot, T., Lee, G.M., Bhattacharyya, S., Gutierrez, P., Denisov, A., Lee, C.H., Cort, J.R., Kozlov, G., Liao, J., Finak, G., Chen, L., Wishart, D., Lee, W., McIntosh, L.P., Gehring, K., Kennedy, M.A., Edwards, A.M., Arrowsmith, C.H., An NMR approach to structural proteomics, *Proc. Natl. Acad. Sci. U. S.*, Feb. 19, 99, 4, 1825–30, 2002.
8. Savchenko, A., Yee, A., Khachatryan, A., Skarina, T., Evdokimova, E., Pavlova, M., Semesi, A., Northey, J., Beasley, S., Lan, N., Das, R., Gerstein, M., Arrowmith, C.H., Edwards, A.M., Strategies for structural proteomics of prokaryotes: Quantifying the advantages of studying orthologous proteins and of using both NMR and X-ray crystallography approaches, *Proteins*, 15, 50, 3, 392–9, 2003.
9. Maxwell, K.L., Bona, D., Liu, C., Arrowsmith, C.H., and Edwards, A.M., Refolding out of guanidine hydrochloride is an effective approach for high-throughput structural studies of small proteins, *Protein Sci.*, Sep., 12, 9, 2073–80, 2003.
10. Zhang, O., Forman-Kay, J.D., Shortle, D., Kay, L.E., Triple-resonance NOESY-based experiments with improved spectral resolution: applications to structural characterization of unfolded, partially folded and folded proteins, *J. Biomol. NMR*, Feb. 9, 2, 181–200, 1997.

11. Xia, Y., Yee, A., Arrowsmith, C.H., and Gao, X., 1H(C) and 1H(N) total NOE correlations in a single 3-D NMR experiment. 15N and 13C time-sharing in t1 and t2 dimensions for simultaneous data acquisition, *J. Biomol. NMR*, Nov. 27, 3, 193–203, 2003.

12. Bax, A., Weak alignment offers new NMR opportunities to study protein structure and dynamics, *Protein Sci.*, Jan. 12, 1, 1–16, 2003.

13. Wishart, D.S. and Sykes, D., The 13C chemical-shift index: a simple method for the identification of protein secondary structure using 13C chemical-shift data, *J. Biomol. NMR*, 4, 171–180, 1994.

14. Cornilescu, G., Delaglio, F., and Bax, A., Protein backbone angle restraints from searching a database for chemical shift and sequence homology, *J. Biomol. NMR*, Mar. 13, 3, 289–302, 1999.

15. Szyperski, T., Yeh, D.C., Sukumaran, D.K., Moseley, H.N., and Montelione G.T., Reduced-dimensionality NMR spectroscopy for high-throughput protein resonance assignment, *Proc. Natl. Acad. Sci. U.S.*, Jun. 11, 99, 12, 8009–14, 2002.

16. Robert, F., Service Propelled by recent advances, NMR moves into the fast lane, *Science*, Jan. 24, 299, 503, 2003.

17. Medek, A., Olejniczak, E.T., Meadows, R.P., Fesik, S.W., An approach for high-throughput structure determination of proteins by NMR spectroscopy, *J. Biomol. NMR*, 18, 3, 229–38, 2000.

18. Wagner, G., Prospects for NMR of large proteins, *J. Biomol. NMR*, Jul. 3, 4, 375–85, 1993.

19. Pervushin, K., Riek, R., Wider, G., and Wuthrich, K., Attenuated T2 relaxation by mutual cancellation of dipole–dipole coupling and chemical shift anisotropy indicates an anvenue to NMR structures of very large biological macromolecules in solution, *Proc. Natl. Acad. Sci. U.S.*, 94, 12366–12371, 1997.

20. Mulder, F.A., Ayed, A., Yang, D., Arrowsmith, C.H., Kay, L.E., Assignment of 1H(N), 15N, 13C(alpha), 13CO and 13C(beta) resonances in a 67 kDa p53 dimer using 4D-TROSY NMR spectroscopy, *J. Biomol. NMR*, Oct. 18, 2, 173–6, 2000.

21. Fernandez, C., Adeishvili, K., and Wuthrich, K., Transverse relaxation-optimized NMR spectroscopy with the outer membrane protein OmpX in dihexanoyl phosphatidylcholine micelles, *Proc. Natl. Acad. Sci. U.S.*, Feb. 27, 98, 5, 2358–63, 2001.

22. Zahn, R., Spitzfaden, C., Ottiger, M., Wuthrich, K., and Pluckthun, A., Destabilization of the complete protein secondary structure on binding to the chaperone GroEL, Nature, Mar. 17, 368, 6468, 261–5, 1994.

23. Tjandra, N., Omichinski, J.G., Gronenborn, A.M., Clore, G.M., and Bax, A., Use of dipolar 1H-15N and 1H-13C couplings in the structure determination of magnetically oriented macromolecules in solution, *Nat. Struct. Biol.*, Sep., 4, 9, 732–8, 1997.

24. Fowler, C.A., Tian, F., Al-Hashimi, H.M., and Prestegard, J.H., Rapid determination of protein folds using residual dipolar couplings, *J. Mol. Biol.*, Dec. 1, 304, 3, 447–60, 2000.

25. Moseley, H.N.B. and Montelione, G.T., Automated analysis of NMR assignments and structures for proteins, *Curr. Opin. Struct. Biol.*, 9, 635–642, 1999.

26. O'Connell, J.F., Pryor, K.D., Grant, S.K., and Leiting, B., A high quality nuclear magnetic resonance solution structure of peptide deformylase from Escherichia coli: application of an automated assignment strategy using GARANT, *J. Biomol. NMR*, Apr. 13, 4, 311–24, 1999.

27. Guntert, P., Salzmann, M., Braun, D., and Wuthrich, K., Sequence-specific NMR assignment of proteins by global fragment mapping with the program MAPPER, *J. Biomol. NMR*, Oct. 18, 2, 129–37, 2000.

28. Damberg, C.S., Orekhov, V.Y., and Billeter, M., Automated analysis of large sets of heteronuclear correlation spectra in NMR-based drug discovery, *J. Med. Chem.*, Dec. 19, 45, 26, 5649–54, 2002.

29. Nilges, M., Macias, M.C., O'Donoghue, S.I. and Oschkinat, H., Automated NOESY interpretation with ambiguous distance restraints: the refined NMR solution structure of the pleckstrin homology domain from alpha-spectrin, *J. Mol. Biol.*, 269, 408–422, 1997.

30. Mumenthaler, C., Guntert, P., Braun, W., and Wuthrich, K., Automated combined assignment of NOESY spectra and three-dimensional protein structure determination, *J. Biomol. NMR*, Dec. 10, 4, 351–62, 1997.

31. Huang, Y.J., Mosely, H.N.B., Baran, M.C., Arrowsmith, C., Powers, R., Tejero, R., Szyperski, T., Montelione, G.T., An integrated platform for automated analysis of protein NMR structures, *Meth. Enzymol.*, 394, 111–141, 2005.

32. Greenfield, N.J., Huang, Y.J., Palm, T., Swapna, G.V., Monleon, D., Montelione, G.T., and Hitchcock-DeGregori, S.E., Solution NMR structure and folding dynamics of the N terminus of a rat nonmuscle alpha-tropomyosin in an engineered chimeric protein, *J. Mol. Biol.*, Sep. 28, 312, 4, 833–47, 2001.

33. Goto, N.K. and Kay, L.E., New development in isotope labeling strategies for protein solution NMR spectroscopy, *Curr. Opin. Struct. Biol.*, 10, 585–592, 2000.

34. Rosen, M.K., Gardner, K.H., Willis, R.C., Parris, W.E., Pawson, T., and Kay, L.E., Selective methyl group protonation of perdeuterated proteins, *J. Mol. Biol.*, 263, 627–636, 1996.

35. Yamazaki, T., Otomo, T., Oda, N., Kyogoku, Y., Uegaki, K., Ito, N., Ishino, Y., and Nakamura, H., Segmental isotope labeling for protein NMR using peptide splicing, *J. Am. Chem. Soc.*, 120, 5591–5592, 1998.

36. Yabuki, T., Kigawa, T., Dohmae, N., Takio, K., Terada, T., Ito, Y., Laue, E.D., Cooper, J.A., Kainosho, M., and Yokoyama, S. Dual amino acid-selective and site-directed stable-isotope labeling of the human c-Ha-Ras protein by cell-free synthesis, *J. Biomol. NMR*, 11, 3, 295–306, 1998.

37. Kigawa, T., Yabuki, T., Yoshida, Y., Tsutsui, M., Ito, Y., Shibata, T., and Yokoyama, S., Cell-free production and stable-isotope labeling of milligram-quantities of proteins, FEBS Lett., 442, 15–19, 1999.

38. Hajduk, P.J., Betz, S.F., Mack, J., Ruan, X., Towne, D.L., Lerner, C.G., Beutel, B.A., Fesik, S.W., A strategy for high-throughput assay development using leads derived from nuclear magnetic resonance-based screening, *J. Biomol. Screen.*, Oct. 7, 5, 429–32, 2002.

39. Dalvit, C., Ardini, E., Flocco, M., Fogliatto, G.P., Mongelli, N., Veronesi, M., A general NMR method for rapid, efficient, and reliable biochemical screening, *J. Am. Chem. Soc.*, Nov. 26, 125, 47, 14620–5, 2003.

40. Dalvit, C., Fogliatto, G., Stewart, A., Veronesi, M., and Stockman, B., WaterLOGSY as a method for primary NMR screening: practical aspects and range of applicability, *J. Biomol. NMR*, Dec. 21, 4, 349–59, 2001.

41. Atreya, H.S., Eletsky, A., Szyperski, T., Resonance assignment of proteins with high shift degeneracy based on 5D spectal information encoded in GFT NMR experiments. *J. Am. Chem. Soc.*, April 6, 127, 4554–5, 2005.

5 High Throughput Protein Crystallography

Bernhard Rupp

CONTENTS

5.1   BACKGROUND AND RATIONALE

5.1.1   DEFINITION AND SCOPE

This chapter aims to introduce scientists with general interest in proteomics to the methods and techniques of macromolecular crystallography. Special emphasis will be placed on the challenges faced and the progress made in achieving high throughput in this predominantly used method of protein structure determination.

The scope of this chapter is the entire process of a crystallographic structure determination, beginning from the presence of a crystal of suitable size and morphology (but of unknown quality as far as diffraction is concerned) up to initial validation of the refined molecular model. This chapter thus will not include the arguably most crucial part of a structure determination, namely the production of suitable, often engineered, protein targets (Chapter 1), nor will it cover the actual crystal growth experiments (Chapter 3). Included in the discussion will be procedures ranging from crystal harvesting, cryo-protection and derivatization, through data collection and phasing, to the cycles of model building, refinement, and initial validation, which constitutes — or at least should constitute — an integrated part of the process of structure determination. The extensive array of structure validation and analysis based on structural and chemical plausibility and related prior knowledge warrant their own full chapter (Chapter 10.) Figure 5.1 provides an overview of the most important steps and methods in the course of a crystallographic structure determination.

5.1.2   CHOOSING CRYSTALLOGRAPHY FOR HIGH THROUGHPUT
STRUCTURE DETERMINATION

The motivation to use **X-ray** crystallography as the primary means of structure determination (out of 26,500 PDB entries in July 2004, about 22,500 are crystal

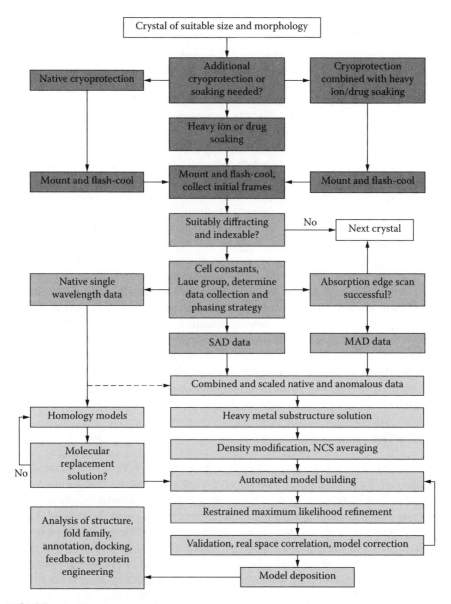

FIGURE 5.1 Flow diagram of the key steps in automated high throughput structure determination, from crystal selection to final structure model. The dark gray shaded boxes indicate steps involved in micromanipulation of crystals (Section 5.2.2). Medium gray shade indicates experimental data collection steps conducted at the X-ray source (5.2.3). In light shade, steps that are conducted exclusively *in silico* (5.2.4).

structures and about 4,000 are nuclear magnetic resonance (NMR) models) lies mainly in the fact that accurate and precise molecular structures, sometimes at near atomic resolution, can be obtained rapidly and reliably via X-ray crystallography.

Elucidating the precise atomic detail of molecular interactions is essential for drug target structures to be useful as leads in rational drug design [1]. Another advantage of crystallographic structure determination is that no principal difficulty limits the accurate description of very large structures and complexes, as evidenced in the nearly 2 MDa structure of the 50S ribosomal subunit [2], determined at 2.4 Å resolution. As the emphasis in proteomics shifts towards obtaining a comprehensive picture of protein interactions [3], the capability to determine large, multi-unit complex structures will become increasingly important.

The price to be paid for obtaining high quality X-ray structures is that a diffracting protein crystal needs to be produced. Crystal growth in itself can prove quite challenging (Chapter 3) and often can only be achieved after substantial protein engineering efforts (Chapter 2). Comparison with NMR data confirms that the core structure of proteins remains unchanged [4], and enzymes packed in crystals even maintain biological activity, often necessitating the design of inactive enzyme substrate analogues in order to dissect the molecular reaction mechanisms. The fact that protein molecules are periodically packed in a crystal places limitations on the direct observation of processes involving large conformational changes, which would destroy the delicate molecular packing arrangement necessary to form a protein crystal. Molecular transport processes or interactions involving extended conformational rearrangements may require multiple, stepwise "snapshot" structure determinations in order to understand such inherently dynamic processes. Time resolved crystallography, although exciting for enzymatic reactions involving limited, local structural changes [5], presently plays no role in high throughput structure determination projects due to substantial technical challenges and limitations in applicability.

5.1.3 WHY HIGH THROUGHPUT?

At this point, it may also be appropriate to recall the motivation behind the desire to achieve high throughput in protein crystallography (as discussed in the wider context of Proteomics in Chapter 1). To a large degree, the driving incentive behind structure determination on a genome scale is the elucidation of the structural basis for molecular function, not just for one enzyme, but for whole functional pathways. Such aspects will become even more important as the classic single gene–single disease targets become increasingly sparse [6]. The goal of producing lead structures suitable for therapeutic drug design requires that such structures are of high quality, providing enough accurate detail for in silico screening or rational compound design [1]. Structures obtained for drug design purposes will naturally have much higher criteria for precision compared to structures determined with the goal to populate the protein fold space, an explicit aim of publicly funded structural genomics projects [7]. In many cases, structures of multiple orthologs, numerous functional mutants, or multiple potential drug-protein complexes have to be determined. The need for high throughput, without compromise in structure quality, becomes quite evident, in particular to take the step from structure-guided lead optimization to true structure-based drug discovery [8,9].

5.1.4 DEFINITION OF HIGH THROUGHPUT PROTEIN CRYSTALLOGRAPHY

The term High Throughput Protein Crystallography (HTPX) is used rather loosely in the scientific literature and comprises work ranging from modest scale, academic efforts focusing on affordable automation [10] to truly industrial scale, commercially driven ventures processing hundreds of crystals a week [9,11–13]. The underlying physical and methodical principles, however, are the same for HTPX as for conventional Low Throughput (LT) protein crystallography. Most of the improved methods and techniques developed for high throughput purposes are also applicable and indeed very useful for LT efforts. It appears fair to say that protein crystallography in general has made significant technical progress through increased public and commercial funding devoted to the development of high throughput technologies since the late 1990s.

5.1.5 HIGH THROUGHPUT VS. LOW THROUGHPUT

The most significant difference separating true HTPX and conventional LT crystallography efforts is the amount of automation and process integration implemented. In contrast to affordable computational prowess, full robotic automation nearly always comes at a substantial cost, and here lies probably the most significant rift between academic low- to medium-throughput and commercial HTPX ventures. Financial, infrastructural, and production related constraints, for example, are very different in academic and commercial environments, as are efficiency considerations. In a high throughput environment it may be neither necessary nor efficient to pursue every recalcitrant target to completion, while in an academic setting careers may depend on determining one specific structure. A review of the most significant differences between academic and industrial high throughput efforts in terms of philosophy and efficiency considerations is provided elsewhere [10].

5.1.6 KEY DEVELOPMENTS

High throughput in crystallography has become possible through two major synergistic developments: A rapid progress in technology, in particular advances in cryo-, synchrotron-, and computational techniques, combined with influx of substantial public and venture capital funding. Practically all true high throughput efforts depend on powerful third generation synchrotron X-ray sources (Table 5.1), largely because of the high brilliance of the X-rays (important for small crystals) and the unique tunability of the wavelength [14]. The capability to choose the wavelength — in contrast to the fixed, characteristic X-ray wavelength defined by the anode material of conventional X-ray sources — is the basis for anomalous phasing techniques that dominate high throughput protein crystallography [15]. Given the potentially enormous rewards of structure guided drug development [16], it comes as no surprise that a substantial number of commercial ventures were able to attract funds to develop and implement high throughput techniques, in particular advanced robotic automation. On the other hand, the public NIH-NIGMS funding of the PSI-I structural genomics pilot projects [7] provided, for the first time on a reasonable scale, the means for the development of nonproprietary high

TABLE 5.1
Major Synchrotron Facilities and Beam Lines Equipped for HTPX

Location	Link	Remarks	Microfocus/Micro-Collimated Beamlines	HTPX Beamlines w. Robotics
ALS: Advanced Light Source, Berkeley, CA, U.S.	bcsb.lbl.gov	DOE facility	—	5.0.1, 5.0.2 and 5.0.3 8.2.1, ALS design[a]
APS: Advanced Photon Source, Argonne, IL, U.S.	www.aps.anl.gov/aps cars9.uchicago.edu/biocars www.sbc.anl.gov	DOE facility, excellent user facilities on site	19ID (40 μm, ribosome 50S, 30S), 19BM (100μm), 14BM, 14ID	19BM, 19ID (December 2003), 14ID (planned for 2005) SGX 31 (Mar Robot)
SSRL/SPEAR: Stanford Synchrotron Radiation Laboratory, Stanford, CA, U.S.	smb.slac.stanford.edu	DOE facility, DOE/NIH upgrade to SPEAR 3 January 2004	50 μm min beam size on all beam lines	1-5, 9-1, 9-2, 11-1, 11-3 (Jan 2004)
CHESS: Cornel High Energy Synchrotron Source, Ithaca, NY, U.S.	www.macchess.cornell.edu	Non-DOE facility, NSF and NIH funding	F1 (special request only)	F1 (Fall 2003)
NSLS: National Synchrotron Light Source, Brookhaven, NY, U.S.	www.px.nsls.bnl.gov	DOE facility	—	X12B (Fall 2003)
ESRF: European Synchrotron Radiation Facility, Grenoble, France	www.esrf.fr/UsersAndScience/Experiments/MX	Multinational funding	BM14 (100 μm) ID29 (40 μm) ID13 (down to 5 μm) ID23 (Fall 2003)	BM14 (EMBL design, Feb 2004) On all ID beam lines summer 2003-2004 BM30 (in-house)

SPring-8: Super Photon ring 8GeV, Hyogo, Japan	www.spring8.or.jp	JASRI and RIKKEN	BL41XU (100 μm)	BL26B1, BL26B2
DESY: Deutsches Elektronen Synchrotron, Hamburg, FRG	www.embl-hamburg.de	EMBL outstation	(2008 PetraIII upgrade)	X-13 (Mar Robot, 2004) BW-7B (in-house, EMBL design)

Note: For a general listing of present and planned worldwide synchrotron facilities, see [129]. Beam line end stations are rapidly added or upgraded with robotics, thus an online search for new capabilities at the beam lines should be conducted to augment this table. Good portals are http://www-ssrl.slac.stanford.edu/sr_sources.html and the Structural Biology Synchrotron Users Organization, http://biosync.sdsc.edu/, which provide periodically updated listings of all macromolecular crystallography beamlines. DOE: U.S. Department of Energy. NIH: National Institutes of Health.

[a]G Snell, C Cork, R Nordmeyer, E Cornell, G Meigs, D Yegian, J Jaklevic, J Jin, RC Stevens, and TE Earnest, *Structure*, 12, 1–12, 2004.

throughput structure determination methods, which have benefited not only the funded projects, but practically every structural biology effort [17]. The next funding round of NIH PSI-II centers, commencing in 2005, will extend methods- and technology-driven development further.

5.1.7 LITERATURE

Excellent monographs, series, and journal special issues reflecting the progress in crystallographic techniques and their implication for structural genomics in recent years are available; some of them are listed in the resource list of general references at the end of this chapter. Many citations throughout the text refer to other reviews or review-like articles that contain further specific technical primary references. This chapter emphasizes aspects of particular relevance to automation and HT protein crystallography. The development in this field is extraordinarily rapid, and to obtain a complete and current picture it will be necessary to supplement the information provided in this article with electronic literature and website searches.

5.2 METHODS OF HIGH THROUGHPUT PROTEIN CRYSTALLOGRAPHY

5.2.1 OVERVIEW

Crystallographic protein structure determination centers around a conceptually quite simple diffraction experiment: A cryo-cooled crystal is placed on a goniostat and exposed to intense and collimated X-rays. The goniostat allows the crystal to be rotated in small increments, and for each orientation a diffraction pattern is collected. The images are indexed, integrated and scaled, and unit cell and space group are determined. A reduced and hopefully complete data set, essentially representing a periodically sampled reciprocal space transform of the molecules forming the crystal, is obtained. Due to the lack of phase information in the diffraction patterns, direct reconstruction of the electron density of the molecules via Fourier transforms is not generally possible for proteins (known as the Phase Problem in Crystallography, Figure 5.6). In the absence of a suitable known structure model, additional data sets of isomorphous derivative crystals and/or anomalous datasets, at suitable wavelengths, need to be collected to allow the determination of phases (detailed in section 5.2.4). Overall, approximately three fourths of all HTPX structures are solved by molecular replacement techniques that exploit a homologous structure or model as a source of initial phases, and the *de novo* phasing* of the remaining one fourth relies heavily upon anomalous techniques (Table 5.2).

The actual phase calculations, electron density reconstruction, model building, and structure refinement are conducted *in silico* with computer programs. The procedure generally begins *de novo* with determination of a heavy (marker) atom substructure, calculation of initial phases, phase improvement by density modification techniques,

---

* I follow the notation of '*de novo*' phasing if no previous protein structure model has been used (thus excluding molecular replacement techniques or difference map techniques). *Ab initio* phasing refers to the use of (direct) methods that derive the phases solely from the intensities of a diffraction data set.

and model building and refinement. The last two steps are generally used in iteration, with improvements until convergence is achieved. Nearly all of the computational methods have been highly automated, and they are currently being integrated into fully automated structure determination packages (Table 5.3). Although challenges remain, particularly in model building at low resolution, improved computational methods are continuously developed and tend to perform well.

With respect to high throughput requirements, the major steps in a crystallographic structure determination can be grouped as indicated by the different shading in Figure 5.1:

- The first group includes processes and steps involving manipulation of the fragile raw crystals, such as harvesting, cryo-protection, soaking, and actual mounting of the crystals on the diffractometer (section 5.2.2).
- In the second group, processes from initial evaluation of the now cryo-protected crystals to completion of data collection are included (section 5.2.3).
- Group three comprises the entirely computational steps from phasing to the analysis of the final structure model. No more manipulation of crystals is necessary (section 5.2.4).

The technical challenges faced in automated mechanical manipulation of crystals are fundamentally different from the computational requirements later in the structure determination process. Robotic micromanipulation, particularly harvesting, soaking, and cryo-cooling of the fragile protein crystals is very expensive to automate and remains a hurdle for full process automation. As crystals are becoming more plentiful, cryo-mounting may eventually develop into a rate limiting step, and demonstrated success of high throughput crystallography at that point may well justify further substantial investment in high throughput robotic crystal harvesting and in-situ diffraction screening techniques.

At the other extreme, the continuous increase in computational power — still roughly doubling every 18 months according to Moore's law (18) — at decreasing cost — and the public funding of development efforts for powerful software packages (Table 5.3) have made automation of the final computational part of structure determination quite successful, although considerable challenges remain in the area of data collection expert systems and automated model building and completion.

5.2.2 PROCESSES INVOLVING CRYSTAL MANIPULATION

Micromanipulations during harvesting, mounting, derivative soaking, and cryo-protection present serious challenges for full robotic automation and for achievement of sustained high throughput. Although crystal harvesting in suitable cryo-loops with magnetic bases has become an inexpensive and reliable de facto standard in cryo-crystallography [19], the selection and capture of the crystals from the crystallization drop under a microscope, as well as the micromanipulations during cryo-protection and soaking sweeps, are still performed manually. However, the advent of precisely fabricated and stable cryo-loop designs [20,21], in combination with micro-manip-

TABLE 5.2
Phasing Methods in High Throughput Protein Crystallography (HTPX)

Phasing Method	Phasing Marker	Derivatization Method	Remarks	Suitability for HTPX
SAD via sulfur atoms (S-SAD)	S in Met, Cys, residues, combined with solvent density modification	None, native protein	Requires highly redundant data collection	Becoming established, increasing use, also in-house
MAD/SAD via naturally bound metals	Naturally bound metal ion, cofactor	None, native protein		Selected cases only
MAD via Se	Se in Se-Met residues	Incorporated during expression in met- cells or via metabolic starvation	1 Se phases 100–200 residues	Reliable standard, generally applicable, few exceptions
MAD via isomorphous metals	Heavy metal ion specifically bound	Soaking or co-crystallization	Hg, Pt, Au, etc Strong signal on L-edges due to XAS white lines	Reliable, generally applicable
SIR(AS) via isomorphous metals	Heavy metal ion specifically bound, density modification	Soaking or co-crystallization	Phasing power proportional to z back soaking necessary	Reliable, nearly always in combination with anomalous method
MIR(AS) via isomorphous metals	Heavy metal ion specifically bound	Soaking or co-crystallization	Multiple derivatives needed back soaking necessary	Reliable, often in combination with anomalous method
SIR(AS) via anions	Heavy anion specifically bound, Br-, I-, I3-	Mostly brief soaking, or co-crystallization	I derivatives also suitable for Cu source, possibly back soaking	Not fully established in HTPX, probably increasing use

SIR(AS) via noble gas	Noble gas specifically bound, Xe, Kr	Pressure apparatus	Xe XAS edge unsuitable for most MAD beam lines	Isolated cases so far
MR via model structure	None	None	Needs homology model with close coordinate r.m.s.d.	70% of cases, increasing. Subject to model bias, particularly at low resolution
Direct Methods	None	None	Atomic resolution, small size	Few cases, but important in metal substructure solution

Notes: SAD: Single-wavelength Anomalous Diffraction; MAD: Multi-wavelength Anomalous Diffraction; SIR: Single Isomorphous Replacement; MIR: Multiple Isomorphous Replacement; (AS): with Anomalous Scattering.

TABLE 5.3
Computer Programs and Program Packages Commonly Used in or Developed for High Throughput Protein Structure Determination

Program	Reference, Website	Coverage	Remarks	Suitability for HTPX
CCP4 Program Suite	[131–133] www.ccp4.ac.uk	Data collection, data processing, phasing, ML refinement, model building, validation. No MD refinement, but the only TLS refinement program	Current version 5.0.2. Graphical interface CCP4i. Multi-author collaboration	Most common program package. No expert system. Can be scripted, many of its program modules found in and interface with other packages
PHENIX	[104,134] www.phenix-online.org	In final version complete from data collection to model building, currently parts from phasing to model building	Author team includes XPLOR/CNS experts. Open Python source, industrial consortium members	Still under development, should be well suited for automation/expert system. Will include MD and simulated annealing refinement
XPLOR/CNS	[135] cns.csb.yale.edu	Program pioneering SA and MD in refinement of X-ray and NMR data. Complete package including phasing and MR	HTML interface. CNX commercial version via MSI/Accelrys	Academic development continued under PHENIX project
MOSFLM	[44] www.mrc-lmb.cam.ac.uk/harry/mosflm	Data image processing, integration, reduction, and scaling via CCP4. Available on most beam lines	Reliable indexing and data collection, freely available via CCP4	Developments under way to integrate expert system, fully automate data collection
HKL2000/DENZO	[121] www.hkl-xray.com	Data collection, integration, reduction, scaling (SCALEPACK module). Commercial indexing/processing service available	License fee also for academics. Interfaces with CCP4	Next to MOSFLM most popular data collection software suite
SOLVE/RESOLVE	[65,81] www.solve.lanl.gov	Combined ML HA solution, phasing, reciprocal space density modification, NCS, and model building program. Interfaces with CCP4	Pseudo-SIRAS Patterson approach to substructure solution, ML HA refinement	Easy to use and to automate via scripts. Interfaced with PHENIX

Program	Reference / URL	Description	Availability	Comments
SHARP, AUTOSHARP	[136] www.globalphasing.com	ML HA refinement, excellent phasing, density modification. Interfaces with CCP4	Modern Server/Client architecture, Bayesian ML methods. Newer versions get faster	Slower but more powerful phasing algorithm. Automated, links to ARP/wARP for model building
ARP/wARP, warpNtrace	[87,137] www.embl-hamburg.de/ARP	Map improvement via dummy atom refinement and model building. Works best at higher resolution	Fully interfaced with CCP4/CCP4i. No source	Fully automated model building, plans exist also for automated ligand building
XPREP,SHELXD, SHELXE	[68] shelx.uni-ac.gwdg.de	HA data processing (XPREP), HA substructure solution by combined Patterson and direct methods, simpler but very fast phasing and density modification for maps	SHELXD and SHELXE publicly available, XPREP, XM, XE commercial (Bruker XAS). No source	Very reliable, fast, most often used as front-end for HA substructure solution for subsequent HA ML refinement and phasing programs. Well updated
Shake and Bake	[67] www.hwi.buffalo.edu/SnB	Direct methods full structure or HA substructure solution via reciprocal-direct scape cycling	Has also been used for complete small protein structure solution	Stand alone, interfaces via HA file to SOLVE/RESOLVE
TEXTAL	[138] textal.tamu.edu:12321	Low resolution automated model building based on pattern recognition	Better performance at low resolution than dummy atom based methods	Incorporated in PHENIX, actively developed, target resolution of as low as 3.5 Å
MAID	[88] www.msi.umn.edu/~levitt	Relatively new model building program, evaluated in Badger, 2003	Combination of building techniques, real space torsion angle dynamics	No GUI, suitable for integration and automation. Works also at low resolution
QUANTA	[139] Accelrys www.accelrys.com/quanta	Commercial descendent from Biosym/Xsight and MSI programs, now Accelrys	Monolithic, expensive, includes ligand building XLIGAND	Well integrated package that delivers most of the functionality needed for structure determination
AUTOSOLVE	Astex Pharmaceuticals www.astex-technology.co.uk/autosolve.html	Complete package including ligand placing and refinement	Not publicly available	Highly automated

TABLE 5.3
Computer Programs and Program Packages Commonly Used in or Developed for High Throughput Protein Structure Determination (continued)

Program	Reference, Website	Coverage	Remarks	Suitability for HTPX
ELVES	Holton J. ucxray.berkeley.edu/~jamesh/elves	Clever UNIX scripting searching local installation for program components and combining them into a very basic expert system	One of the first attempts towards expert systems	Limited to the performance of the other available programs
O	[140] alpha2.bmc.uu.se/~alwyn/o_related .html	Descendant of first generation of pioneering graphic modeling programs	Many associated additional programs and utilities, partly included in CCP4	GUI, scriptable, steeper learning curve than X-fit. Interfaces with CCP4. Well supported user group
XtalView/Xfit	[141] www.sdsc.edu/CCMS/Packages/X TALVIEW	Complete package for basic data processing, brute force HA location, phasing, and semi-automated model building (XFIT)	XFIT module allows easy model building and correction. Also Windows version MI-Fit available (Molecular Images)	GUI, intended for fast manual building and correction of models. Easy to learn. Fast FFT support. Supported by CCP4i
EPMR	[101,102] ftp.agouron.com/pub/epmr/	Evolutionary algorithm for molecular replacement, fast FFT allows 6d search	Good convergence, automate search for multiple copies.	Well suited for multiple automated searches
Shake&wARP	[108] tuna.tamu.edu	Automated MR and bias removal program based on EPMR (101) and dummy atom placement/refinement via CCP4 programs	Data preparation, MR, and multiple map averaging, bias minimized real space correlation. Source available	Implemented as web service intended for automated structure validation. Slow but excellent, averaged bias minimized maps

Notes: HA: Heavy Atom; ML: Maximum Likelihood; MD: Molecular Dynamics; SA: Simulated Annealing; TLS : Torsion, Libration, Screw. Additional compilation of specific programs is available in (130).

ulation actuators for robots and advanced real-time machine vision tools will likely lead to the development of harvesting robots within a few years.

Once a crystal has been flash-cooled to cryogenic temperatures, further manipulation of the cryo-pins and the actual placing of the rather sturdy pins onto the goniostat can be performed quite fast and reliably by robotic arms with grippers.

5.2.2.1 Selection and Harvesting of Crystals

Crystals are grown using a variety of crystallization techniques, and not all techniques are equally well suited for crystal harvesting. Practically all high throughput crystallization experiments are set up in some Society of Biological Screening (SBS) standard compliant multi-well format with 96 to 1536 wells. Initial crystallization screening against many different conditions is usually performed with the objective of minimizing material usage, and suitable micro-batch screening methods under oil [22] or free interface diffusion experiments in micro-chips [23] are not necessarily designed with ease of harvesting in mind. In optimization experiments, where minor variations around successful crystallization conditions are set up, vapor diffusion sitting drops of 1 μl to 50 nl suitable for harvesting are most commonly used. Nanoliter drop sizes not only determine the maximum size of crystals that can be obtained, but also significantly affect nucleation and growth kinetics [24]. The use of nanoliter drop technology was the subject of an infringement dispute involving the patent holder Syrrx in San Diego and Oculus Pharmaceuticals (U.S. Patent 6,296,673, "Methods and apparatus for performing array microcrystallizations").

For harvesting, crystals are selected under the microscope and lifted from the drop with a small, suitably sized cryo-loop. A number of research groups have developed plate scanning and crystal recognition software, which can reliably detect crystals (at least those of reasonably defined shape) and the methods are expected to improve further [25–27]. Crystals are judged by size and appearance (often deceiving), and crystals with isotropic dimensions in the range of 100 μm are considered most desirable. The growing availability of powerful and automated micro-focus beam lines on third generation synchrotron sources allows, in ideal cases, very small crystals approaching 10 to 20 μm in the smallest dimension to be used, thus also permitting successful data collection on highly anisotropic crystal needles or plates [28,29]. Below μm size, intensity issues and line broadening due to limited periodic sampling, as well as radiation damage in the protein crystal, generally become limiting factors [30].

5.2.2.2 Soaking and Derivatization of Crystals

For *de novo* phasing methods based on the determination of a marker atom substructure, some atoms in the protein must act as sources of isomorphous and/or anomalous differences (See Table 5.2, Phasing Methods). Such marker atoms can be natively present heavy atoms, such as Fe, Zn, or Cu. In sulfur single-wavelength anomalous diffraction phasing (S-SAD), the sulfur atoms of Cys or Met residues act as marker atoms. In those cases of native marker atoms, no derivatization or soaking is necessary. The advantage of no need for soaking also holds for Se-Met labeled proteins, and no marker atoms are needed when Molecular Replacement (MR) phasing with homologous structure models will be attempted.

Nonnative marker ions can either be co-crystallized (i.e., added *a priori* to the crystallization cocktail), or the native crystals can be soaked for minutes to hours in mother liquor with a metal ion added in mM concentration, followed by short, optional back-soaking to remove the unbound ions from the crystal's solvent channels [31,32]. Soaking and co-crystallization techniques are also used to incorporate ligands, cofactors, inhibitors, or drugs into the crystals.

Heavy metal ions, or halide anions such as bromide and iodide [33,34] can also be introduced during brief sweeps in combined heavy-ion cryo-buffers. Due to the location of metal or iodine X-ray absorption L-edges or even uranium M-edges [35] not too far below the characteristic Cu-K wavelength (8keV), SAD/SIRAS phasing should become an increasingly interesting (in-house) alternative to synchrotron based (multi-wavelength) methods [33].

5.2.2.3 Cryo-Protection and Loop Mounting

Rapid cooling of crystals to cryogenic temperatures (quenching or flash-cooling[*]) has become a standard procedure in macromolecular crystallography [36]. The foremost reason for cryo-cooling is the drastic reduction of radiation damage [30], eliminating the need for multiple crystals with the associated data merging errors; and secondly increasing the resolution due to reduced thermal vibrations. An additional benefit is that, once the crystals are cryo-protected and safely embedded in the solid amorphous mother liquor of their loops, they can be handled quite easily and reliably by high throughput mounting robots.

Successful cryo-cooling depends on a number of factors, only few of them well established under controlled conditions. If the crystallization cocktail does not *a priori* contain high enough concentrations of reagents like PEG, MPD, or glycerol to prevent freezing (i.e., the formation of ice destroying the crystal) and disorder, the mounted crystal needs to be swiped through a cryoprotectant before being flash cooled. Cryoprotectants primarily prevent ice formation in the mother liquor, which can be established by quenching a swiped loop without crystal in liquid nitrogen or the diffractometer nitrogen cold stream and checking the diffraction pattern for the absence of ice rings (a diffraction pattern displaying typical ice rings is shown in [37]). For reasons not entirely clear, even rapid quenching and amorphous state of mother liquor do not guarantee successful cryo-cooling. Excessive increase in mosaicity and loss of resolution are common mishaps. A number of annealing procedures occasionally reducing mosaicity have been reported [38,39] and studied [40], but have not yet become established standard procedure in HTPX pipelines. Although a common recipe for cryoprotectants is to spike mother liquor with glycerol, PEG, MPD or other additives, few systematic studies of generally applicable procedures for high throughput efforts appear in the literature. Clearly, these methods are of great relevance to the objectives of high throughput crystallization, but they are time

---

[*] The term *freezing* should be avoided as it encompasses the definition of ice formation, which is detrimental to the crystals.

consuming and risky for the crystals, with little systematic or automated procedures developed so far. Problems during soaking and cryo-protection are probably the single most significant source of loss of crystals in both high and low throughput crystallography.

5.2.2.4 Robotic Sample Mounting

Automated mounting of the cryo-pins on the diffractometer greatly enhances utilization of valuable synchrotron beam time. Practically every major synchrotron facility and larger biotech companies have developed mounting robots for their HTPX beam lines (see for example [41,42], Table 5.1). Commercial systems, which are also suitable for in-house lab sources, are available (Mar Research, Rigaku-MSC, Bruker-Nonius). Under the premise that every crystal deserves screening, fast and reliable storage and mounting procedures are needed to achieve high-throughput data collection. The sample transport and storage system developed at the Advanced Light Source (ALS) Macromolecular Crystallization Facility together with the Engineering Division of Lawrence Berkeley National Laboratory [43] may serve as an example for an easy-to-use and quite practical development. The basic handling unit, a cylindrical, puck-shaped cassette containing 16 cryo-pins, also serves as an integral part of a complete, automated cryogenic sample alignment and mounting system, which has been routinely operating since 2002 at the ALS protein crystallography beam line 5.0.3 (Figure 5.2).

Seven puck cassettes, each holding 16 Hampton-style, magnetic base cryo-pins, fit into a standard dry shipping dewar. The pucks are loaded at the crystallization lab with the crystals on cryo-pins, and four pucks are transferred into the robot-hutch liquid nitrogen vessel. The mounting robot can randomly access any sample with a cooled, robotic gripper that transfers the sample to the diffractometer within seconds, maintaining the crystal temperature below 110 K. Crystals are centered automatically on the computer-controlled goniometer head. In addition to saving valuable beam time, mounting robotics also allow the safe removal and re-storage of a sample, should the initial analysis of diffraction snapshots cast doubt on the crystal quality. Potentially better crystals can be mounted and examined without risk of losing the best one found so far.

5.2.3 DATA COLLECTION

Data collection is in fact the last physical experiment that is conducted in the long process of a structure determination, and it deserves full attention. Considering the constant loss and thus increasing value of targets throughout the steps of expression, purification, and crystallization, failure to obtain useful data in the final experiment is most costly. Diffraction data quality largely (and without mercy) determines the quality, and hence usefulness and value, of the final protein structure model.

5.2.3.1 High Throughput Considerations

In high throughput mode it may not be worthwhile, except in special cases, to collect data sets with resolution worse than 2.3–2.5 Å, but better to pursue additional

FIGURE 5.2 Prototype of ALS developed automated sample mounting system. Top: Overall view of sample mounting robot in hutch of beam line 5.0.3. at the Advanced Light Source (ALS) in Berkeley, CA, U.S.A. [43]. Bottom left: detail view of pucks, 4 of each contained in the Dewar visible at the bottom of top panel. Bottom right: detail view of the pneumatically operated, cryo-cooled sample gripper, which retrieves the magnetic base sample pin from the Dewar and mounts them on the goniostat. The crystals in the sample loops are automatically centered on a motorized goniometer head (left side of instrument in top panel). (From B Rupp, BW Segelke, H Krupka, T Lekin, J Schafer, A Zemla, D Toppani, G Snell, and T Earnest, *Acta Crystallogr.*, D58, 2002. Reproduced with permission from the IUCr.)

crystallization optimization or protein engineering. Overall throughput might well be higher when adopting a high resolution strategy, particularly in view of the increased difficulty to accurately build and refine low resolution models, and considering the reduced information content in low resolution models. A decision whether to pursue a low resolution structure will be influenced by whether a structure serves the purpose of fold determination, or must satisfy the more stringent quality criteria for a drug target structure.

5.2.3.2 Data Collection as a Multi-Level Decision Process

Once the cryo-pins with the crystals are transferred to the robot-hutch dewar, the remaining steps of the crystal structure determination can principally proceed in fully automated mode. At present, however, there are still weaknesses and substantial off-line processing in the data collection stage. At several points, strategic decisions need to be made whether to accept a given level of data quality (and hence, a certain probability of failure in the structure determination), or to proceed to the next crystal. Clearly, reliable robotics provides the advantage of safely un-mounting and storing an acceptable but not optimal crystal for later use, and to proceed to evaluate hopefully better ones.

Any expert system handling the chain of decisions during initial crystal assessment and data collection must be able to evaluate a significant number of parameters at its decision points [44]. Problems and irregularities can occur at several stages during data collection, and often show up and become critical only later. At present, data collection expert systems are not yet developed to completely handle the entire decision process. Such a system requires tight interfacing with the data indexing, integration, and reduction programs, and with the beam line hardware and robotics control software (reference [45] exemplifies the substantial complexities involved).

5.2.3.3 Initial Assessment of Crystal Quality and Indexing

The most prevalent data collection technique in protein crystallography, practically exclusively used in HTPX efforts, is the rotation method. During each exposure, the crystal is successively rotated by a small increment (usually 0.2 to 1.0 deg) around a single axis, until sufficient coverage of the reciprocal (diffraction) space unit cell is attained.

Once the crystals are centered on the goniostat, a first diffraction pattern is recorded. The crystal is exposed to a collimated, fine beam of X-rays, for a few seconds on a powerful synchrotron, and up to several minutes on weaker or laboratory X-ray sources. The diffracted X-rays are recorded mostly on CCD area detectors (longer read-out times disfavor image plate detectors for high throughput use on synchrotrons), and recorded as an image of diffraction spots, also referred to as a (rotation or oscillation) frame. The first frame immediately shows the extent to which the crystal diffracts. Good diffraction implies single, resolved, and strong spots, extending far out in diffraction angle to high resolution. The relation between diffraction, resolution, and structure detail is shown in Figure 5.3. The first diffraction snapshot also can reveal the presence of ice rings [46]. Although icing affects the reflections in proximity to the ice ring, frames with not too excessive ice rings can be processed with little difficulty [36,47].

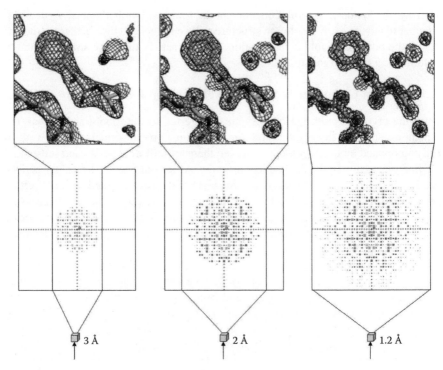

FIGURE 5.3 Diffraction and electron density at increasing resolution. The crystals (bottom) diffract X-rays increasingly better (diffraction limit or resolution of 3.0, 2.0 and 1.2 Å from left to right). Increasing diffraction (corresponding to finer sampling) produces many more reflections in the diffraction pattern (center row of figure) and hence, a more detailed reconstruction of the electron density map (contour grid) and building of a more accurate model is possible (top row). The modeled protein and ligand are shown in a ball and stick representation. Figures prepared using Raster3D (EA Merritt and DJ Bacon, *Meth. Enzymol.*, 277, 1997) and XtalView (DE McRee, *J. Mol. Graph.*, 10, 44–46, 1992.)

Depending on the quality of the data and the indexing algorithm used, it can be possible to index the diffraction pattern based on a single frame or snapshot [47]. Indexing means the assignment of a consistent set of three reciprocal basis vectors, which span the reciprocal lattice represented by the diffraction spots. The corresponding direct vectors (a,b,c) and angles between them (α,β,γ) define the crystal unit cell. In practice, more than one frame is used for indexing, for several reasons: Crystals may not diffract isotropically, and snapshots in different orientations assure that anisotropy does not cause unacceptably low resolution in certain directions/orientations of the crystal. A single frame also may not contain enough reflections to allow reliable determination of the internal Laue symmetry of the diffraction pattern, which again determines the possible Laue group and crystal system of the crystal.[*]

---

[*] It is a common misconception that the crystal system is determined by the cell constants and angles. The internal symmetry overrides the apparent symmetry deduced from the cell constants. It is possible, for example, that an apparently orthorhombic cell (a $\neq$ b $\neq$ c, $\alpha = \beta = \gamma = 90$ deg) is monoclinic, with $\beta = 90$ deg.

Several space groups may belong to one Laue group, and it is not always possible to unambiguously determine the crystal's space group at this early stage from systematic absences of reflections alone. Proper determination of the Laue group is necessary to develop a strategy to collect a complete set of diffraction data. The data collection strategy also depends on the selected phasing strategy, as discussed below.

Typical difficulties arise during indexing, when it is not possible to find a consistent unit cell for the crystal. Large mosaic spread, large spot size, streaking and overlap, multiple or satellite spots due to macroscopic twinning, and excessive ice rings can cause problems. Spot overlap due to large unit cell dimensions should automatically trigger a re-evaluation at larger detector distances, and a strategy with multiple sweeps at increasing detector offset angles may have to be generated (discussed also under ultra high resolution strategies). After indexing, the choice of Laue symmetry may not be unambiguous, and if in doubt, a lower symmetry must be selected in developing the data collection strategy, depending on the planned phasing technique. Even after successful data reduction and space group determination, the possibility of microscopic twinning [48], not recognizable from the appearance of the diffraction pattern, exists in certain space groups, and should be automatically evaluated [49].

Increased frame exposure time in pursuit of high resolution tends to lead to low resolution detector pixel saturation. Automatic detection of saturation should routinely trigger collection of a second, faster low resolution data collection sweep. The need for good low resolution data for any phasing method (including MR) has been pointed out repeatedly [50].

5.2.3.4 Data Collection Strategies for Phasing

As indicated in Figure 5.1, each phasing technique requires a suitable data collection strategy to obtain the necessary coverage of the reciprocal (diffraction) space. It is seldom a disadvantage to collect as much redundant data as possible, and given the high throughput capabilities of synchrotrons, a few general strategies suffice to cover most standard phasing techniques [51,52].

The simplest case of data collection is a single wavelength data set without consideration of the anomalous signal. Although anomalous contributions from all atoms in the crystal are present at varying degrees at all wavelengths, they are miniscule for the light elements (H,C,N,O) comprising most of the scattering matter in native proteins, and special techniques described later are employed to utilize the minute anomalous signal of sulfur for phasing. A single wavelength data set covering the reciprocal space unit cell contains the data necessary for structure solution by Molecular Replacement (MR). Except in cases of special orientation of a crystal axis nearly parallel to the rotation axis in uniaxial systems, and at very high resolution, a practically complete set of data can be collected in a single sweep of successive frames [51]. The extent of the necessary rotation range is calculated by a strategy generator from the crystal orientation matrix, instrument parameters, and the Laue symmetry. As always, excessive pixel saturation of intense low resolution reflections may require a second sweep with shorter exposure times.

5.2.3.5   Data Collection for High Resolution Structures

In fortunate cases, crystals diffract to very high resolution. The term is loosely used and shall indicate in our case crystals diffracting better than about 1.5 Å, with exceeding 1.2 Å denoted as the onset of atomic resolution, or ultra high resolution. As illustrated in Figure 5.3, high resolution permits us to discern fine details in the structure, which can be understood as a manifestation of tighter sampling intervals or slices throughout the crystal. As the number of reflection increases with the volume of sampling space, even a numerically less impressive increase in resolution leads to a substantial increase in recorded data (Figure 5.3), thus drastically improving the accuracy and precision of the subsequent structure refinement. Given the (rare) case of very strong data with resolution of at least 1.2 Å (Sheldrick's Rule, [53,54]) and small protein size, structures can be determined *ab initio* via Direct Methods (section 5.2.4.2, [55]).

Additional effort is required to collect complete data sets at very high resolution. In certain crystal orientations, it is principally impossible for geometrical reasons to record all reflections. A part of the diffraction pattern (affectionately called the "apple core") remains unrecorded. While this range is small (a few percent) at normal resolution, it can become large at very high resolution, and suitable hardware that allows movement of the crystal about another axis must be interfaced with the strategy devising program. Despite the large recording area of modern detectors and the short wavelengths used at synchrotrons[*], additional sweeps at larger detector offset angles can become necessary and the data collection strategy quite elaborate, as in the early days of area multi-wire detectors with small solid angle coverage. A finer slicing of the rotation range in frames of about 0.2 deg has certain benefits [56], and an increasing number of data collection programs will probably implement this option in the near future [57].

5.2.3.6   Single Anomalous Diffraction Data and SAD from Sulfur

Anomalous data collection requires that in addition to a unique wedge of data covering the reciprocal space asymmetric unit, the Friedel mates (reflections of inverse indices in the centrosymmetrically related part the diffraction pattern) must be recorded. A most useful difference in intensity between Friedel mates results from wavelength dependent, anomalous scattering contributions, and intensity difference data are the basis for location of the anomalously scattering atoms in the phasing stage (see section 5.2.4).

Anomalous data are commonly recorded in smaller blocks (15–30 deg) of data and their inverse segment. Possible radiation damage or beam decay require this precaution. Splitting the data set into smaller blocks that include the corresponding inverse has the additional benefit that, after recording the first block, the significance of the anomalous difference signal can be determined, and the data collection expert

---

[*] According to Bragg's Law, shorter wavelength (higher energy) compresses the diffraction pattern and more data can be recorded within the same solid angle. Large unit cell dimensions, which require larger crystal-detector distance for spot resolution, can partly eliminate this advantage.

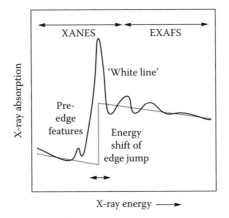

FIGURE 5.4 X-ray absorption edge. An X-ray absorption edge scan for the anomalous marker atom is necessary for optimal choice of wavelengths in MAD experiments. The theoretical position of the absorption edge (saw tooth line) shifts due to varying chemical environment, and the scan is decorated with features stemming from electronic transitions in the X-ray Absorption Near Edge Spectrum (XANES) region and from nearest neighbor scattering in the Extended X-ray Absorption Fine Structure (EXAFS) region. Absorption edges can have white lines, which result from electronic transitions into unoccupied atomic energy levels for elements with np or nd levels with n > 3. White lines contribute to a substantial increase in anomalous peak signal.

system should adjust data collection times accordingly. Alternatively, if no usable anomalous signal can be expected in reasonable time[*], it may be more efficient to abandon data collection and to proceed to another crystal.

Anomalous data collection for single wavelength experiments does not require measurement of an X-ray absorption spectrum (XAS), (Figure 5.4). The experiment must be conducted at or above the absorption edge of the selected marker element, but the exact determination of the absorption edge spectrum, as is necessary to optimize dispersive ratios in multiple anomalous diffraction (MAD) experiments, is not needed. In cases where white lines in the spectrum can be present (elements of the third period and higher), experimentally determining the exact absorption maximum is of advantage for maximizing the anomalous differences.

Special considerations are required when using native sulfur of the Met and Cys residues as anomalous markers. Even the longest practically usable X-ray wavelengths[**] (around 2.5–2 Å) are far above the K absorption edge of sulfur, and the anomalous difference signal is often as low as ~0.5% of the total signal [58]. However, given sufficiently redundant data collection via integration of multiple sweeps covering the reciprocal space unit many times (720 deg and more of rotation with varying crystal orientation), data with S/N ratio sufficiently high to extract

---

[*] Note that to obtain twice (n = 2) the S/N ratio or signal, four times (n^2) as much data collection time is needed, which over-proportionally reduces throughput and hence, process efficiency.

[**] Geometric and X-ray optical constraints of the tunable beam line components, as well as rapidly increasing absorption at longer wavelengths (lower X-ray energies), set a limit to how long a wavelength can be used experimentally.

anomalous intensity differences can be collected. In combination with powerful density modification techniques, the SAS method proposed 20 years ago by Wang [59], has recently been shown to be quite successful [58,60]. Given that no special marker atoms need to be introduced into the protein, the method is likely to gain rapid acceptance in high throughput crystallography (see Matthews [61] for a review about SAD data collection and phasing).

5.2.3.7 Multiple Anomalous Diffraction Data

MAD phasing [14,62] exploits additional redundancy in anomalous signal by not only using anomalous (Friedel- or Bijvoet-) differences *within* each data set, but also dispersive differences *between* data sets recorded at different wavelengths. To optimize these differences, an accurate experimental absorption edge scan for the phasing element needs to be recorded, and between two and four MAD wavelengths are selected (Figure 5.5). The anomalous differences are largest at the absorption edge maximum (max f''), and the dispersive differences are largest between the other data sets and the data set recorded at the inflection point of the edge jump, corresponding to a minimum in f. The high-remote data are usually recorded several

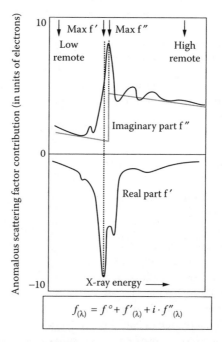

$$f_{(\lambda)} = f^\circ + f'_{(\lambda)} + i \cdot f''_{(\lambda)}$$

FIGURE 5.5 Choice of MAD wavelengths. The normalized X-ray absorption edge scan for the anomalous marker atom defines optimal wavelengths choice in MAD experiments. Top Panel shows the imaginary contribution f'' to the atomic scattering factor of the marker atom (Formula in box). The Kramers-Kronig transform (lower panel) shows a minimum at the inflection point of the edge and is located at the maximum of the real contribution f. Note that the values for max f' and f'' are close together; an exact edge scan is therefore necessary for optimal MAD experiments.

hundred eV above the edge and contain still substantial internal anomalous signal. The least anomalous difference signal, but still dispersive contributions, are expected from the optional low-remote data set, collected several hundred eV below the absorption edge. In view of signal loss through radiation damage, the most common strategy is to obtain peak wavelength data first (still enabling SAD phasing with a reasonable chance), followed by inflection point data (this second data set providing mostly dispersive differences), and the high-remote data set (with redundant internal anomalous differences and large dispersive differences against inflection data). All other basic data collection strategy considerations regarding completeness and S/N ratio discussed in the previous sections apply to MAD data as well. MAD data collection is currently the phasing method of choice in high throughput protein crystallography [14].

5.2.3.8 Raw Data Warehousing

On top of the data collection and decision making involved to this point, the challenge of raw data warehousing is substantial in HTPX efforts. On high throughput beam lines equipped with 3 x 3 module CCD detectors, saving a single image (frame) can require over 10 MB of disk space, and data accumulation rates can approach gigabytes per minute. Online data reduction while rapidly archiving the original image data places substantial demands on the IT infrastructure. Automated harvesting of experimental parameters and data collection statistics practically dictates the use of a powerful data base system for storage and retrieval.

5.2.4 CRYSTALLOGRAPHIC COMPUTING

Once data collection is successfully finished, the remaining steps of the structure solution are carried out *in silico*. Crystallographic computing has made substantial progress, largely due to abundant and cheap high-performance computing. It is now possible to solve and analyze complex crystal structures entirely on $2k laptop computers. Consequently, automation has reached a high level, although many compatibility and integration issues remain. As a result, the actual process of structure solution, although the theoretically most sophisticated part in a structure determination, is commonly not considered a bottleneck in HTPX projects. Given reliable data of decent resolution (~2.5 Å or better) and no overly large or complex molecules, many structures can in fact be solved *de novo* and refined within several hours.

5.2.4.1 Data Reduction and Scaling

The raw data obtained from the data collection program or expert system need to be further merged, sorted and reduced into a unique set according to the Laue group, and if not already known, the possible space groups need to be determined. If multiple data sets are used for phasing, these data sets must be brought onto a common scale as well. Depending on the amount of integration, this may be handled by the data collection experts system or by sequential programs, and partly by phasing programs. At this stage, the possible number of molecular subunits in the asymmetric unit of the crystal can be estimated [63,64], but as in case of the space group, the answer may not be unambiguous and must await the metal substructure solution. An automated HTPX program package or system has to successfully handle the multiple possibilities and provide proper decision branching [44]. For isomorphous data sets from different crystals, additional complications resulting from multiple indexing possibilities also need to be expected and accounted for.

5.2.4.2 Phasing: General Remarks

The core of the phase problem, which makes protein crystallography nontrivial, is the absence of direct phase information in diffraction data. Two quantities per reflection need to be known in order to reconstruct the electron density (Figure 5.6): the magnitude of the scattering vectors (or structure factor), which is proportional to the square root of the measured reflection intensity, and the relative phase angle of each scattering vector, which cannot be directly measured. Unfortunately, the phases dominate the electron density reconstruction, a fact giving rise to the phenomenon of phase- or model-bias. Incorrect phases from a model tend to reproduce incorrect model features despite experimental data from the true structure. Bias minimization will be discussed in the section about molecular replacement.

Practically all macromolecular phasing techniques used in HTPX depend on the presence and solution of a marker atom substructure (Table 5.2). By creating difference intensities between data sets with and without the contributions from the marker atoms, the initial problem is reduced to solving a substructure of a few to a few hundred, versus many thousands to ten thousands of atoms (Figure 5.7). The intensity differences can arise between absence (native) and presence (derivative) of heavy atoms, which forms the basis of isomorphous replacement techniques. Dif-

$$\rho_{(x,y,z)} = \frac{1}{V} \sum_{-h}^{h} \sum_{-k}^{k} \sum_{-l}^{l} |F_{hkl}| \cdot \exp[-2\pi i(hx + ky + lz - \alpha_{hkl})]$$

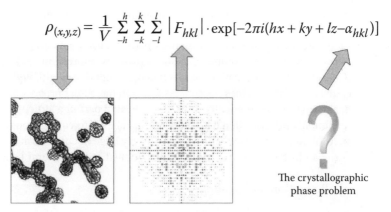

The crystallographic phase problem

FIGURE 5.6 The nature of the crystallographic phase problem. Reconstruction of electron density $_{(x,y,z)}$ via Fourier transformation (formula) requires two values for each reflection: The structure factor amplitude $|F_{hkl}|$, which is proportional to the square root of the measured reflection intensity and readily available, and the phase angle $_{hkl}$, which is unknown. Additional phasing experiments need to be carried out to obtain the missing phases. The need to determine phases by other means than direct measurement is referred to as the Crystallographic Phase Problem.

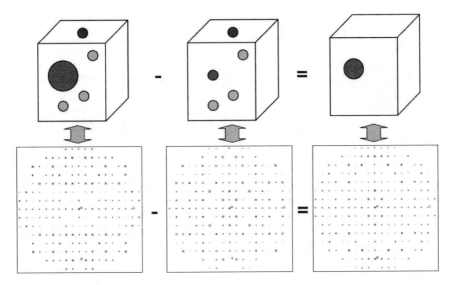

FIGURE 5.7 The principle of crystallographic difference methods. Top row presents the real space scenario, showing how differences simplify the solution of the isomorphous marker substructure (big atom). Left crystal, heavy marker atom derivative; middle crystal, native; and right side, fictitious "difference crystal." The diffraction patterns (bottom row) are the reciprocal space representation of the real space scenario described above with crystals. In a similar way, anomalously scattering marker atoms creates anomalous differences *within* a diffraction data set, and dispersive differences *between* data sets recorded at different wavelengths.

ferences also arise from different anomalous intensities originating from native (S, Fe, Cu), engineered (Se), or soaked heavy metal derivative anomalous scatterers at a single wavelength; or additionally from dispersive differences between data sets recorded at different wavelengths (MAD)[*]. In anomalous methods, all data are preferably collected from one single crystal and are then perfectly isomorphous. Combined with highly redundant data collection, excellent experimental phases can be obtained even for weaker high resolution reflections *via* anomalous phasing techniques. Consequently, anomalous methods are the workhorse of high throughput phasing. Combinations of various phasing techniques are also possible, providing higher redundancy in the phase angle determination.

5.2.4.3 Substructure Solution

The solution of the marker substructure is the first step in determination of *de novo* phases. Three-dimensional maps containing peaks at the position of interatomic distance vectors between the marker atoms can be created from difference intensity data without the use of phases. Such difference Patterson maps (Figure 5.8) contain

---

[*] The pseudo-SIRAS-like treatment of MAD data as presented here and used for example in SOLVE (Terwilliger et al., 2001), is different in details from the explicit solution of the MAD equations originally used by Hendrickson (Hendrickson, 1991).

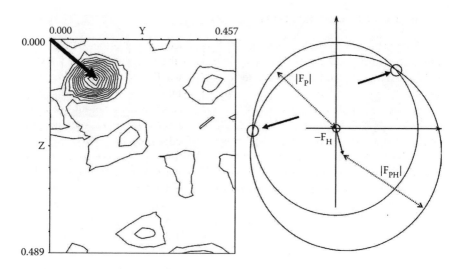

FIGURE 5.8 Phasing via metal substructure. Left panel: Harker section of a Patterson map created from the difference dataset in Figure 5.7. The marker atom positions are derived from analysis of distance vectors leading from the origin to the Patterson peaks. Right panel: Once the positions of the marker atoms are refined (represented by substructure vector F_H), the phase equations are solved and two possible solutions for the phase angle are obtained (circled). Degeneracy of the phase angle solution is resolved by using additional derivative structures, anomalous data, and/or density modification techniques.

strong peaks in certain sections and along certain directions, and the correct marker atom positions giving rise to a consistent peak pattern can be determined. A number of software packages used in HTPX use Patterson techniques in varying flavors to find consistent solutions in one or more difference maps [65]. Due to the centrosymmetry of the Patterson space, the handedness of the metal substructure cannot be determined by Patterson methods alone. Anomalous contributions, direct methods (discussed below), or map interpretability yield additional information to break the inherent substructure enantiomer ambiguity.

Direct Methods provide an alternative avenue of solving the metal substructure *ab initio* from intensities reduced to normalized structure factor amplitudes. Statistical inferences about phase relations of improved starting atom sets from Patterson superposition techniques cycled with real space phase expansion [55,56] and dual space methods (SnB, [67]) are particularly successful. Substructure solution is highly automated, and Se substructures containing up to 160 atoms have been successfully solved with both SHELXD [68] and SnB [69]. In rare cases of strong data, atomic resolution (better than 1.2 Å) and modest size (from a few hundred to currently just over one thousand nonhydrogen atoms), complete protein structures can be solved by Direct Methods [55]. Direct Methods can also extract additional information from intensities determining the absolute handedness of the metal substructure [70]. The heavy atom positions are also used to determine the NCS operator needed for subsequent map averaging and density modification (discussed below).

5.2.4.4 Initial Phase Calculation

The phase angles of the reflections can be determined once positions of the marker atoms are refined, and hence, the magnitude and the phase angle of the marker contribution to the total diffraction intensity are known. Figure 5.8B shows a graphic representation (Harker diagram) visualizing the solution of the phasing equations. Two limitations become immediately clear. First, the solution is not unique if only one difference data set is available, leading to the need to break a second phase ambiguity in addition to the metal substructure handedness discussed above. Second, based on errors in both the measured intensities and the marker atom positions, each circle intersection defining a phase angle will contain a certain error. The inherent phase ambiguity can be removed by use of multiple derivatives; adding anomalous signal in single derivative cases; or via multiple wavelength methods. Multiple determinations of each phase angle also increase the probability for the phase angle to be correct, and thus increase the figure of merit for the best phase estimate.

Once a phase angle for each reflection is obtained, a first electron density map can be computed. If the handedness of the metal substructure has not yet been determined (except in pure MIR cases), a map containing anomalous contributions will be not or much less interpretable (lower figure of merit) if the wrong enantiomer of the substructure was used in the phasing calculations. In the case of single anomalous phasing data (SAD), even a proper map will contain density of the molecule, superimposed with noise features. To improve the interpretability of all maps and to compensate for the lack of a unique sign of the phases in the SAD case, iterative density modification and filtering techniques [59] are applied in the next step.

5.2.4.5 Density Modification Techniques

One of the most powerful tools at hand to obtain readily interpretable maps, which is particularly important for automated model building, are (direct space) density modification techniques. Implementations of solvent flattening [71], solvent flipping [72], histogram matching [73] and reciprocal space maximum likelihood methods [74] exploit the fact that protein molecules pack loosely in the lattice. Substantial solvent channels (about 30 to 70% of the crystal volume) are filled with nonordered solvent, giving rise to a uniform density distribution in the solvent region. Setting the solvent electron density to a constant value (flattening), in repeated cycles with adjustments in the solvent mask under extension of resolution [75], leads to drastically reduced phase angle errors, and hence, clearer and better interpretable electron density maps (Figure 5.9). Density modification is also important in permitting phase extension to higher resolution in the frequent case where the derivative or anomalous data used for substructure solution and phasing do not extend to the same high resolution as the native data [76].

Another powerful variation of density modification is map averaging, which is applicable both in presence of quite frequent non crystallographic symmetry (NCS), and in model bias removal techniques based on multiple perturbed models (discussed later). The principle of NCS averaging is that if more than one molecule (for example,

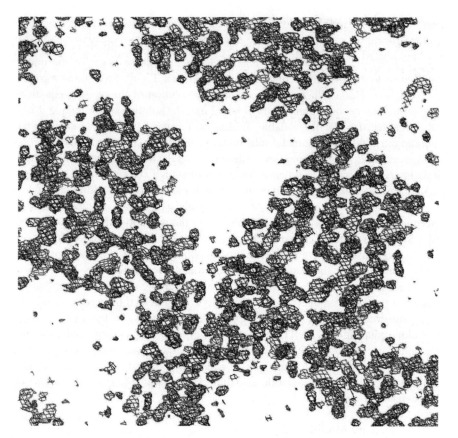

FIGURE 5.9 Experimental electron density map after density modification. The map shows a clear outline of the packed protein molecules, with solvent channels between them. Presence of these solvent channels allows soaking of small molecule ligands or drugs into protein crystals.

a homodimer) is present in the asymmetric unit of the crystal creating additional noncrystallographic symmetry, the diffraction pattern and hence the back-transformed map will contain redundant information. The electron density of the different copies of the molecule can be averaged (consistent features will amplify whereas noise and ambiguous density will be suppressed), and again, a greatly improved electron density map of the molecule results[*]. Map averaging is also possible between different crystal forms of the same protein [77], and in view of the increasing number of different crystal forms obtained via high throughput crystallization efforts, may be attractive to routinely implement.

    Full automation of NCS averaging is not trivial [78]. One subtlety is that the atoms of metal substructure often are found in different asymmetric units and/or in adjacent unit cells. Early determination and refinement of the NCS operators is

---

[*] Density modification by map averaging is in fact so powerful that virus capsid structures, which are highly symmetric and contain up to 60 copies per molecule, can be phased without marker techniques.

necessary for subsequent map averaging, as is determination of the proper molecular envelope (or mask) to submit only the map of one complete molecular subunit for initial automated model building. Increased attention towards automated utilization of NCS [79–81] will eventually lead to stable integrated expert systems handling automated map averaging.

5.2.4.6 Map Interpretation and Model Building

Once an electron density map of best possible quality is obtained from improved experimental phases, a model of the protein structure must be built into the electron density. The process is generally more successful with clean maps and at higher resolution. Traditionally, model building was carried out by hand, using programs that graphically display the electron density and allow placement and manipulation of protein backbone markers and residues, combined with varying real space and geometry refinement tools (Table 5.3). High throughput requires that the interpretation of the map, building, and if possible, refinement and correction steps are carried out automatically by specialized programs without the need for graphical user interaction. The programs generally follow a procedure similar to what is used in manual model building, with the benefit of fast and automated library searches. In the first step, Cα backbone atoms are placed at recognizable branching points of main and side chain density, and the longest contiguous chain is sought [82,83]. Search of fragment libraries [71] and use of preferred rotamers [84] can improve the model quality already in the first building cycles. At higher resolution, the electron density shapes of the residues are also less degenerate than at low resolution, aiding proper feature recognition and faster sequence alignment. Sequence anchoring either on stretches of distinct residues or at marker atom sites facilitates tracing of the chains properly as well as recognition and correction of branching errors. Interestingly enough, no automated model building program currently appears to take advantage of the simultaneous positional and sequential marking by Se atoms. One automated program, RESOLVE [81,85], uses iterative model building in combination with maximum likelihood density modification, and ARP/wARP is based on cycled dummy atom and fragment building and refinement [86,87]. Table 5.3 contains a summary of additional public and commercial programs. A brief direct comparison of RESOLVE, MAID [88] and ARP/wARP has been compiled recently [89].

5.2.4.7 Refinement

The initial raw model built into the electron density must be refined and, by phase combination of experimental and model phases, an improved map is obtained. The model building step is repeated, additional model parts are built into the improved density and necessary corrections are made to the model. The refinement itself consists of adjusting general scaling parameters to match observed and calculated data overall, and the atomic coordinates of the initial model are refined so that the differences between observed and calculated data are minimized [90]. The common global measure of this agreement is the R-value, often expressed in percent:

$$R = \frac{\displaystyle\sum_{hkl} \left\| F_{obs} \right| - k \left| F_{cal} \right\|}{\displaystyle\sum_{hkl} \left| F_{obs} \right|}$$

With improving model quality, individual atomic temperature factors (a measure for positional displacement either via thermal motion or through disorder) are also refined, and in cases of atomic resolution, it may be possible to refine ansiotropic temperature factors. Even at modest resolution, however, grouping and refining parts of the molecule through torsion, libration, screw (TLS) motion modes improves refinement and model quality [91]. Bulk solvent model corrections are also applied during the refinement [92].

A general problem in protein structure refinement is the low data/parameter ratio. Such refinements are not stable against experimental data alone, and additional restraint terms creating penalties for deviations from geometry target values are introduced (93). The poor data/parameter ratio not only requires special refinement techniques, but also implementation of safeguards against overfitting and introduction of model bias. Strict crossvalidation against a small subset of unused (free) data, monitored by the R value for the free data set [94] is standard practice. Properly used, this free R prevents overfitting, i.e., introduction or variation of parameters that do not contribute to model improvement but only reduce the fitting residual (an example would be excessive solvent building). Maximum Likelihood (ML) refinement target functions [95] allowing for error in the model based on the sigmaA map coefficients [96] are now universally implemented and reduce, but not eliminate, susceptibility to model bias. Refinements using ML targets and torsion angle refinement combined with simulated annealing techniques [97,98] have a large convergence radius and are especially useful for refinement of initial models far from the correct one, as is often the case in Molecular Replacement (MR, discussed below). Despite substantial progress in automated model building, a surprisingly large number of final adjustments and repair of housekeeping errors still need to be made manually to fine-tune the structural model. Integration and cycling of refinement and building with real-time validation, including local validation via real space fit correlation and chemical plausibility, will further improve the quality of automatically built and refined models.

5.2.4.8 Automated Molecular Replacement

The principle of MR is to use a homologous search model, sufficiently close to the unknown structure, and to properly replace — in the sense of repositioning — the molecule(s) in the unit cell until good correlation between the observed diffraction data and the data calculated using the replaced search model indicates a solution. Once the correct position of the search model is determined, initial — but highly biased — phases for the unknown structure can be obtained. About 75% of the structures in high throughput efforts are currently solved using MR, particularly in repeated structural screening of the same protein co-crystallized with different drugs. Rapid data collection and automated molecular replacement routines are the backbone of high throughput structure based drug screening [1].

A single native data set only is necessary for MR phasing, provided a homology model within no more than a few Å backbone coordinate r.m.s.d. is found in the Protein Data Bank (PDB), [99], or built by computer modeling [100]. Generally, accuracy of the model appears to be more important than completeness for convergence of the MR search [101]. Conventional search programs perform separate rotational and translational searches to find the proper position. Innovations in the method such as full 6-dimensional, fast evolutionary searches [102], or combinations with new maximum likelihood based approaches increasing the radius of convergence of the searches, are becoming established [98,103–105]. The process is easily automated, and given the anticipated rise in coverage of structural folds available in the public databases due to the structural genomics efforts, MR sees constantly increasing use.

A general strategy for automated search model building is to identify a set of possible template structures with sequence alignment tools and to retrieve them automatically from the protein structure database. Search models are built from each template either directly or obtained via homology modeling, and target side chains can be automatically built. Parallel molecular replacement searches for each of the highest scoring models are branched to a computer cluster and the models are evaluated according to their correlation coefficient to observed data. Fold recognition models, although steadily increasing in quality [100], still may not produce successful MR probes. The immediate feedback possible through evaluation of the model against experimental data, however, should allow for adaptive correction of the model building algorithms in response to MR scoring. Model completion techniques such as loop building and gap filling are likely to benefit from such experimental restraints.

A drawback of the MR method is its high susceptibility to model (phase) bias recreating the model's features in the electron density, and the bias minimization techniques described in the refinement section must be rigorously applied. Effects of model bias can be insidious [106] and are not easily recognized by commonly used global structure quality descriptors such as R and free R [107]. A fully automated model bias removal protocol (Shake&wARP, [108], available as a web service), based on a modified dummy atom placement and refinement and protocol [109], uses a combination of model perturbation, omit techniques, and maximum likelihood refinement together with multiple map averaging techniques to effectively minimize phase bias.

5.2.4.9 Initial Model Validation

The refined final model is subjected to an array of validation techniques, ranging from a ranking via global validators and geometry deviations against other structures, to detailed chemical and folding plausibility checks based on prior knowledge (Chapter 10). Global indicators such as R and even the cross-validated free R value cannot be specific as far as the local quality of a structure model is concerned. Similar considerations hold for average deviations from geometry targets, which largely reflect the weight of the restraints chosen in refinement. While limited local errors may not be a great concern for a structure solved to populate fold space, a drug target structure needs to be of high quality around the specific drug or ligand binding site. Local methods of assessment evaluate the correlation of the model against a bias minimized electron density map [108,110,111], or check local geometry descriptors for outliers on a per residue basis (PROCHECK [112], WHAT-CHECK [113], see also Chapter 10). Extended stretches of consistently high deviations in either case are indicative of serious problem zones within the model. An adaptive response by the model building program to such analysis would be desirable in fully automated structure determination packages. Presently, the validation programs are largely self-contained and standalone programs, and model corrections are still made off-line and *a posteri*. Correction of nuisance or housekeeping errors such as stray atoms, nomenclature violations, sequence inconsistencies, etc. could be easily automated. Currently, most of these nuisance errors are corrected only at the off-line, PDB deposition level (Autodep, ADIT, Chapter 10). It should be expected that such corrections will be automated to a much higher level, up to fully automated data harvesting and deposition/annotation procedures.

In the author's experience, automatically solved and built structures do not seem to be any less reliable (or more wrong) than conventionally determined structures. The automated phasing programs fail to a similar degree as those scripted by a human operator, but automation provides the benefit that a much more rigorous and consistent pursuit of multiple options can be implemented. Model building programs manage fairly well at the beginner's level with good maps. They tend to fail in borderline cases when a skilled crystallographer might still be able, with serious effort (and fantasy), to successfully bootstrap a build. On the other hand, building programs are also devoid of any desire to salvage an abysmal project, which avoids a number of bias issues *a priori* (see [106] or [114] for illustrative examples). The author feels that the danger of flooding the structural data bases with low quality models from automated HTPX structure determination projects is overstated.

5.3 SUMMARY OF PROGRESS AND CHALLENGES

Development of high throughput techniques for crystallographic structure determination has been fairly rapid, with a much clearer picture of the more challenging areas now emerging. While the technical advances of public efforts are relatively easy to track [17], proprietary developments are much less accessible, and their throughput figures tend towards the optimistic side [115].

5.3.1 SYNCHROTRON X-RAY SOURCES

The intensity and brilliance of third generation synchrotron light sources has reached limiting radiation dose levels, where further increase in brute intensity will not substantially increase throughput (ESRF ID13, APS 19ID). Significant radiation damage to the crystals [116], and recognizable modifications of the protein molecules by high radiation [117] already require attenuation of the most powerful X-ray beams. Upgrades of older synchrotron sources like SSRL's SPEAR to third generation output levels, however, will add significantly to the available synchrotron capacity. Tunable micro-focus sources based on Compton scattering have potential [118], and future developments may lead to viable instruments and new techniques such as broad bandwidth SAD phasing. The much-discussed free electron lasers will probably not readily contribute to future structural genomics efforts, even if the considerable technical problems can be overcome [119,120].

5.3.2 ROBOTIC CRYSTAL HARVESTING AND MOUNTING

Increasing the capacity of existing beam lines by efficient use of beam time is a key benefit of rapid automated mounting methods. A few $100k invested in robotics can substantially increase the throughput capacity of $10M beam lines. In contrast, crystal harvesting, cryo-protection, soaking, and loop-mounting are still largely off-line procedures. The (mis)handling of crystals during these steps is probably the single most significant point of failure in the structure determination process. Whether the manual mounting steps will turn into bona fide throughput-limiting bottlenecks will depend on what level of sustained throughput capacity protein production and crystallization can achieve.

5.3.3 FULLY AUTOMATED DATA COLLECTION SYSTEMS

Data collection, through its intimate connection of direct experiment control and computation, and with key decisions necessary relatively late in the process, presents a challenge for full automation. Increasing sophistication and automation of the upcoming versions of data collection suites [44,56,121] and their seamless interfacing into the subsequent phasing programs will likely reduce the number of abandoned data collections and data sets, and hence increase throughput.

5.3.4 AUTOMATED PHASING, MODEL BUILDING, REFINEMENT, AND DEPOSITION

Cooperation between publicly sponsored program developer teams such as CCP4, SOLVE or PHENIX, as well as improvement and interfacing of independently developed suites such as SHARP or MAID (Table 5.3) have already greatly increased the ease of the computational parts of structure solution. The next logical step would be to interface tightly with the data collection suites acting as a front end for the structure solution programs, and to incorporate feedback from independent validation programs into model building, refinement, and deposition [122] of validated models. Although in each of these areas excellent programs exist, seamless integration is still missing.

5.3.5 INTERFACING WITH STRUCTURAL BIOINFORMATICS

As already discussed in section 5.1.3, structures are determined in HTPX efforts with mainly two objectives in mind: Use of the structure to determine a potentially new fold of unknown function, or as a lead structure for drug design. Although the former is probably not the real high throughput driver, direct interfacing during build with fold comparison programs such as DALI [123] and further fold analysis including active site searches (see Chapter 19 of [124]) would be desirable. Data harvesting, automated annotation [125], and interfacing to laboratory management systems are just some aspects of integration within the realm of structural bioinformatics [124].

As structure guided drug design is perhaps the major driver for HTPX [1,9], automated ligand building and docking have been the focus of development in commercial software (for example, Accelrys, Astex, Molsoft), and these approaches are currently being implemented in at least one publicly available model building program (ARP/wARP, Victor Lamzin, EMBL Hamburg, personal communication). Interfacing to Virtual Ligand Screening, automated lead optimization, and in-silico ADME/T property prediction programs would be the next step towards fully automated drug HTPX guided target structure analysis, up to structure based drug discovery [8].

5.3.6 THROUGHPUT ESTIMATES

A quick analysis of the PDB deposition data reveals that the common notation of an exponential increase in PDB structure depositions per year may not be maintained. After a brief, nearly overexponential surge in the early 1990s (perhaps largely to the credit of Hendrickson's MAD technique) the number of depositions per year has increased less rapidly, and the curve has flattened considerably since the late 1990s. If the deposition rate indeed reflects the impact of new technology, then one would expect a similar deposition surge in the near future, similar to that occurring in the early 1990s, as a result of the HTPX efforts. The question arises to what degree the synchrotron sources (and improved anomalous in-house phasing techniques) then can satisfy the need for more and more beam time. Assuming a conservative 8 hrs data collection time for complete 4-wavelengths MAD data, and 150 full operating days a year, about 8 such beam lines could produce the data for all structures deposited in 1 year. As there are approximately 10 to 15 times as many PX beam lines available throughout the world, a shortage of beam time cannot be easily explained by a lack of hardware. Even accounting for industrial efforts and additional data sets, long waiting times for beam time appear surprising. Suboptimal use of beam time, dropped crystals, aborted scans, unprocessible datasets, and unsolved phasing probably account for most of the discrepancy. In nearly all of these instances of failure, intelligent automation (which is not necessarily an oxymoron) will greatly enhance the success rates and efficiency, and allow for a further, manifold increase in structure determinations and depositions. Nonetheless, a "killer application" in protein production and crystallization could prove this extrapolation dreadfully wrong.

ACKNOWLEDGMENTS

The following individuals have provided comments, insight, information or updates helpful during the preparation of this chapter: Gerry McDermott, Thomas Earnest, Harry Powell, Gerard Bricogne, Victor Lamzin, Clemens Vonrhein, Dirk Kostrewa, Tim Harris, Aled Edwards, Andrzej Joachimiak, Ehmke Pohl, Martin Walsh, Sean McSweeney, and Katherine Kantardjieff. The author thanks James C. Sacchettini, Texas A&M University, for support of his sabbatical leave at Texas A&M University.

Preparation of the manuscript at the Texas A&M University was funded by NIH P50 GM62410 (TB Structural Genomics) center grant and the Robert A. Welch Foundation. LLNL is operated by University of California for the US DOE under contract W-7405-ENG-48.

RESOURCE LIST: GENERAL REFERENCES, CRYSTALLOGRAPHIC TECHNIQUES

G Rhodes, *Crystallography Made Crystal Clear*, 2nd ed., Academic Press, London, UK, 2000. Easy introductory reading for the user of protein structure models.

J Drenth, *Principles of Protein X-ray Crystallography*, 2nd ed., Springer Advanced Texts in Chemistry, New York, NY, Springer, 1999. More detailed derivations requiring some mathematical background.

C Jones, B Mulloy, and M Sanderson, eds., *Crystallographic Methods and Protocols*, Humana Press, Totowa, NJ, 1996. Selected chapters for a quick overview on specific topics.

C Giacovazzo, H Monaco, D Viterbo, F Scordari, G Gilli, G Zanotti, and M Catti, *Fundamentals of Crystallography*, 2nd ed., IUCr Texts on Crystallography, Vol. 7. Oxford Science Publications, Oxford, UK, 2002. General crystallographic treatise with emphasis on computational crystallography.

M Rossmann and E Arnold, *Crystallography of Biological Macromolecules. International Tables for Crystallography*, Vol. F, Kluwer Academic Publishing, Dordrecht, NL, 2001. Concise advanced reading from protein expression to model analysis. Several chapters from this volume are explicitly cited in the Reference list.

C Carter and R Sweet, eds., *Macromolecular Crystallography. Methods in Enzymology*, Vol. 276, 277, 368, Academic Press, London, UK, 1997. Collection of in-depth chapters detailing key topics.

H Wyckoff, C Hirs, and S Timasheff, eds., *Diffraction Methods for Biological Macromolecules. Methods in Enzymology*, Vol. 114, 115, Academic Press, London, UK, 1985. Overview of state of the art 20 years ago, but still valuable reading for selected chapters.

COMPLETE JOURNAL SPECIAL ISSUES

Nature Structural Biology, 5, August 1998, Synchrotron Supplement.

Nature Structural Biology, 7, November 2000, Structural Genomics Supplement.

Accounts of Chemical Research, 36, March 2003, Special Issue on Structural Genomics.

Journal of Structural Biology, 142, April 2003, Macromolecular Crystallization in the Structural Genomics Era.

J. Synchrotron Radiation, 11(1), 2004, Proceedings of International Symposium on Diffraction Structural Biology, Tsukuba, Japan.

Acta Crystallographica, D59(11), 2003, Proceedings of CCP4 Study Weekend on Experimental Phasing.

Acta Crystallographica, D58(11), 2002, Proceedings of CCP4 Study Weekend on High Throughput Structure Determination.

Acta Crystallographica, D57(10), 2001, Proceedings of CCP4 Study Weekend on Molecular Replacement and its Relatives.

Acta Crystallographica, D55(10), 1999, Proceedings of CCP4 Study Weekend on Data Collection and Processing.

REFERENCES

1. TL Blundell, H Jhoti, and C Abell. High-throughput crystallography for lead discovery in drug design, *Nature Rev. Drug Discovery*, 1, 45–54, 2001.

2. N Ban, P Nissen, J Hansen, PB Moore, and TA Steitz, The complete atomic structure of the large ribosomal subunit at 2.4 Å resolution, *Science*, 289, 905–20, 2000.

3. I Xenarios, D Eisenberg, Protein interaction databases, *Curr. Opin. Struct. Biol.*, 12, 334–339, 2001.

4. M Etter, NMR and X-ray crystallography: interfaces and challenges, *ACA Trans.*, 24, 1988.

5. K Moffat, Ultrafast time resolved crystallography, *Nature Struct. Biol.*, Suppl., 5, 641–642, 1998.

6. J Knowles and G Gromo, Target selection in drug discovery, *Nature Rev. Drug Discovery*, 2, 63–69, 2003.

7. JC Norvell and A Zapp-Machalek, Structural genomics programs at the US National Institute of General Medical Sciences, *Nature Struct. Biol.*, Suppl 7, 931, 2000.

8. J Yon and H Jhoti. High-throughput structural genomics and proteomics: where are we now? *Targets*, 2(5), 201–207, 2003.

9. S Burley, The FAST and the curios, *Mod. Drug Discovery*, 7(5), 53–56, 2004.

10. B Rupp, High throughput crystallography at an affordable cost: The TB Structural Genomics Consortium crystallization facility, *Acc. Chem. Res.*, 36, 173–181, 2003.

11. S Dry, S McCarthy, and T Harris, Structural genomics in the biotechnology sector, *Nature Struct. Biol.*, Suppl. 7, 946–949, 2000.

12. KE Goodwill, MG Tennant, and RC Stevens, High-throughput X-ray crystallography for structure based drug design, *Drug Discovery Today*, 6, 15, S113–S118, 2001.

13. U Heinemann, G Illing, and H Oschkinat, High-throughput three-dimensional protein structure determination, *Curr. Opin. Biotechnol.*, 12, 4, 348–354, 2001.

14. WA Hendrickson, Synchrotron Crystallography, *Trends Biochem. Sci.*, 25, 637–643, 2000.

15. CM Ogata, MAD phasing grows up, *Nature Struct. Biol.*, Suppl. 5, 638–640, 1998.

16. T Harris, The commercial use of structural genomics, *Drug Discovery Today*, 6(22), 1148, 2001.

17. PE Bourne, The status of structural genomics, *Targets*, 2, 5, 181–182, 2003.

18. GE Moore, Cramming more components onto integrated circuits, *Electronics*, 8, 15, 2–5, 1965.

19. DW Rodgers, Cryocrystallography techniques and devices, *Int. Tables for Crystallography F*, 202–208, 2001.

20. RE Thorne, Z Stum, J Kmetko, K O'Neill, and R Gillilan, Microfabricated mounts for high-throughput macromolecular cryocrystallography, *J. Appl. Crystallogr.*, 36, 6, 1455–1460, 2003.

21. A Sanjo and RE Cacheu, New Microfabricated Device Technologies for High Throughput and High Quality Protein Crystallization, *ICCBM9 Book of Abstracts*, O.I.2, 2002.

22. A D'Arcy, A MacSweeney, M Stihle, and A Haber, The advantages of using a modified microbatch method for rapid screening of protein crystallization conditions, *Acta Crystallogr.*, D59, 396–399, 2003.

23. C Hansen, E Skordalakes, J Berger, and S Quake, A robust and scalable microfluidic metering method that allows protein crystal growth by free interface diffusion, *Proc. Natl. Acad. Sci.*, USA, 99, 16531–16536, 2002.

24. E Bodenstaff, F Hoedemaker, E Kuil, H deVrind, and J Abrahams, The prospects of nanocrystallography, Acta Crystallogr., D59, 1901–1906, 2002.

25. G Spraggon, SA Lesley, A Kreusch, and JP Priestle, Computational analysis of crystallization trials, Acta Crystallogr., D58(11), 1915–1923, 2002.

26. JR Luft, RJ Collins, NA Fehrman, AM Lauricella, CK Veatch, and GT DeTitta, A deliberate approach to screening for initial crystallization conditions of biological macromolecules, J. Struct. Biol., 142, 1, 170–179, 2003.

27. J Wilson, Towards the automated evaluation of crystallization trials, Acta Crystallogr., D58, 11, 1907–1914, 2002.

28. S Cusack, H Belrhali, A Bram, M Burghammer, A Perrakis, and C Riekel, Small is beautiful: protein micro-crystallography, Nat. Struct. Biol., Suppl. 5, 634–447, 1998.

29. C Riekel, Recent developments in microdiffraction on protein crystals, J. Synchrotron Rad., 11,1, 4–6, 2004.

30. E Garman and C Nave, Radiation damage to crystalline biological molecules: current view, J. Synchrotron Rad., 9, 327–328, 2002.

31. SA Islam, D Carvin, MJE Sternberg, and TL Blundell, HAD, a Data Bank of Heavy-Atom Binding Sites in Protein Crystals: a resource for use in multiple isomorphous replacement and anomalous scattering, Acta Crystallogr., D54, 1199–1206, 1998.

32. TJ Boggon and L Shapiro, Screening for phasing atoms in protein crystallography, Structure, 7, 8, R143–R149, 2000.

33. G Evans and G Bricogne, Triiodide derivatization and combinatorial counter-ion replacement: two methods for enhancing phasing signal using laboratory Cu Ka X-ray equipment, Acta Crystallogr., D58, 976–991, 2002.

34. RA Nagem, Z Dauter, and I Polikarpov, Protein crystal structure solution by fast incorporation of negatively and positively charged anomalous scatterers, Acta Crystallogr., D57, 996–1002, 2001.

35. Y Liu, CM Ogata, and W Hendrickson, Multiwavelength anomalous diffraction analysis at the M absorption edges of uranium, Proc. Natl. Acad. Sci., USA 98, 19, 10648–10653, 2001.

36. E Garman, Cool data: quantity AND quality, Acta Crystallogr., D55, 1641–1653, 1999.

37. R Sweet, The technology that enables synchrotron structural biology, Nature Struct. Biol. Suppl. 5, 654–656, 2000.

38. S Kriminski, CL Caylor, MC Nonato, KD Finkelstein, and RE Thorne, Flash-cooling and annealing of protein crystals, Acta Crystallogr., D58, 459–471, 2002.

39. LB Hanson, CA Schall, and GJ Bunick, New techniques in macromolecular cryo-crystallography: macromolecular crystal annealing and cryogenic helium, J. Struct. Biol., 142(1), 77–87, 2003.

40. DH Juers and BW Matthews, The role of solvent transport in cryo-annealing of macromolecular crystals, Acta Crystallogr., D60(3), 412–421, 2004.

41. WI Karain, GP Bourenkov, H Blume, and HD Bartunik, Automated mounting, centering and screening of crystals for high-throughput protein crystallography, Acta Crystallogr., D58, 1519–22, 2002.

42. SW Muchmore, J Olson, R Jones, J Pan, M Blum, J Greer, SM Merrick, P Magdalinos, and VL Nienaber, Automated crystal mounting and data collection for protein crystallography, Structure, 8, 12, 243–246, 2000.

43. G Snell, C Cork, R Nordmeyer, E Cornell, G Meigs, D Yegian, J Jaklevic, J Jin, RC Stevens, and TE Earnest, Automatic sample mounting and alignment system for biological crystallography at a synchrotron source, Structure, 12, 1–12, 2004.

44. A Leslie, H Powell, G Winter, O Svensson, D Spruce, S McSweeney, D Love, S Kinder, E Duke, and C Nave, Automation of the collection and processing of X-ray diffraction data — a generic approach, *Acta Crystallogr.*, D58, 1924–1928, 2002.

45. TM McPhillips, SE McPhillips, H-J Chiu, AE Cohen, AM Deacon, PJ Ellis, E Garman, A Gonzalez, NK Sauter, RP Phizackerley, SM Soltis, and P Kuhn, Blu-Ice and the Distributed Control System: software for data acquisition and instrument control at macromolecular crystallography beamlines, *J. Synchrotron Rad.*, 9, 401–406, 2002.

46. RM Sweet, The technology that enables synchrotron structural biology, *Nature Struct. Biol.*, Suppl. 5, 654–656, 1998.

47. HR Powell, The Rossmann Fourier autoindexing algorithm in MOSFLM, *Acta Crystallogr.*, D55, 10690–1695, 1999.

48. S Parsons, Introduction to twinning, *Acta Crystallogr.*, D 59, 11, 1995–2003, 2003.

49. TO Yeates, Detecting and overcoming crystal twinning, *Meth. Enzymol.*, 276, 344–58, 1997.

50. Z Dauter and KS Wilson, Principles of monochromatic data collection, *Int. Tables For Crystallography F*, 177–195, 2001.

51. Z Dauter, Data-collection strategies, *Acta Crystallogr.*, D55, 1703–1717, 1999.

52. W Minor, D Tomchick, and Z Otwinowski, Strategies for macromolecular synchrotron crystallography, *Struct. Fold. Des.*, 8, 5, 105–110, 2000.

53. GM Sheldrick, Phase annealing in SHELX-90: direct methods for larger structures, *Acta Crystallogr.*, A46, 467–473, 1990.

54. RJ Morris and G Bricogne, Sheldrick's 1.2Å rule and beyond, *Acta Crystallogr.*, D59, 615–617, 2003.

55. G Sheldrick, H Hauptman, C Weeks, R Miller, and I Uson, Ab initio phasing, *Int. Tables for Crystallography F*, 333–354, 2001.

56. W Pflugrath, The finer things in X-ray diffraction data collection, *Acta Crystallogr.*, D55, 1718–1725, 1999.

57. AGW Leslie, Integration of macromolecular diffraction data, *Acta Crystallogr.*, D55, 1969–1702, 1999.

58. UA Ramagopal, M Dauter, and Z Dauter, Phasing on anomalous signal of sulfurs: what is the limit? *Acta Crystallogr.*, D59, 1020–1027, 2003.

59. BC Wang, Resolution of Phase Ambiguity in Macromolecular Crystallography, *Meth. Enzymol.*, 115, 90–112, 1985.

60. Z Dauter, M Dauter, and ED Dodson, Jolly SAD, *Acta Crystallogr.*, D58, 496–508, 2002.

61. BW Matthews. Heavy atom location and phase determination with single wavelength diffraction data. International Tables for Crystallography F, 293–298, 2001.

62. WA Hendrickson, Determination of macromolecular structures from anomalous diffraction of synchrotron radiation, *Science*, 254, 51–58, 1991.

63. KA Kantardjieff and B Rupp, Matthews coefficient probabilities: Improved estimates for unit cell contents of proteins, DNA, and protein-nucleic acid complex crystals, *Prot. Sci.*, 12, 1865–1871, 2003.

64. BW Matthews, Solvent content of protein crystals, *J. Mol. Biol.*, 33, 491–497, 1968.

65. TC Terwilliger and J Berendsen, Automated MAD and MIR structure solution, *Int. Tables for Crystallography F*, 303–309, 2001.

66. I Uson and GM Sheldrick, Advances in direct methods for protein crystallography, *Curr. Opin. Struct. Biol.*, 9, 642–648, 1999.

67. CM Weeks and R Miller, The design and implementation of SnB v2-0, *J. Appl. Crystallogr.*, 32, 120–124, 1999.

68. TR Schneider and GM Sheldrick, Substructure solution with SHELXD, *Acta Crystallogr.*, D58, 1772–1779, 2002.
69. F vonDelft and TL Blundell, The 160 selenium atom substructure of KMPHT, *Acta Crystallogr.*, A58, C239, 2002.
70. Q Hao, YX Gu, CD Zheng, and HF Fan, OASIS: a computer program for breaking phase ambiguity in one-wavelength anomalous scattering or single isomorphous substitution (replacement) data, *J. Appl. Crystallogr.*, 33, 980–981, 2000.
71. KD Cowtan, Modified phased translation functions and their application to molecular-fragment location, *Acta Crystallogr.*, D54, 750–756, 1998.
72. JL Abrahams and AGW Leslie, Methods used in the structure determination of the bovine mitochondiral F1 ATPase, *Acta Crystallogr.*, D52, 30–42, 1996.
73. KYJ Zhang, SQUASH — combining constraints for macromolecular phase refinement and extension, *Acta Crystallogr.*, D49, 213–222, 1993.
74. TC Terwilliger, Reciprocal space solvent flattening, *Acta Crystallogr.*, D55, 1863–71, 1999.
75. KD Cowtan and P Main, Phase combination and cross validation in iterated density-modification calculations, *Acta Crystallogr.*, D52, 43–48, 1996.
76. KYJ Zhang, KD Cowtan, and P Main, Phase improvement by iterative density modification, *Int. Tables for Crystallography F*, 311–324, 2001.
77. G Taylor, The phase problem, *Acta Crystallogr.*, D 59(11), 1881–1890, 2003.
78. RW Grosse-Kunstleve and PD Adams, On symmetries of substructures, *Acta Crystallogr.*, D 59(11), 1974–1977, 2003.
79. TC Terwilliger, Statistical density modification with non-crystallographic symmetry, *Acta Crystallogr.*, D58, 2082–2086, 2002.
80. C Vonrhein and GE Schultz, Locating proper non-crystallographic symmetry in low-resolution electron-density maps with the program GETAX, *Acta* Crystallogr., D55, 225–229, 1998.
81. TC Terwilliger, SOLVE and RESOLVE: automated structure solution, density modification and model building, *J. Synchrotron. Rad.*, 11, 1, 49–52, 2004.
82. TR Ioerger and JC Sacchettini, Automatic modeling of protein backbones in electron density maps via prediction of C? coordinates, *Acta Crystallogr.*, D58, 2043–2054, 2002.
83. T Oldfield, Automated tracing of electron density maps of proteins, *Acta Crystallogr.*, D59, 483–491, 2003.
84. RL Dunbrack, Jr., Rotamer libraries in the 21st century, *Curr. Opin. Struct. Biol.*, 12, 431–440, 2002.
85. TC Terwilliger. Maximum likelihood density modification, *Acta Crystallogr.*, D56, 965–972, 2000.
86. R Morris, A Perrakis, and V Lamzin, ARP/wARP's model-building algorithms. I. The main chain, *Acta Crystallogr.*, D58, 968–75, 2002.
87. R Morris, P Zwart, S Cohen, F Fernandez, M Kakaris, O Kirillova, C Vonrhein, A Perrakis, and V Lamzin, Breaking good resolutions with ARP/wARP, *J. Synchrotron Rad.*, 11, 1, 56–59, 2004.
88. DG Levitt, A new software routine that automates the fitting of protein X-ray crystallographic electron-density maps, *Acta Crystallogr.*, D57, 1013–1019, 2001.
89. J Badger, An evaluation of automated model-building procedures for protein crystallography, *Acta Crystallogr.*, D59, 823–827, 2003.
90. LF TenEyck and K Watenpaugh, Introduction to refinement, *Int. Tables for Crystallography F*, 369–374, 2001.

91. MD Winn, MN Isupov, and GN Murshudov, Use of TLS parameters to model aniso-tropic displacements in macromolecular refinement, *Acta Crystallogr.*, D57, 122–223, 2001.

92. D Kostrewa, Bulk Solvent Correction: Practical Application and Effects in Reciprocal and Real Space, *CCP4 Newsletter Protein Crystallogr.*, 34, 9–22, 1997.

93. JH Konnert and WA Hendrickson, A restrained-parameter thermal-factor refinement procedure, *Acta Crystallogr.*, A36, 110–119, 1980.

94. AT Brünger, Free R value: A novel statistical quantity for assessing the accuracy of crystal structures, *Nature*, 355, 472–475, 1992.

95. GN Murshudov, AA Vagin, and ED Dodson, Refinement of Macromolecular Struc-tures by the Maximum-Likelihood Method, *Acta Crystallogr.*, D53, 240–255, 1997.

96. RJ Read, Model Phases: Probabilities, bias and maps, *Int. Tables for Crystallography F*, 325–331, 2002.

97. AT Brünger, J Kuryan, and M Karplus, Crystallographic R Factor refinement by molecular dynamics, *Science*, 235, 458–460, 1987.

98. PD Adams, NS Panu, RJ Read, and AT Brünger, Extending the limits of molecular replacement through combined simulated annealing and maximum-likelihood refine-ment, *Acta Crystallogr.*, D55, 181–190, 1999.

99. HM Berman, J Westbrook, Z Feng, G Gilliland, TN Bhat, H Weissig, IN Shindyalov, and PE Bourne, The Protein Data Bank, *Nucleic Acids Res.*, 28, 235–242, 2000.

100. DT Jones, Evaluating the potential of using fold-recognition models for molecular replacement, *Acta Crystallogr.*, D57, 1428–1434, 2001.

101. CR Kissinger, DK Gehlhaar, BA Smith, and D Bouzida, Molecular replacement by evolutionary search, *Acta Crystallogr.*, D57, 10, 1474–1479, 2001.

102. CR Kissinger, DK Gelhaar, and DB Fogel, Rapid automated molecular replacement by evolutionary search, *Acta Crystallogr.*, D55, 484–491, 1999.

103. RJ Read, Pushing the boundaries of molecular replacement with maximum likelihood, *Acta Crystallogr.*, D57, 1373–1382, 2001.

104. PD Adams, K Gopal, RW Grosse-Kunstleve, LW Hung, TR Ioerger, AJ McCoy, NW Moriarty, RK Pai, RJ Read, TD Romo, JC Sacchettini, NK Sauter, LC Storoni, and TC Terwilliger, Recent developments in the PHENIX software for automated crys-tallographic structure determination, *J. Synchrotron Rad.*, 11, Part 1, 53–5, 2004.

105. LC Storoni, AJ McCoy, and RJ Read, Likelihood-enhanced fast rotation functions, *Acta Crystallogr.*, D60, 432–438, 2004.

106. B Rupp and BW Segelke, Questions about the structure of the botulinum neurotoxin B light chain in complex with a target peptide, *Nature Struct. Biol.*, 8, 643–664, 2001.

107. GJ Kleywegt and TA Jones, Model building and refinement practice, *Meth. Enzymol.*, 277, 208–230, 1997.

108. V Reddy, S Swanson, JC Sacchettini, KA Kantardjieff, B Segelke, and B Rupp, Effective electron density map improvement and structure validation on a Linux multi-CPU web cluster: The TB Structural Genomics Consortium Bias Removal Web Service, *Acta Crystallogr.*, D59, 2200–2210, 2003.

109. A Perrakis, TK Sixma, KS Wilson, and VS Lamzin, wARP: Improvement and Exten-sion of Crystallographic Phases by Weighted Averaging of Multiple-Refined Dummy Atomic Models, *Acta Crystallogr.*, D53, 448–455, 1997.

110. CI Branden and TA Jones, Between objectivity and subjectivity, *Nature*, 343, 687–689, 1990.

111. AA Vaguine, J Richelle, and SJ Wodak, SFCHECK: a unified set of procedures for evaluating the quality of macromolecular structure-factor data and their agreement with the atomic model, *Acta Crystallogr.*, D55, 191–20, 1999.

112. RA Laskowski, MW MacArthur, DS Moss, and JM Thornton, PROCHECK: a program to check the stereochemical quality of protein structures, *J. Appl. Crystallogr.*, 26, 2, 283–291, 1993.

113. RRW Hoft, G Vriend, C Sander, and EE Albola, Errors in protein structures, *Nature*, 381, 272–272, 1996.

114. GJ Kleywegt and TA Jones, Where freedom is given, liberties are taken, *Structure*, 3, 535–540, 1995.

115. M Paris, Vapornomics, *Nature Biotechnol.*, 19, 301, 2001.

116. WP Burmeister, Structural changes in a cryo-cooled protein crystal owing to radiation damage, *Acta Crystallogr.*, D56, 328–341, 2000.

117. M Weik, RBG Ravelli, G Kryger, S McSweeney, ML Raves, M Harel, P Gros, I Silman, J Kroon, and JL Sussman, Specific chemical and structural damage to proteins produced by synchrotron radiation, *Proc. Natl. Acad. Sci.*, USA, 97, 623–628, 2001.

118. VF Hartemann, HA Baldis, AK Kerman, A Le Foll, NC Luhmann, Jr, and B Rupp, Three-dimensional theory of emittance in compton scattering and X-ray protein crystallography, *Phys. Rev.*, E64, 16501-1-16501-26, 2000.

119. R Henderson, Excitement over X-ray lasers is excessive, *Nature*, 415, 833, 2002.

120. JR Helliwell, Overview and new developments in softer X-ray (2Å <l< 5Å) protein crystallography, *J. Synchrotron Rad.*, 11, 1, 1–3, 2004.

121. Z Otwinowski and W Minor, Processing of X-ray diffraction data collected in oszillation mode, *Meth. Enzymol.*, 267, 307–326, 1997.

122. J Westbrook, Z Feng, L Chen, H Yang, and HM Berman, The Protein Data Bank and structural genomics, *Nucleic Acids Res.*, 31, 489–91, 2003.

123. L Holm and C Sander, Protein structure comparison by alignment of distance matrices, *J. Mol. Biol.*, 233, 123–138, 1993.

124. PE Bourne and H Weissig, *Structural Bioinformatics*, NY, Wiley-Liss, Hoboken, NY, 2003.

125. T Peat, E deLaFortelle, C Culpepper, and J Newman, From information management to protein annotation: preparing protein structures for drug discovery, *Acta Crystallogr.*, D58, 1968–1970, 2002.

126. B Rupp, BW Segelke, H Krupka, T Lekin, J Schafer, A Zemla, D Toppani, G Snell, and T Earnest, The TB structural genomics consortium crystallization facility: towards automation from protein to electron density, *Acta Crystallogr.*, D58, 1514–1518, 2002.

127. EA Merritt and DJ Bacon, Raster3D: Photorealistic molecular graphics, *Meth. Enzymol.*, 277, 505–524, 1997.

128. DE McRee, A visual protein crystallographic software system for X11/Xview, *J. Mol. Graph.*, 10, 44–46, 1992.

129. JR Helliwell, Synchrotron radiation facilities, *Nature Struct. Biol.*, 7, 614–617, 1998.

130. VS Lamzin and A Perrakis, Current state of automated crystallographic data analysis, *Nature Struct. Biol.*, 7, 979–981, 2000.

131. CCP4, The CCP4 Suite: Programs for Protein Crystallography, *Acta Crystallogr.*, D50, 760–763, 1994.

132. MD Winn, AW Ashton, PJ Briggs, CC Ballard, and P Patel, Ongoing developments in CCP4 for high-throughput structure determination, *Acta Crystallogr.*, D58(11), 1929–1936, 2002.

133. E Potterton, PJ Briggs, M Turkenberg, and ED Dodson, A graphical user interface to the CCP4 program suite, *Acta Crystallogr.*, D59, 1131–1137, 2003.

134. PD Adams, RW Grosse-Kunstleve, LW Hung, TR Ioerger, AJ McCoy, NW Moriarty, RJ Read, JC Sacchettini, NK Sauter, and TC Terwilliger, PHENIX: building new software for automated crystallographic structure determination, *Acta Crystallogr.*, D58, 1948–54, 2002.

135. AT Brünger, PD Adams, GM Clore, WL DeLano, P Gros, RW Grosse-Kunstleve, JS Jiang, J Kuszewski, M Nilges, NS Pannu, RJ Read, LM Rice, T Simonson, and GL Warren, Crystallography and NMR system: a new software suite for macromolecular structure determination, *Acta Crystallogr.*, D54, 905–921, 1998.

136. E Fortelle and G Bricogne, Maximum-Likelihood Heavy-Atom parameter refinement for multiple isomorphous replacement and multi-wavelength anomalous diffraction methods, *Meth. Enzymol.*, 276, 472–494, 1997.

137. A Perrakis, R Morris, and VS Lamzin, Automated protein model building combined with iterative structure refinement. *Nature Struct. Biol.*, 6, 458–463, 1999.

138. T Holton, TR Ioerger, JA Christopher, and JC Sacchettini, Determining protein structure from electron density maps using pattern matching, *Acta Crystallogr.*, D56, 722–724, 2000.

139. T Oldfield, X-LIGAND: an application for the automated addition of flexible ligands into electron density, *Acta Crystallogr.*, D57, 696–705, 2001.

140. TA Jones, JY Zou, SW Cowan, and M Kjeldgaard, Improved methods for the building of protein models in electron density maps and the location of errors in these models, *Acta Crystallogr.*, A47, 110–119, 1991.

141. DE McRee, XtalView/Xfit — a versatile program for manipulating atomic coordinates and electron density, *J. Struct. Biol.*, 125, 156–165, 1999.

6 From Sequence to Function

Martin Norin

CONTENTS

6.1 INTRODUCTION

The major challenge for the postgenomics efforts is to functionally assign and validate a large number of novel target genes and their corresponding proteins. Although this volume is focusing on structural proteomics and how analyses of protein structures may contribute to functionally annotate genes, other methods based on genetics, protein expression patterns, and gene/protein sequences are widely and efficiently applied to complement structural analysis from experimentally derived structures.

The two dominating experimental technologies to find functional traits in genomes are based on genetics and gene expression data, respectively. Genetic analyses look at variations of DNA sequences to establish associations to phenotypic traits. In the analyses of gene expression patterns, associations between separate genes and physical end-points are established by structuring patterns of how genes are being switched on and off. Recently Schadt et al. [1] have shown how these two methods may be synergistically combined to distinguish causal from reactive genes in traits dependent on complex multigenetic biological systems. These new approaches have coined a new term named *Systems Biology*.

Although these new Systems Biology approaches may create physical maps of gene associations, they require large resources and alone give little insight into the detailed mechanisms of action of individual gene products. Gene and protein sequence comparison methods along with protein structure analyses remain important to fill these information gaps. Especially when combining them with other experimental data. Marcotte et al. [2] were able to functionally annotate 2557 previously uncharacterized yeast proteins using an algorithm combining links derived from:

- Experimental protein interaction data
- Homologues with related metabolic function
- Related phylogenetic profiles
- Homologues that are fused into single genes in another organism
- Correlated mRNA expression profiles

The remainder of this chapter is intended to give a brief overview on functional predictions based on analyses of gene/protein sequence comparisons, and to serve as a background and reference to subsequent chapters focusing on theoretical methods to predict structures and functional sites from three-dimensional structural data of proteins.

6.2 GENE AND PROTEIN SEQUENCE DATABASES

Sequence analysis is often the first guide for the prediction of residues in a protein family that may have functional significance. These are independent of any direct information of protein conformation. In general these methods only use DNA or protein amino-acid sequences as input data and have the advantage of being straightforward and fast. To date, this is the most common method for automatic annotations of large DNA sequence data sets and entire genomes. In the initial analysis of the human genome [3,4] the research groups were able to computationally assign functional categories for 40 to 60% of the predicted genes.

The basis for any sequence comparison method is the availability of comprehensive annotated databases of genes and proteins. The most general databases for nucleotide sequences are the

- EMBL nucleotide sequence database (http://www.ebi.ac.uk/embl)
- Genbank database http://www.ncbi.nlm.nih.gov/Genbank)
- DNA Database of Japan (DDBJ, http://www.ddbj.nig.ac.jp/)

The main sources of entries in these databases are submissions from individual research groups, large genome projects and patent applications. Fortunately, the three databases are synchronized, and updated database entries are exchanged between the groups on a daily basis. Therefore, it is rather a matter of preferred database organization and flavor in selecting which database to use for a particular application. In addition to these general databases, the genome projects maintain databases of specific genomes. For example the Sanger Institute provides databases (http://www.sanger.ac.uk/DataSearch/) of ensembles of human, mouse, rat, worm, fish, and insect genomes.

The SWISSPROT database (http://www.expasy.ch/sprot/) contains highly annotated protein amino-acid sequences. The entries in SWISSPROT are curated and annotated by a number of experts in different protein families. The TrEMBL database supplements SWISSPROT with computer-annotated translations of EMBL nucleotide sequence entries that have not yet been integrated into SWISSPROT. A common web interface, SRS (http://us.expasy.org/srs5), provides a fast sequence retrieval system to both databases.

It is important to keep in mind that entries in all these databases contain errors emanating from limitations in the experimental data, computational methods and human interpretations. Karlin and co-workers [5] also illustrate this problem, where annotated sequences in the *Drosophila* genome were compared with the corresponding entries in the SWISSPROT database. The study revealed that sequence discrepancies of more than 1% were likely to exist for close to 50% of the genes.

6.3 SEQUENCE COMPARISON METHODS

The simplest way of gaining functional insights of novel protein sequences is to find evolutionary relationships by searching in databases for homologous proteins of known function. Evolutionary conserved functional sites can be revealed by comparing protein sequences to the PROSITE database (http://us.expasy.org/prosite). During the last couple of years other advanced sequence fingerprint methods have been shown to be very powerful. Two examples of these are PSI-BLAST [6] (http://www.ncbi.nlm.nih.gov/BLAST) and hidden Markov models [7]. These methods are also useful in predicting functional and structural organization of proteins, e.g., the hidden Markov model based topology prediction method (TMHMM), which was used to correctly predict 97–98% of transmembrane helices [8].

A number of publicly available sequence fingerprint and clustering databases exist, and the optimal database for specific applications varies depending on differences in their underlying methodology. The development of the InterPro database (http://www.ebi.ac.uk/interpro) has added significant value to each of the most commonly used databases by integrating the information from several important databases into an analysis and documentation resource to retrieve combined and complementary data.

6.4 SUMMARY

Genetic and gene expression methods may give systemic insights into the association of genes to variations in phenotypes as well as how proteins relate to each other in their biological pathways. A deeper mechanistic and functional understanding of each protein may be reached by thorough analysis of its amino-acid sequence. Taking into account the three-dimensional structures of proteins may further enhance this understanding. This is especially important in gaining knowledge about the specificity between similar proteins. By combining structural and sequence data with biological knowledge, we are well prepared to make use of biotechnology and chemistry to select and prioritize between potential target proteins, to modulate biological systems in a rational way.

REFERENCES

1. Schadt, E.E., Monks, S.A., Drake, T.A., Lusis, A.J., Che, N., Colinayo, V., Ruff, T.G., Milligan, S.B., Lamb, J.R., Cavet, G., Linsley, P.S., Mao, M., Stoughton, R.B., and Friend, S.H., *Nature*, 442, 297–302, 2003.

2. Marcotte, E., Pellegrini, M., Thompson, M., Yeates, T., Eisenberg, D., *Nature*, 402, 83–86, 1999.
3. International Human Genome Consortium, *Nature*, 409, 860–921, 2001.
4. Venter, J., et al., *Science*, 291, 1304–1351, 2001.
5. Karlin, S., Bergman, A., Gentles, A.J., *Nature*, 411, 259–260, 2001.
6. Altschul, S., Madden, T., Schaffen, A., Zhang, J., Miller, W., Lipman, D., *Nucleic Acid Res.*, 25, 3389–3402, 1997.
7. Eddy, S., *Bioinformatics*, 14, 755–763, 1998.
8. Krogh, A., Larsson, B., von Heijne, G., Sonnhammer, E.L., *J. Mol. Biol.*, 305, 567–580, 2001.

7 Comparative Modeling and Structural Proteomics

Guoli Wang, J. Michael Sauder, and Roland L. Dunbrack, Jr.

CONTENTS

7.1 INTRODUCTION

A central premise of structural proteomics is that comparative modeling will significantly broaden its impact by providing structural models of many thousands of proteins. The goal is described either as obtaining enough structures to represent "fold space," that is at least one structure of every fold, or alternatively as obtaining a structure so that every protein of interest is within modeling distance of a protein of known structure [1]. The latter description requires many more structures since large superfamilies will require multiple structures to cover all member sequences within the frequently mentioned modeling distance of at least 30% sequence identity.

The purpose of comparative modeling in the context of structural proteomics is not the modeling of all proteins homologous to the solved structures. Rather it depends on the interest in the modeling target and its relationship to the template structure from structural proteomics. The purposes of comparative modeling in structural proteomics are not substantially different than outside these projects. They can be roughly categorized as follows:

1. Analysis of sequence-structure-function relationships of orthologues and paralogues of the structural proteomics template. These studies might include, for instance, active site identification by locating conserved residues within the structures. Experimental information may be available for the orthologue of the structural template and not the template itself. Because one of the goals of structural proteomics is to determine structures that represent large superfamilies of proteins, many families homologous to the template may be of unknown function. In this case, comparative modeling may be used to provide a basis for predicting functional differences within a superfamily. For instance, a structural proteomics structure may have a known substrate, but its homologues of unknown structure may act on unknown substrates, or vice versa.

2. Inhibitor design, when the structural proteomics target is homologous to a human protein associated with disease or to a pathogen protein that is a good drug target. Comparative modeling may also be used to assist drug discovery by providing models of off-target proteins, including metabolizing enzymes or proteins that might also bind a drug, causing undesirable side effects.

3. Structure-function relationships of mutations, especially those caused by polymorphisms in the human genome that might relate to disease susceptibility, drug responsiveness, or side effect profiles.

4. Target selection. Comparative modeling depends, in part, on bioinformatics analysis of protein families and superfamilies, including database searching, multiple sequence alignment, sequence clustering, and prediction of transmembrane regions, coiled coils, and unstructured regions. All of these methods are needed to determine targets in structural proteomics that will expand the number of proteins in genomes of interest that can be modeled effectively. Explicit modeling may be used in some cases to determine whether an active site or binding site of an important protein or family of proteins can be well modeled by an existing structure, or whether a new structure should be determined.

5. Modeling used to facilitate structure determination in structural proteomics. If a homologue to a structural proteomics target is available, it may be used to build a model of the target for use in determining the X-ray crystallographic phases using molecular replacement. Modeling might

also be used to improve phase determination in the process of structure refinement in low-resolution or incomplete structures or to guide protein engineering to improve solubility or crystallizability of structural proteomics targets.

In this chapter, we discuss the state-of-the-art in comparative modeling methods and recent developments. We emphasize publicly available programs, rather than proprietary methods or ones for which no software is available. We use the categories of modeling projects described above to illustrate the potential impact of structural proteomics through comparative modeling. Finally, we discuss future prospects.

7.2 METHODS OF COMPARATIVE MODELING

Since it was first recognized that proteins can share similar structures [2], computational methods have been developed to build models of proteins of unknown structure based on related proteins of known structure [3]. Most such modeling efforts, referred to as *homology modeling* or *comparative modeling*, follow a basic protocol laid out by Greer [4,5]:

1. Template identification and target-template sequence alignment. The first step in modeling is to identify a template structure related to the target sequence, and align the target sequence to the template sequence and structure. This may be performed by sequence alignment methods that provide database searching with some form of significance testing, or by structure-based methods such as threading, that judge the compatibility of the target sequence to known structures. Once a template is identified, any number of methods may be used to determine the best sequence alignment.

2. Backbone modeling. For core secondary structures and all well-conserved parts of the alignment, backbone coordinates can be borrowed from the template according to the sequence alignment of the target and template. For segments of the target sequence for which coordinates cannot be borrowed from the template because of insertions and deletions in the alignment (usually in loop regions of the protein), these segments can be built using some *ab initio* or database method.

3. Side-chain modeling. The sequence alignment is used to build the side chains of the target sequence onto the backbone model. Since backbone and side-chain positions must be mutually compatible, Steps 2 and 3 may be performed simultaneously, or cyclically, in which the backbone is constructed, followed by side chains, followed by adjustment of the backbone, etc.

4. Refinement. The model may be refined by energy minimization or molecular dynamics, although these methods are not yet proven to improve the model by moving the atoms closer to the positions seen in the true structure.

An alternative strategy has been developed by Blundell and colleagues, based on averaging a number of parent structures, if these exist, rather than using a single structure [6–8]. More complex procedures based on reconstructing structures (rather than perturbing a starting structure) by satisfying spatial restraints derived from the template or templates using distance geometry [9] and molecular dynamics simulations [10,11] have also been developed. In particular, Andrej Sali's Modeller program is widely used.

Many methods have been developed for template identification, sequence alignment, and structure modeling. There is an interesting dichotomy, however, between the identification/alignment programs and the structure modeling programs. In recent years, it has become the norm for identification/alignment programs to be publicly available as web servers or downloadable computer programs. Moreover, the fold identification abilities and sequence alignment accuracy of these programs have been tested and compared extensively, either by the developers themselves, or by benchmarking systems, such as EVA [12], Livebench [13], and CASA [14]. By contrast, only a few loop and side-chain modeling programs are publicly available. In most cases, they have not been extensively benchmarked, either by the developers themselves, or more importantly by disinterested parties. The recent Critical Assessment of Structure Prediction Methods (CASP) Meeting in December 2002 yielded only a little discussion of side-chain modeling and almost none on loop modeling. This is partly for procedural reasons, in that participants submit either coordinates or alignments but not both, making it difficult to determine which loops were borrowed from the template and which were constructed in the modeling process.

We now review the commonly used methods and recent developments in each step of the modeling process in turn.

7.2.1 TEMPLATE IDENTIFICATION AND SEQUENCE ALIGNMENT

The methods necessary for this step depend in part on what is available in the Protein Data Bank (PDB) [15] (or other proprietary sources of structures, if available). That is, a simple pairwise BLAST search [16] of the PDB may readily provide a template with high significance. If the sequence identity is high (over 35%) and there are few gaps, the sequence alignment is likely to be quite accurate [17]. Some manual adjustment might be made based on observation of the structure, by moving gaps into loop regions where a change in sequence length is least likely to disturb the structure. If BLAST fails, then PSI-BLAST might be used next to search the PDB. This is most easily accomplished by iteratively searching the nonredundant protein sequence database available from NCBI [18] with the target sequence as query. The first round of PSI-BLAST is the same as BLAST. PSI-BLAST uses the multiple sequence alignment generated by this search to construct a position-specific scoring matrix (PSSM) or profile of the sequence family related to the query. This takes the form of a $20XL$ matrix, where L is the length of the sequence, containing the log-odds scores of finding each of the 20 amino acids in each column of the multiple sequence alignment. Gaps in the query or in the hits in the alignment are ignored in the PSI-BLAST PSSM. This matrix can now be used to search the nonredundant database again, but this time scoring each alignment with the PSSM instead of a substitution matrix such as BLOSUM used in the first round. This procedure can run for any number of rounds. We find it most convenient to take the matrix that results from each round of searching *nr* (which can be output from PSI-BLAST) to search a separate database containing only sequences in the PDB. If the alignment covers the full structure and there are only a few well-placed gaps, this alignment (with some manual adjustment) may be suitable for modeling. The use of PSI-BLAST can also be extended by so-called intermediate sequence search methods, in which homology relationships are inferred transitively between target sequence and template sequence via other sequences related to both within some level of statistical significance [17.19]. One implementation of this is to begin new PSI-BLAST searches based on a sample of the hits identified from the search initiated with the target sequence [20]. This process can continue until no new hits are identified in the protein superfamily of the target sequence. Hits of intermediate sequences to sequences in the PDB can then be used to make target-template alignments for modeling.

A number of more powerful methods are available, mostly as web servers. Hidden Markov models (HMMs) [21–24] are generalizations of profiles that provide more flexible treatment of gaps within the profile than PSI-BLAST. This enables more distant relationships to be identified and provides more accurate sequence alignments [25]. A further generalization is to align profiles to one another, rather than comparing a single profile to individual sequences in the PDB [26–29]. In this case, it is necessary to construct a database of profiles for proteins of known structure and to search this database with a profile based on the target sequence family. This method performs better than either PSI-BLAST or HMMs based on profile-sequence alignment.

Structural information may also be used to improve or remove homologue identification and alignment. This can be done in a one-dimensional way, such that information is added to each position in the sequence or profile based on secondary structure or surface accessibility [30]. This can be in the form of secondary-structure-dependent gap penalties [31] or scoring matrices based on similarity of predicted and observed secondary structure. Threading involves moving the target sequence through known structures, and scoring pairwise amino acid interactions [32–36]. As the sequence-based methods have become more powerful, threading has declined in popularity.

The recent CASP meeting demonstrated the power of the so-called meta-servers [37–39]. These are web-based programs that submit a query sequence to several other publicly available fold recognition and sequence alignment servers (including PSI-BLAST, HMMs, threading, etc), and then use the results to produce a consensus alignment and model of the protein. These meta-servers are able to increase the signal-to-noise ratio over that of the individual programs that they exploit.

While the goal of structural proteomics is eventually to have a structure within easy modeling distance of any protein of interest, this will not be achieved any time soon. In many cases, the methods for remote homologue detection and alignment just described will be crucial for exploiting the results of structural proteomics.

7.2.2 STRUCTURE MODELING

Some basic insight into sequence-structure-function relationships can always be obtained simply from the sequence alignment to a known structure. That is, one may readily identify conserved residue positions in the active site or binding site of a protein, or that mutations of buried hydrophobic amino acids to charged ones are deleterious. A crude model can be built easily and quickly by using side-chain prediction on the template backbone according to the target-template sequence alignment. For instance, 3D-PSSM [30] uses the SCWRL program [40] to build simple models from a structure-derived profile alignment, without loop modeling or further refinement.

There are a number of publicly available programs designed solely for side-chain prediction onto a fixed backbone. The most popular among these, as demonstrated at the recent CASP meeting, is the SCWRL program [40]. SCWRL was designed with comparative modeling in mind. It can take an input backbone model and a sequence (potentially different from the input backbone) and build side chains very quickly. SCWRL preserves the residue numbering and chain identifiers, in contrast to some other side-chain prediction programs. SCWRL uses a backbone-dependent rotamer library [41–43] to build an initial model of the protein side chains. Clashes among clusters of interacting residues are resolved using a backtracking algorithm.

The SCAP program builds side chains onto a fixed backbone by cyclically placing side chains from a Cartesian-coordinate-based rotamer library [44]. In this way, variation in bond lengths, bond angles, and dihedral angles is accounted for, rather than relying on fixed bond lengths and angles. Classical rotamer-library-based methods usually ignore this kind of variation, and for some side chains the effect may be large (for instance, large buried aromatic residues). SCAP works on one side chain at a time, finding the energy minimum over the Cartesian rotamers in its library, with the current model of all other side chains as a framework. It works through all the side chains cyclically in linear order until convergence. This method is not guaranteed to find a global energy minimum, but nevertheless works well in practice.

There has been much research into search strategies that are guaranteed to find a global minimum of an energy function. There are a number of variations on the dead-end elimination algorithm that seek to remove some rotamers from consideration that can be proven not to be part of the global minimum configuration [45–48]. The most recent work is that of Looger and Hellinga, who present a schedule of DEE steps for the sequence design problem [49]. Their program is not publicly available. Recently, Canutescu et al. have presented a graph-theory algorithm, implemented in SCWRL3.0, that when combined with the simplest form of the DEE algorithm is faster than published DEE methods [50].

Also recently, the energy functions used in side-chain prediction have received some attention. Petrella et al. have demonstrated that the CHARMM energy function can find the correct conformation for 93% of side chains, when the other side chains are in their X-ray crystallographic positions [51]. Liang and Grishin developed an energy function based on probabilities from the backbone-dependent rotamer library and volume overlap and surface contact potentials [52]. They achieve higher accuracy rates than any other method previously published with a Monte Carlo search strategy, although the method is very slow.

When there are insertions and deletions in the alignment, new backbone segments must be built to substitute for the template backbone. Unless there are major changes in secondary structure, this modeling entails removing all or part of a loop from the template backbone and building a new loop of longer or shorter length. There are many loop-building methods published, but very few standalone programs that in our experience are suitable for comparative modeling purposes. One exception is the Loopy program of Xiang and Honig, which uses an energy function that contains an entropy term (so-called *colony energy*) that favors loop conformations in wide energy basins [53]. Fiser and Sali have incorporated a loop-building module within Modeller that starts by building the atoms of the loop in a straight line joining the anchor positions in the neighboring secondary structures [54]. The energy of the loop is then minimized according to the CHARMM energy potential [55]. In contrast to most published methods, the test set used to evaluate this module was fairly large, consisting of 560 loops. The program was not tested in direct comparative modeling situations, but rather in X-ray structures perturbed via molecular dynamics simulations. Blundell and colleagues recently presented a promising loop modeling method based on sampling from a library of fine-grained ϕ, ψ conformations for each residue type [56,57]. Methods used to predict protein structures *ab initio* (i.e., with no structural template) also have promise in being used in the more constrained situation of modeling loops in comparative modeling. The Rosetta method is very successful at *ab initio* structure prediction [58,59], and was used by the research group of David Baker in comparative modeling in the most recent CASP experiment. Methods such as Rosetta are likely to have their largest impact in modeling long insertions that have elements of regular secondary structure.

Apart from programs that perform individual aspects of the modeling process (alignment, loops, side chains), relatively few programs are designed to perform the entire process. The Modeller program is widely used for building comparative models from target-template sequence alignments [10,11,54]. Modeller uses the template structure or structures and the alignment to derive spatial restraints such as atom distances and dihedral angles and then builds a model of the protein by minimizing the CHARMM-based energy[55] with forces placed on the restraint variables. One advantage of this kind of method is that it is readily able to consider multiple templates, since this only adds additional restraints to the list. Sequence information can be used to weight one template over another in regions of the protein where that template has a higher sequence identity to the target.

Xiang and Honig have recently developed the Jackal suite of programs presented at the CASP5 meeting that perform complete homology modeling, based on multiple structure alignments, loop [53] and side-chain [44] modeling, and refinement.

7.2.3 Current Challenges

While sequence alignments and template identification have improved tremendously in recent years, alignment at very low sequence identity (below 15 to 20%) remains a great challenge. In part, this is because at such great evolutionary distances, two related proteins have changed structure quite significantly, potentially adding and subtracting units of regular secondary structure, changing loop lengths significantly, and significantly redesigning the hydrophobic core of the protein. At very low identity (less than 10%), two structure alignment methods, such as Dali [60,61] and CE [62,63], will agree with each other only over 40% of the alignment [17]. In such cases, it is difficult to define what correct means [64,65]. Modeling from alignments at low identity is difficult when large segments of backbone need to be constructed or remodeled. Success in this area is likely to come from *ab initio* methods that are able to build segments of secondary structure and search the conformational space significantly to place them properly.

As more and more structures are solved, the availability of multiple templates for a given target is increasingly common. Modeller handles these by deriving restraints weighted by local sequence similarity. 3d-Jigsaw developed by Paul Bates is also an automated protocol for using multiple templates [66] with a self-consistent mean-field approach for selecting segments from the available templates. One difficult aspect of using multiple templates is choosing which templates to use in an automated way. Clearly, using low-resolution structures or those much more distantly related to the target than the best template is likely to result in poorer results than using the single best template. Also, some templates may have ligands that result in conformational changes, and the appropriate choices are still almost always made manually. But the best way to make these decisions automatically is not immediately evident.

One of the greatest challenges is in refinement. Once an alignment is obtained and a model built, it is very difficult to move this model closer to the target structure (if it were known) than the backbone model implied by the alignment. In fact, at CASP, relatively few models were an improvement on the best alignment to the template structure, and there is no clear method for doing this consistently. This problem occurs at all levels of sequence identity. It is particularly difficult at low sequence identity, or when there are large insertions, which are very difficult to model accurately.

Finally, protein–protein docking based on two comparative models is a significant challenge, because of the approximate nature of the structures and possible conformational changes upon binding [67–70]. In most cases, one needs to know the approximate site of docking before an accurate model can be made of the complex. Information from mapping sequence conservation onto the structure is one way of obtaining such information [71]. Experimental approaches such as chemical cross-linking and protease digestion studies can narrow the search to specific regions of the protein surfaces, so that computational docking can more easily identify probable structures for the complex [72].

7.3 IMPACT OF STRUCTURAL GENOMICS TARGETS ON MODELING OF GENOME SEQUENCES

While structural proteomics projects are in their early phases, we decided to explore the impact of the 323 structures currently available in the PDB (as of April 12, 2003). Because the nonredundant protein sequence database has many duplicated sequences of the same gene products with only slight variations, it is difficult to get a true estimate of impact using this database. Using all completed genomes is complicated by the predominance of bacterial genomes. Instead, we chose a set of seven organisms that have well-studied genomes: *Mycoplasma genitalium*[73], *Archaeoglobus fulgidus* (an archaebacterium) [74], *Escherichia coli* (strain CFT073) [75], *Drosophila melanogaster, Homo sapiens*[76], *Caenorhabditis elegans* [77], and *Saccharomyces cerevisiae* [78]. While four of these organisms are eukaryotes, they are from diverse phyla of this branch of life.

Of the 323 structural proteomics targets in the PDB, 84 of them (26%) are on hold and the structures are not yet actually available. An additional 13 do not yet have PDB accession codes, leaving 226 available structures, comprising 219 unique sequences. It is not clear to us what the justification for this situation is, given the extensive public financing of these projects. A handful of these, mostly from the *Haemophilus influenzae* structural genomics project [79], have been on hold for well over a year. We expect the structures will be available in the near future.

We used a straightforward procedure for determining what genome sequences are homologous to either structural-proteomics targets or other proteins in the PDB. We used PSI-BLAST [16] to build position-specific scoring matrices, or profiles for each sequence in the PDB, including those from structural proteomics. We already had on hand a profile for each sequence whose structure is released for use in our PISCES server (http://dunbrack.fccc.edu/pisces), which provides lists of PDB sequences with sequence identity and structural quality cutoffs. Each sequence was used as a query against the nonredundant protein sequence database (*nr*, containing 1.4 million sequences) for 3 rounds, with E-value cutoff of 0.0001 for inclusion in the profile matrix. We used a word-score cutoff of 18 (default is 22), which provides a more sensitive search in each round. Each profile was then used to search each genome, generating alignments to genome sequences homologous to sequences with known structures. The E-value cutoff used was 1.0, with an effective database size set to that of the *nr*. Low-complexity regions in *nr* were masked with the seg program [80].

In Table 7.1, we show the percentages of sequences in each genome covered in the PDB as a whole and by structural proteomics targets. The percentage of sequences that can be modeled at least in part ranges from 38% for *C. elegans* to 60% for *M. genitalium.* The percentage of residues that can be modeled is of course lower, especially for eukaryotic genomes that possess many multi-domain protein sequences and large regions of low-complexity. We analyzed the number of genome sequences homologous to structural proteomics targets in the PDB by whether: 1) the structural proteomics target could be a modeling template for a genome sequence or domain; 2) if it was the *best* template in the PDB for the genome sequence; or 3) if it was the *only* template. Of the 90,945 sequences in these 7 genomes, 41,428 or 45.5% have domains that can be modeled on the basis of some structure in the PDB (including the structural proteomics targets), 6,224 or 6.8% can be modeled based on a structural proteomics target in the PDB, 1,295 or 1.4% are *best* modeled by a structural proteomics target, and 495 or 0.5% can *only* be modeled based on structural proteomics.

The number of sequences that can be modeled based on structural proteomics targets varies substantially across the 219 currently available structures. In Figure 7.1, the number of genome sequences modelable is plotted for the 219 targets (along the x-axis, sorted by most to fewest homologous sequences in the 7 genomes). As

TABLE 7.1
Modeling Genomes with the PDB and Structural Proteomics Targets

Genome	N Sequences	N Residues (in thousands)	%Residues modelable with PDB	%Sequences modelable with PDB	%Sequences modelable with SPT	%Sequences modelable with SPT best	%Sequences modelable with SPT only
MG	484	176	46.3	60.5	18.2	6.6	3.5
AF	2420	670	42.4	46.3	13.5	5.1	1.8
EC	5379	1576	42.0	43.3	11.4	3.9	1.5
YS	6304	2967	25.9	43.2	8.4	1.7	0.6
DM	17670	9385	22.5	42.5	7.4	1.4	0.5
CE	21341	9443	23.0	37.8	5.2	0.9	0.4
HU	37347	17069	28.9	51.3	6.0	1.0	0.4
Total	90945	41287	26.7	45.5	6.8	1.4	0.5

Notes: Modelable sequences were determined with PSI-BLAST as described in the text. Abbreviations: SPT: Structural proteomics targets available in the PDB; MG: *Mycoplasma genitalium*; AF: *Archaeoglobus fulgidus*; EC: *Escherichia coli CFT073*; YS: *Saccharomyces cerevisiae*; DM: *Drosophila melanogaster*; CE: *Caenorhabditis elegans*; HU: *Homo sapiens*.

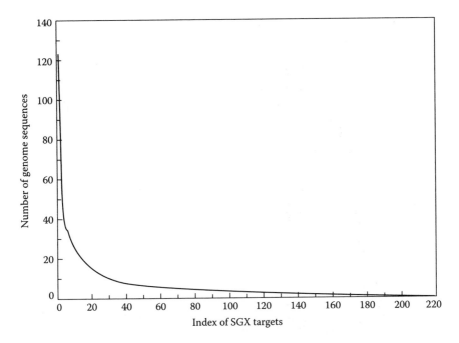

FIGURE 7.1 Distribution of number of modelable genome sequences (out of seven genomes, as in Table 7.1) for structural proteomics targets in the PDB. The number of genome sequences homologous to each structural proteomics target is shown in the vertical axis vs. an index that runs over the list of targets sorted by the number of genome sequences that are modelable from each structural proteomics template.

with all structures in the PDB, most structural proteomics targets are actually distantly related to sequences in the genomes. In Figure 7.2, the fraction of sequences that can best be modeled with structural proteomics template at each sequence identity in bins of 5% is shown for all seven genomes. The curves demonstrate that a majority of genome sequences are modelable only at sequence identities less than 30%. The results are not substantially different for the PDB as a whole (not shown) or for those structural-proteomics-only targets in the genomes.

The scientific or medical impact of a single structure cannot easily be measured, and is essentially subjective depending on one's interests. It depends on the role of the protein and its homologues in experimental organisms or in the human genome, and not only on the number of sequences that can now be modeled based on the structure as the best or only template available. Nevertheless, it is interesting to examine some of the structural proteomics structures available that allow new or improved modeling of the largest number of sequences in these genomes. In Table 7.2, we list the top 10 in terms of either whether the structural proteomics structure is the best template (first 10 in list) or the only template (second 10 in list). The RNA binding domain in PDB entry 1D8Z is the best template for 123 sequences in the 7 genomes. However, because there are many other RNA binding domains in the PDB available as modeling templates for these sequences, it does not appear at

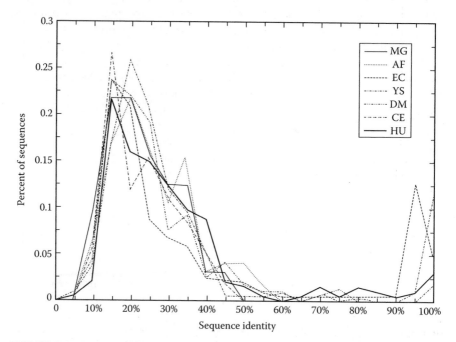

FIGURE 7.2 Sequence identities from alignments of genome sequences to structural pro-teomics targets currently in the PDB, where the structural proteomics structure is the best template available. Height of each curve indicates the percentage of sequences in the seven genomes that can be aligned to a structural proteomics target at each level of sequence identity in bins of 5% (0–5, 5–10, etc.) along the x-axis.

all in the only top 10. While there are many methyltransferase structures in the PDB, they have diverged so significantly that PDB entry 1DUS is the top only template in these genomes. This structure has been used to analyze several families of methyltransferases [81].

7.4 PURPOSES OF COMPARATIVE MODELING FROM STRUCTURAL PROTEOMICS TARGETS

In this section, we examine several categories of comparative modeling and give published examples where possible, using structures from structural proteomics.

7.4.1 ANALYSIS OF SEQUENCE-STRUCTURE-FUNCTION RELATIONSHIPS OF ORTHOLOGUES AND PARALOGUES OF THE STRUCTURAL PROTEOMICS TEMPLATE

A very general paradigm in biology is the analysis of sequence-structure-function relationships in proteins. In the context of structural proteomics where the emphasis is on high-throughput approaches, the application of this paradigm often involves predictions of many protein structures homologous to the template. These homologues may be restricted to the orthologues of the structural proteomics template, or may in fact include most or all known paralogues of the template, in an effort to understand the full range of structures and biological functions represented in the family or superfamily. In the latter vein, some structural scaffolds have evolved numerous functions over many duplications through evolutionary history. Koonin and colleagues [82] describe a number of such promiscuous folds, including the α, β-barrel, the RRM-like fold, the Rossmann fold, and the β-propeller fold. Many of these domains apparently duplicated in the last universal common ancestor of all modern life forms, prior to the divergence of bacteria, archaea, and eukaryotes. Others experienced large numbers of duplications much later in evolution, such as the kinases in eukaryotes, but nevertheless exhibit great diversity in structure and function within a single genome. Folds that have multiple representatives in any single genome of interest are obviously some of the most interesting structural proteomics targets. For example, Hurley et al. have recently reviewed structural genomics of signaling domains, such as PX, TULP, and GAF domains, that have many paralogues in the human genome [83].

The New York Structural Genomics Research Consortium (NYSGRC) has been very active in using new templates from structural proteomics to model as many proteins as possible. New structures are used immediately to populate MODBASE, a database of comparative models constructed with the MODELLER program and the MODPIPE protocol [84]. NYSGRC has used MODBASE to determine the impact of its new structures, finding that on average, models of 100 proteins from the nonredundant protein sequence database without any prior structural characterization in MODBASE can be made for each new template structure solved. While such models are not characterized in terms of structure-function relationships, they are immediately available for analysis by biologists with interest in a certain protein or family. For instance, Chakravarty and Varadarajan used MODBASE to compare a set of orthologues containing 300 thermophilic and 900 mesophilic proteins [85]. They found several features likely to be responsible for the increased thermostability of proteins in thermophiles, including surface-exposed salt-bridges, shorter loops between regular secondary structures, and increased cation-π interactions. Other observations that could not have been made simply through analysis of multiple sequence alignments include locations of altered charge distribution and the surface exposure of amino acid substitutions.

There are only a few published studies in which structures arising from structural proteomics have been used for extensive structure-function analysis and annotation through comparative modeling. One example [86] describes two nucleotide-binding proteins in *E. coli*, YajQ and YnaF, identified through affinity chromatography. YnaF is homologous to a structural proteomics target from *M. jannaschii*, MJ0577, which was the first structural proteomics structure published [87]. YnaF is also more distantly related to UspA, whose structure is known [88]. MJ0577 was used as a template for building a model of YnaF with MODELLER, and it was determined that the ATP-binding site and dimeric structure of MJ0577 and UspA are likely to be preserved in YnaF. This is very likely to be true for the many other homologues of these proteins in bacteria. Because no template was available for YajQ and its

TABLE 7.2
Modeling Impact of Top 10 Structural Proteomics Targets

	PDB	SPT Name	Project	Protein Name	Function	Nseq best	%ID	%Cov
1	1D8Z	1D8Z	RSGI	ELAVL3	HuC 1st RNA-binding dom.	123	30.0	24.6
2	1DUS	BSGCAIR30382	BSGC	MJ0882	Methyltransferase	100	18.0	44.5
3	1HZD	1HZD	RSGI	AUH	RNA-binding; enoyl-CoA hydratase	42	30.4	70.2
4	1HLV	1HLV	RSGI	CENP-B	Centromere DNA binding domain	36	27.9	25.3
5	1OOW	282968	JCSG	TM1102, RNC	Ribonuclease III	35	23.9	24.4
6	1OIZ	283478	JCSG	TM1621	Glycerophosphoryl diester phosphodiesterase homologue	35	19.8	62.1
7	1LPL	F53F4.3	SECSG	F53F4.3	Cytoskeleton-associated protein (CAP-GLY)	31	42.3	13.4
8	1LNZ	NYSGRC-T131	NYSGRC	P20964	SpoOB-assoc. GTP-binding	28	29.0	62.8
9	1D9A	1D9A	RSGI	ELV3_HUMAN	HuC 2nd RNA-binding dom.	28	21.5	18.4
10	1MI1	HC3	NESG	neurobeachin	PH-Beach domain; PKA-anchoring	26	56.9	25.4

	PDB	SPT Name	Project	Protein Name	Function	Nseq only	%ID	%Cov
1	1OIZ	283478	JCSG	TM1621	Glycerophosphoryl diester phosphodiesterase homologue	34	19.4	63.7
2	1DUS	BSGCAIR30382	BSGC	MJ0882	Methyltransferase	30	16.6	44.7
3	1MI1	HC3	NESG	neurobeachin	PH-Beach domain; PKA-anchoring	26	56.9	25.4
4	1HLV	1HLV	RSGI	CENP-B	Centromere DNA binding domain	23	23.5	23.3
5	1KSK	RSUA_ECOLI	BSGI	RsuA	16S rRNA pseudouridine synthase	23	21.5	55.9
6	1LPL	F53F4.3	SECSG	F53F4.3	Cytoskeleton-associated protein (CAP-GLY)	18	39.1	9.4
7	1IW5	mmk001000564	RSGI	BolA-like protein	BolA-like Protein	15	45.8	82.5
8	1J6O	282539	JCSG	TM0667	Deoxyribonuclease	14	24.7	82.8
9	1NXU	ER82	NESG	ER82	Oxidoreductase	12	29.3	88.8
10	1NY1	SR127	NESG	SR127	Deacetylase	12	18.3	27.1

Notes: *Nseq best* is the number of sequences in the 7 genomes listed in Table 7.1 that are best modeled by the structural proteomics target structure (SPT) listed in the second column by PDB code.

Nseq only is the number of sequences in the 7 genomes listed in Table 7.1 that can only be modeled by the SPT structure listed in the second column by PDB code.

%ID gives the average sequence identity between the SPTs and the modelable genome sequences. %Cov gives the percentage of residues in these targets that can be modeled from the SPT.

many homologues, Saveanu et al. performed preliminary NMR experiments to identify secondary structure units and distance constraints from NOEs for use in structural modeling using MODELLER. Structures of RRMs were also used as constraints in MODELLER, because of similar secondary structure and nucleotide binding properties. The model showed extensive positive electrostatic potential, and RNA binding was confirmed with additional NMR experiments.

Bonanno et al. from the NYSGRC have determined two structures in the sterol/isoprenoid biosynthesis pathway, and they used these structures to model other proteins in the pathway that are paralogues of the structural proteomics targets [89]. These two proteins are mevalonate-5-diphosphate decarboxylase (MDD) and isopentenyl diphosphate isomerase (IDI). These structures were used to model 379 other proteins, including two other enzymes in the pathway of isopentenyl diphosphate synthesis — mevalonate kinase and phosphomevalonate kinase, as well as other small molecule kinases, including galactokinases and homoserine kinases. MDD phosphorylates its substrate before elimination of both the phosphoryl group and the carboxylate, while the other enzymes are kinases alone. Models of 21 proteins falling into 4 sequence families without annotation were constructed within MODPIPE. These models can presumably be used to predict function for these proteins, especially if the active sites turn out to be similar to those in proteins of known substrates. Similarly, the IDI structure was used to model a large number of proteins with various annotations, but also 42 proteins without annotated enzymatic activity. Again, these models might be used to predict function, in line with the stated goal of structural proteomics. However, in practice, this is quite difficult to do, leaving the stated promise of structural proteomics combined with comparative modeling in this case as yet unfulfilled.

There are a number of recent studies not based on structural proteomics targets that demonstrate the potential power of high-throughput comparative modeling and sequence-structure-function analysis. These are the kinds of studies that, when applied to new folds, are likely to be extremely informative in diverse areas of biology. The first example is a study of the PII nitrogen regulatory proteins by Kinch and Grishin [20]. This diverse set of proteins is involved in sensing and regulating intracellular carbon and nitrogen status in prokaryotes and eukaryotic organelles by binding small metabolically important molecules such as 2-ketoglutarate and glutamine. These proteins have a fold similar to that of ferredoxin. Using a transitive BLAST procedure, in which hits found in the nonredundant protein database on an initial search are used as queries for additional searches, five groups of orthologues, as defined in the COGs database [90], were identified in the PII superfamily. These include families with no functional annotation, and no previously identified structural information. Structure prediction on one of the groups indicates a potential metal binding site in place of the ATP binding site in PII proteins. If these metal binding proteins form the trimer seen in PII structures, then the authors speculate that they may act as channels. Indeed, loss of family member CutA in *E. coli* is associated with sensitivity to divalent metal ions. This study presents an excellent example of leveraging a single structure (or small number of structures) to predict functions of large numbers of previously unannotated proteins in many genomes. Aloy et al. have used structure alignments to link very distantly related families in a similar manner, identifying numerous unexpected homologies, with important implications for structural genomics [91].

An example that involves more explicit modeling, in addition to sequence alignment, is a recent study of ferredoxins [92]. A set of 179 ferredoxin sequences was analyzed and divided into three groups based on their patterns of cysteine positions. Structures of those proteins with the pattern $CX_4CX_2CX_{29}C$ (88 sequences) were predicted with the MODELLER program. A number of conclusions were drawn. First, modeling of ferredoxins of known structure demonstrated that using multiple templates in MODELLER produced better models than using a single template, when the templates were approximately the same evolutionary distance away from the modeling target. Second, the secondary structures across this large set of proteins are well maintained throughout the superfamily, and hydrophobic residues in the core are well conserved as well. Third, conserved exposed residues cluster in the ferredoxin interface responsible for binding ferredoxin-dependent enzymes. These residues form a negatively charged patch, which in some ferredoxins has a positive charge at its center, apparently compensated for in the binding partner (e.g., ferredoxin reductase). This approach of mapping sequence conservation and variation within and between protein families onto protein structure surfaces has been generalized and formalized as the evolutionary trace (ET) method in a series of papers by Olivier Lichtarge and colleagues [71,93,94].

7.4.2 INHIBITOR DESIGN AND OFF-TARGET MODELING

Structure-based drug design is an increasingly powerful method of optimizing small molecule binding and selectivity. Examples of FDA-approved drugs whose development was assisted through the use of protein structure include HIV protease inhibitors [95] and influenza neuraminidase inhibitors [96]. The success of companies such as Vertex, and the advent of second-generation structure-based companies such as Affinium, Astex, Plexxikon, Structural GenomiX, and Syrrx, confirms the growing interest in using co-complex structures of lead compounds bound to their protein targets to accelerate drug discovery. Despite the huge investment in public and private structural genomics, however, there are still many challenges with experimental structure determination. Computer-aided drug design with comparative models of validated targets is therefore necessary when experimental structures are not available.

In addressing the issue of drug-binding specificity, one can use a combination of sequence analysis and structure modeling to predict or evaluate putative off-targets that may also bind the compound and lead to undesirable side effects. This approach has the potential to be most useful in large protein families with similar binding sites. A recent example is the modeling of three cysteine proteases in the malaria parasite, *Plasmodium falciparum* [97]. While drug development had produced inhibitors to falcipain-1, genome sequencing indicated the presence of two other very similar proteases in the parasite's genome. These proteases are in fact the primary hemoglobinases of *P. falciparum* and thus important targets for drug design. Comparative modeling was used to produce structures for use in computational drug design and identification of new lead compounds. Another example can be found in the 518 kinases [98] in the human genome that are also targets for genome-wide comparative modeling of a protein superfamily in the service of inhibitor design [99]. A number of human kinases have been modeled for drug design [100–102]. It is likely that such genome-wide approaches will become commonplace in drug development.

7.4.3 STRUCTURE-FUNCTION RELATIONSHIPS OF MUTATIONS

A major goal of the Human Genome Project is to identify DNA sequence variations among individuals and populations [103]. While some genome sequence variations in noncoding regions may alter the expression levels of genes, many variations in the coding region will alter the amino acid sequence of expressed proteins. The most frequent mutations in coding regions are *missense* mutations that alter a single amino acid in an expressed protein. Many missense mutations, some of which are polymorphisms present in at least 1% of the population, have been associated with disease states, including cancer [104–111], heart disease [112], diabetes [113–115], cystic fibrosis [116], and Alzheimer's disease [117–119]. The Human Gene Mutation Database presently includes 1,367 genes in which there are 19,746 missense and nonsense mutations associated with disease [120]. It has been estimated that there are up to 250,000 missense polymorphisms in the human genome [121–123].

Establishing inherited risk of a disease based on a known sequence variation requires two steps: 1) a plausible biological basis that the proposed genetic difference plays a role in the disease process; and 2) evidence of epidemiological association of the genetic difference with increased susceptibility of the disease. Without a plausible biological basis, it is always possible that the genetic association is caused by linkage to nearby genetic differences and not the variation under study. Data from the Human Genome Project and other sources will provide an enormous number of genetic differences. It is therefore a priority to develop methods for identifying which missense mutations in human genes are likely to have adverse consequences in terms of protein function and disease, and which may be benign variations with little or no consequences. Such methods would have a strong impact on focusing epidemiological and clinical work, as well as experimental study of gene function.

One way of doing this is through comparative modeling of human proteins based on known structures. There has not been significant published utilization of structural proteomics templates for this purpose to date. One example is a model of the structure of human galactokinase [124], built from a template of homoserine kinase from *M. jannaschii* [125]. Five mutations in human galactokinase result in galactosemia and cataract formation in newborns. Three of these mutations were determined to lie in the hydrophobic core, leading to likely destabilization of the protein fold. The remaining two were in the binding site for galactose and ATP, and are predicted to interfere with substrate binding or catalysis.

More systematic studies of comparative modeling of disease and nondisease-associated missense mutations have been presented in a nonstructural proteomics context [126-132]. These studies attempt to distinguish deleterious mutations from nondeleterious mutations using multiple sequence alignments, amino acid substitution matrices, and position within protein structure models. The structural models, for instance, are used to identify whether mutations of hydrophobic amino acids to polar or charged residues are buried, and therefore likely to reduce structural stability. These studies indicate that as much as 25% of missense mutations produced by single nucleotide polymorphisms are deleterious [127]. As more structures belonging to new folds and new families are solved in the structural proteomics projects, these analyses will approach full coverage of the human genome.

7.4.4 TARGET SELECTION

Target selection has been one of the most important issues in successful structural proteomics projects. The role of comparative modeling in target selection is to help prioritize targets, since one of the goals of structural proteomics is to determine enough structures such that nearly all proteins are within modeling distance of a protein of known structure. Many of the tools of bioinformatics used in comparative modeling and related sequence analysis are needed for effective target selection. Prediction of transmembrane segments [133], coiled-coil regions [134], long unstructured regions [135], and regular secondary structure [136] are all used in comparative modeling contexts, and are now used to analyze genomes in preparation for target selection in structural proteomics.

Besides eliminating potentially problematic targets (transmembrane proteins, nonglobular proteins, etc.), target selection is performed by clustering of protein sequences across genomes and the nonredundant protein sequence database. Numerous such clusterings of the database have been presented, including Pfam [137], ProDom [138], ProtoMap [139], COGs [90], Picasso [140], and Pedant [141], and reviewed by Linial and Yona [142]. While explicit modeling is not always performed in the process of target selection, the capabilities of comparative modeling in part determine what structures would be most useful. Because interest in a particular model may be primarily focused on an active site, some templates even if distantly related to a modeling target may be sufficiently similar in the active site, so that a new structure may not be a high priority. This is especially true of peptide binding proteins, such as proteases, where existing structures may be sufficient for examining specificity. For instance, we produced a model of the BACE protease associated with Alzheimer's disease, based on the distantly related structure of pepsin, readily explaining much of the observed specificity of BACE for amyloid precursor protein sites [143]. Substrate/enzyme contacts predicted by the model were later verified when an experimental structure [144] and mutation data [145] became available.

7.4.5 MODELING USED TO FACILITATE STRUCTURE DETERMINATION
IN STRUCTURAL PROTEOMICS

Comparative modeling may also be used for the purposes of facilitating structure determination in structural proteomics. While many structural proteomics targets are thought to be new folds, at least some are known in advance to be homologous to proteins of known structure. Such targets may be only distantly related to known structure, and therefore a new structure will improve the modeling prospects of many proteins more closely related to the target than the existing structure. In these cases, determining the structure may be accomplished by molecular replacement, where a model of the structural proteomics target based on the known structure is used to determine initial phases for calculating the electron density from the experimentally measured structure factor intensities. The TB structural genomics group solved its first structure, dTDP-4-dehydrorhamnose 3,5-epimerase, using an automated molecular replacement methodology [146,147], based on molecular modeling with SCWRL [40], molecular dynamics with CNS [148], and molecular replacement with a six-dimensional evolutionary search program EPMR [149] and Shake&wARP [150].

Comparative modeling can be used to improve low-resolution structures or to fill out incomplete structures due to missing experimental data. For instance, the Southeast Collaboratory for Structural Genomics (SEGSC) used residual dipolar coupling data and chemical shift data to determine a backbone model for rubredoxin as a test case [151]. Given a reasonably accurate backbone model, side-chain prediction programs can achieve 80 to 85% correct χ_1 rotamer predictions, with most errors confined to the protein surface [50,52]. The SESGC also uses residual dipolar couplings to verify ab initio or homology predicted structures of larger proteins. Similarly, loop modeling may be quite useful in completing membrane protein structures, which typically have disordered surface loop regions between transmembrane helices. The National Institute of Advanced Industrial Science and Technology in Japan has used loop modeling within its structural proteomics project on membrane proteins [152].

In addition to using modeling for structure determination (i.e., obtaining phase information for molecular replacement and subsequent refinement of loops or side chains), it can be helpful for protein engineering to optimize protein solubility or crystallizability. A comparative model greatly facilitates the identification of surface exposed amino acids that may make unfavorable interactions or those that may be involved in crystal packing interactions.

There are numerous examples in the literature where a single point mutation has been shown to have a dramatic effect on the solubility of a protein or its ability to crystallize [153]. While data from optimizing mutations is still too sparse to be strongly predictive, some general trends are beginning to emerge that serve as starting points for designing mutants. Changes that may increase the likelihood of crystallization include mutation of glutamate and lysine residues to alanine to reduce conformational flexibility [154,155] or substitution of lysine to other charged amino acids [156]. There have been numerous studies analyzing intermolecular interactions within with crystal lattice [157–159], but mutagenesis to improve crystallization is still very much hit-or-miss. As the size of the Protein Data Bank grows through structural genomics efforts, rational mutagenesis methods combined with comparative modeling may greatly facilitate crystallization of difficult proteins.

Extending the protein engineering possibilities beyond point mutants, it is possible to use models to facilitate design decisions when deleting flexible loops or choosing linker length when making fusions or chimeras. In the protein kinase family, there are several regions that may have significant insertions, including the activation loop, which may have dozens or hundreds of additional residues (compared to the average of less than 30 residues), which complicate crystallization. These regions of high conformational entropy can often be deleted without affecting kinase activity, while the protein is much more amenable to crystallization. A good example is KDR/VEGFR2 where deletion of 50 amino acids was necessary in order to obtain crystals and consequent structure determination [160].

7.5 INDUSTRY EXPERIENCE

The role of comparative modeling within biotech and pharmaceutical companies varies significantly, covering the whole spectrum from inflated, unrealistic promises within some biotech companies, to distrust or ignorance within some large well-established companies. A reasonable middle ground would suggest that a healthy dose of skepticism is good when interpreting any model, be it structural or otherwise. However, until there are numerous cases where model-based predictions have been validated experimentally and led to time and cost savings during drug development, the use of models will not be widespread enough to furnish this proof. To some degree, the burden of proof has fallen to certain biotech companies that claim to use models for functional analysis or virtual screening and lead discovery, and to computer-aided drug design groups within big pharmaceutical companies.

Ultimately, the utility of protein structure models in drug discovery is for generating hypotheses that can be confirmed experimentally. The decision is one of trust and resources — how much trust is justified in predictions made by expert analysis of comparative models, and how much in the way of resources should be expended to prove or disprove the model. The last few years have demonstrated that selling protein comparative models is not a commercially viable business strategy, nor is selling experimental protein structures. Biotech companies founded on these precepts have begun using their unique strengths to accelerate aspects of more traditional drug discovery.

For example, San Diego-based Structural GenomiX, Inc. (SGX) was founded on the premise of developing a high-throughput protein X-ray structure-determination pipeline that could eventually be used for drug discovery. During the early development of the company's pipeline, most targets were conserved bacterial proteins with no close homologues in the Protein Data Bank. Despite rapid success in solving large numbers of bacterial structures, no major deals targeting anti-bacterials were completed until the New York Structural Genomics Research Consortium (NYSGRC) harnessed the protein production capability of SGX by transferring oversight and a portion of NIH funding to the company. However, the main focus of SGX is now on the synergy between high-throughput co-crystallization and chemical synthesis for iterative lead optimization. Comparative modeling plays a role at both ends of the pipeline. Models are sometimes used when predicting domain construct boundaries for primer design, and are frequently used when engineering loop deletions or predicting amino acid substitutions that may enhance solubility or crystallizability. At the other end of the process, models are used for molecular replacement solutions and refinement, as well as providing templates for virtual screening.

7.6 FUTURE PROSPECTS

The promise of comparative modeling is in part a central motivation for structural proteomics, as a means of leveraging the investment in these projects on a much wider scale than the solved targets themselves. At the same time, research on methods of comparative modeling has been spurred by structural proteomics, since a great deal is now riding on the accuracy of the methods developed in the computational community. In recent years, a new emphasis on large-scale benchmarking of modeling methods has developed. This has primarily occurred for sequence alignment and homology detection but in the future should extend to benchmarking of loop modeling, side-chain modeling, and structure refinement.

Many modeling methods depend on statistical analysis of existing structures or on borrowing parts such as loops from known structures. As the size of the structure database increases both from structural proteomics structures and other structures, these methods will improve as a matter of course. Building models depends on accurate sequence alignments, which in practice depend on large multiple sequence alignments of whole families and superfamilies of proteins. With many more sequenced genomes in coming years, these alignments will also improve. This is especially true for rare proteins that only exist in certain phyla that have been previously underrepresented in the sequence databases.

Some components of the comparative modeling process, such as refinement of models by molecular dynamics or other methods, are heavily dependent on available computational power. Increases in computer speed and inexpensive Linux-based clusters have already had a strong impact on modeling methods, and will continue to do so. More complete conformational searches and more complex energy functions will be performed with increasing computational power. Another major benefit for comparative modeling from structural proteomics is the likely availability of multiple templates for any modeling target. Multiple templates allow greater certainty for identifying core structural elements and their positions, and indicate more limited regions that can be searched more extensively via simulations.

The greatest impact of combining structural proteomics and comparative modeling will be obtained from widespread use of structure in all areas of biology, from microbiology and cell biology to immunology, virology, and all areas of medicine. With more structures, the distance of any protein of interest to an existing structure will become closer, just as with improved methods of modeling the acceptable modeling distance will become longer, until nearly any protein can be modeled to a desired accuracy for most purposes. With structural models of all of the individual components of any biological systems, biologists will concentrate on the most fascinating questions of how these components interact in the complex dynamic systems of living cells and organisms.

ACKNOWLEDGMENTS

Funding from NIH grants R01 HG-02302 (RLD) and CA-06927 (Fox Chase Cancer Center) is gratefully acknowledged.

REFERENCES

1. R Sanchez, U Pieper, F Melo, N Eswar, MA Marti-Renom, MS Madhusudhan, N Mirkovic, and A Sali, *Nat. Struct. Biol.*, 7, Suppl. 986–990, 2000.
2. MF Perutz, JC Kendrew, and HC Watson, *J. Molecular Biol.*, 13, 669–678, 1965.
3. WJ Browne, AC North, and DC Phillips, *J. Molecular Biol.*, 42, 65–86, 1969.
4. J Greer, *Proc. Natl. Acad. Sci. U.S.*, 77, 3393–3397, 1980.
5. J Greer, *Proteins*, 7, 317–334, 1990.
6. TL Blundell, BL Sibanda, MJE Sternberg, and JM Thornton, *Nature*, 326, 347–352, 1987.
7. MJ Sutcliffe, I Haneef, D Carney, and TL Blundell, *Protein Eng.*, 5, 377–384, 1987.
8. MJ Sutcliffe, FR Hayes, and TL Blundell, *Protein Eng.*, 1, 385–392, 1987.
9. TF Havel and ME Snow, *J. Molecular Biol.*, 217, 1–7, 1991.
10. A Sali and TL Blundell, *J. Molecular Biol.*, 234, 779–815, 1993.
11. R Sanchez and A Sali, *Proteins*, Suppl. 1, 50–58, 1997.
12. VA Eyrich, MA Marti-Renom, D Przybylski, MS Madhusudhan, A Fiser, F Pozos, A Valencia, A Sali, and B Rost, *Bioinformatics*, 17, 1242–1243, 2001.
13. JM Bujnicki, A Elofsson, D Fischer, and L Rychlewski, *Protein Sci.*, 10, 352–361, 2001.
14. RY Kahsay, G Wang, N Dongre, G Gao, and RL Dunbrack, Jr., *Bioinformatics*, 18, 496–497, 2002.
15. HM Berman, J Westbrook, Z Feng, G Gilliland, TN Bhat, H Weissig, IN Shindyalov, and PE Bourne, *Nucleic Acids Res.*, 28, 235–242, 2000.
16. SF Altschul, TL Madden, AA Schäffer, J Zhang, Z Zhang, W Miller, and DJ Lipman, *Nucleic Acids Res.*, 25, 3389–3402, 1997.
17. JM Sauder, JW Arthur, and RL Dunbrack, Jr., *Proteins*, 40, 6–22, 2000.
18. DL Wheeler, DM Church, S Federhen, AE Lash, TL Madden, JU Pontius, GD Schuler, LM Schriml, E Sequeira, TA Tatusova, and L Wagner, *Nucleic Acids Res.*, 31, 28–33, 2003.
19. M Gerstein, *Bioinformatics*, 14, 707–714, 1998.
20. LN Kinch and NV Grishin, *Proteins*, 48, 75–84, 2002.
21. SR Eddy, G Mitchison, and R Durbin, *J. Comput. Biol.*, 2, 9–23, 1995.
22. A Krogh, M Brown, IS Mian, K Sjolander, and D Haussler, *J. Mol. Biol.*, 235, 1501–1531, 1994.
23. R Hughey and A Krogh, *Comput. Appl. Biosci.*, 12, 95–107, 1996.
24. K Karplus, C Barrett, and R Hughey, *Bioinformatics*, 14, 846–856, 1998.
25. K Karplus and B Hu, *Bioinformatics*, 17, 713–720, 2001.
26. S Pietrokovski, *Nucleic Acids Res.*, 24, 3836–3845, 1996.
27. L Rychlewski, L Jaroszewski, W Li, and A Godzik, *Protein Sci.*, 9, 232–241, 2000.
28. G Yona and M Levitt, *J. Molecular Biol.*, 315, 1257–1275, 2002.
29. R Sadreyev and N Grishin, *J. Molecular Biol.*, 326, 317–336, 2003.
30. LA Kelley, RM MacCallum, and MJ Sternberg, *J. Molecular Biol.*, 299, 499–520, 2000.
31. RF Smith and TF Smith, *Protein Eng.*, 5, 35–41, 1992.
32. DT Jones, WR Taylor, and JM Thornton, *Nature*, 358, 86–89, 1992.
33. SH Bryant and CE Lawrence, *Proteins: Structure, Function and Genetics*, 16, 92–112, 1993.
34. B Rost, R Schneider, and C Sander, *J. Molecular Biol.*, 270, 471–480, 1997.
35. L Jaroszewski, L Rychlewski, B Zhang, and A Godzik, *Protein Sci.*, 7, 1431–1440, 1998.

36. DT Jones, M Tress, K Bryson, and C Hadley, *Proteins*, Suppl. 104–111, 1999.
37. JM Bujnicki, A Elofsson, D Fischer, and L Rychlewski, *Bioinformatics*, 17, 750–751, 2001.
38. K Ginalski, A Elofsson, D Fischer, and L Rychlewski, *Bioinformatics*, 19, 1015–1018, 2003.
39. J Lundstrom, L Rychlewski, J Bujnicki, and A Elofsson, *Protein Sci.*, 10, 2354–2362, 2001.
40. MJ Bower, FE Cohen, and RL Dunbrack, Jr., *J. Molecular Biol.*, 267, 1268–1282, 1997.
41. RL Dunbrack, Jr. and M Karplus, *J. Molecular Biol.*, 230, 543–574, 1993.
42. RL Dunbrack, Jr. and FE Cohen, *Protein Sci.*, 6, 1661–1681, 1997.
43. RL Dunbrack, Jr., *Curr. Opinions Struct. Biol.*, 12, 431–440, 2002.
44. Z Xiang and B Honig, *J. Molecular Biol.*, 311, 421–430, 2001.
45. J Desmet, M De Maeyer, B Hazes, and I Lasters, *Nature*, 356, 539–542, 1992.
46. RF Goldstein, *Biophys. J.*, 66, 1335–40, 1994.
47. J Desmet, M De Maeyer, and I Lasters, *Pac. Symp, Biocomput.*, 122–133, 1997.
48. M De Maeyer, J Desmet, and I Lasters, *Methods Mol. Biol.*, 143, 265–304, 2000.
49. LL Looger and HW Hellinga, *J. Molecular Biol.*, 307, 429–445, 2001.
50. AA Canutescu, AA Shelenkov, and RL Dunbrack, Jr. *Protein Sci.,* 12, 2001–2014, 2003.
51. RJ Petrella and M Karplus, *J. Molecular Biol.*, 312, 1161–1175, 2001.
52. S Liang and NV Grishin, *Protein Sci.*, 11, 322–331, 2002.
53. Z Xiang, CS Soto, and B Honig, *Proc. Natl. Acad. Sci. U.S.*, 99, 7432–7437, 2002.
54. A Fiser, RK Do, and A Sali, *Protein Sci.*, 9, 1753–73, 2000.
55. AD MacKerell, Jr., D Bashford, M Bellott, RL Dunbrack, Jr., J Evanseck, S Fischer, J Gao, H Guo, S Ha, D Joseph-McCarthy, L Kuchnir, K Kuczera, FTK Lau, C Mattos, S Michnick, T Ngo, DT Nguyen, B Prodhom, WE Reiher, III, B Roux, M Schlenkrich, J Smith, R Stote, J Straub, M Watanabe, J Wiórkiewicz-Kuczera, D Yin, and M Karplus, *J. Phys. Chem.*, B102, 3586–3616, 1998.
56. PI de Bakker, MA DePristo, DF Burke, and TL Blundell, *Proteins*, 51, 21–40, 2003.
57. MA DePristo, PI de Bakker, SC Lovell, and TL Blundell, *Proteins*, 51, 41–55, 2003.
58. KT Simons, R Bonneau, I Ruczinski, and D Baker, *Proteins*, 37, 171–176, 1999.
59. R Bonneau, J Tsai, I Ruczinski, D Chivian, C Rohl, CE Strauss, and D Baker, *Proteins*, Suppl. 5, 119–126, 2001.
60. L Holm and C Sander, *Trends Biochem. Sci.*, 20, 478–480, 1995.
61. L Holm and C Sander, *J. Molecular Biol.*, 233, 123–138, 1993.
62. IN Shindyalov and PE Bourne, *Protein Eng.*, 11, 739–747, 1998.
63. IN Shindyalov and PE Bourne, *Nucleic Acids Res.*, 29, 228–229, 2001.
64. A Godzik, *Protein Sci.*, 5, 1325–1338, 1996.
65. Z Feng and MJ Sippl, *Folding & Design*, 1, 123–132, 1996.
66. PA Bates, LA Kelley, RM MacCallum, and MJ Sternberg, *Proteins*, Suppl. 5, 39–46, 2001.
67. M Zacharias, *Protein Sci.*, 12, 1271–1282, 2003.
68. T Wang and RC Wade, *Proteins*, 50, 158–169, 2003.
69. GM Clore and CD Schwieters, *J. Am. Chem. Soc.*, 125, 2902–2912, 2003.
70. J Fernandez-Recio, M Totrov, R Abagyan, *Protein Sci.*, 11, 280–291, 2002.
71. O Lichtarge, HR Bourne, and FE Cohen, *J. Molecular Biol.*, 257, 342–358, 1996.
72. J Yi, H Cheng, MD Andrake, RL Dunbrack, Jr., H Roder, and AM Skalka, *J. Biol. Chem.*, 277, 12164–12174, 2002.

73. CM Fraser, JD Gocayne, O White, MD Adams, RA Clayton, RD Fleischmann, CJ Bult, AR Kerlavage, G Sutton, JM Kelley, et al., *Science*, 270, 397–403, 1995.

74. HP Klenk, et al., *Nature*, 390, 364–370, 1997.

75. RA Welch, V Burland, G Plunkett III, P Redford, P Roesch, D Rasko, EL Buckles, SR Liou, A Boutin, J Hackett, D Stroud, GF Mayhew, DJ Rose, S Zhou, DC Schwartz, NT Perna, HL Mobley, MS Donnenberg, and FR Blattner, *Proc. Natl. Acad. Sci.*, U.S., 99, 17020–17024, 2002.

76. ES Lander, et al., *Nature*, 409, 860–921, 2001.

77. The C. elegans Sequencing Consortium, *Science*, 282, 2012–2018, 1998.

78. A Goffeau, BG Barrell, H Bussey, RW Davis, B Dujon, H Feldmann, F Galibert, JD Hoheisel, C Jacq, M Johnston, EJ Louis, HW Mewes, Y Murakami, P Philippsen, H Tettelin, and SG Oliver, *Science*, 274, 546, 563–567, 1996.

79. E Eisenstein, GL Gilliland, O Herzberg, J Moult, J Orban, RJ Poljak, L Banerjei, D Richardson, and AJ Howard, *Curr. Opinions Biotechnol.*, 11, 25–30, 2000.

80. JC Wootton, *Computational Chem.*, 18, 269–285, 1994.

81. M Feder, J Pas, LS Wyrwicz, and JM Bujnicki, *Gene*, 302, 129–138, 2003.

82. V Anantharaman, L Aravind, and EV Koonin, *Curr. Opinions Chem. Biol.*, 7, 12–20, 2003.

83. JH Hurley, DE Anderson, B Beach, B Canagarajah, YS Ho, E Jones, G Miller, S Misra, M Pearson, L Saidi, S Suer, R Trievel, and Y Tsujishita, *Trends Biochem. Sci.*, 27, 48–53, 2002.

84. U Pieper, N Eswar, AC Stuart, VA Ilyin, and A Sali, *Nucleic Acids Res.*, 30, 255–259, 2002.

85. S Chakravarty and R Varadarajan, *Biochemistry*, 41, 8152–8161, 2002.

86. C Saveanu, S Miron, T Borza, CT Craescu, G Labesse, C Gagyi, A Popescu, F Schaeffer, A Namane, C Laurent-Winter, O Barzu, and AM Gilles, *Protein Sci.*, 11, 2551–2560, 2002.

87. TI Zarembinski, LW Hung, HJ Mueller-Dieckmann, KK Kim, H Yokota, R Kim, and SH Kim, *Proc. Natl. Acad. Sci. U.S.*, 95, 15189–15193, 1998.

88. MC Sousa and DB McKay, *Structure*, (Camb) 9, 1135–1141, 2001.

89. JB Bonanno, C Edo, N Eswar, U Pieper, MJ Romanowski, V Ilyin, SE Gerchman, H Kycia, FW Studier, A Sali, and SK Burley, *Proc. Natl. Acad. Sci. U.S.*, 98, 12896–12901, 2001.

90. RL Tatusov, MY Galperin, DA Natale, and EV Koonin, *Nucleic Acids Res.*, 28, 33–36, 2000.

91. P Aloy, B Oliva, E Querol, FX Aviles, and RB Russell, *Protein Sci.*, 11, 1101–1116, 2002.

92. I Bertini, C Luchinat, A Provenzani, A Rosato, and PR Vasos, *Proteins*, 46, 110–27, 2002.

93. S Madabushi, H Yao, M Marsh, DM Kristensen, A Philippi, ME Sowa, O Lichtarge, *J. Molecular Biol.*, 316, 139–154, 2002.

94. H Yao, DM Kristensen, I Mihalek, ME Sowa, C Shaw, M Kimmel, L Kavraki, and O Lichtarge, *J. Molecular Biol.*, 326, 255–261, 2003.

95. J Greer, JW Erickson, JJ Baldwin, and MD Varney, *J. Med. Chem.*, 37, 1035–1054, 1994.

96. JN Varghese, *Drug Dev. Res.*, 46, 176–196, 1999.

97. MP Joachimiak, C Chang, PJ Rosenthal, and FE Cohen, *Mol. Med.*, 7, 698–710, 2001.

98. G Manning, DB Whyte, R Martinez, T Hunter, and S Sudarsanam, *Science*, 298, 1912–1934, 2002.

99. SG Buchanan, JM Sauder, and T Harris, *Curr. Pharm. Des.*, 8, 1173–1188, 2002.

100. S Ghosh, RK Narla, Y Zheng, XP Liu, X Jun, C Mao, EA Sudbeck, and FM Uckun, *Anticancer Drug Des.*, 14, 403–410, 1999.

101. S Mahajan, S Ghosh, EA Sudbeck, Y Zheng, S Downs, M Hupke, and FM Uckun, *J. Biol. Chem.*, 274, 9587–9599, 1999.

102. EA Sudbeck, XP Liu, RK Narla, S Mahajan, S Ghosh, C Mao, and FM Uckun, *Clin. Cancer Res.*, 5, 1569–1582, 1999.

103. FS Collins, A Patrinos, E Jordan, A Chakravarti, R Gesteland, and L Walters, *Science*, 282, 682–689, 1998.

104. KM Carlson, S Dou, D Chi, N Scavarda, K Toshima, CE Jackson, SA Wells, Jr., PJ Goodfellow, and H Donis-Keller, *Proc. Natl. Acad. Sci. U.S.*, 91, 1579–1583, 1994.

105. L Garrigue-Antar, T Munoz-Antonia, SJ Antonia, J Gesmonde, VF Vellucci, and M Reiss, *Cancer Res.*, 55, 3982–3987, 1995.

106. CA Marchese, F Bertolino, B Ceccopieri, M Vanzetti, D Scaglione, L Locatelli, M Montera, L Romio, N Resta, A Stella, G Guanti, and C Mareni, *Scand. J, Gastroenterol.*, 31, 917–920, 1996.

107. Y Wang, W Friedl, C Lamberti, C Ruelfs, R Kruse, and P Propping, *Hum. Genet.*, 100, 362–364, 1997.

108. JW Luca, LC Strong, and MF Hansen, *Hum. Mutat.*, Suppl. S58–61, 1998.

109. L Izatt, J Greenman, S Hodgson, D Ellis, S Watts, G Scott, C Jacobs, R Liebmann, MJ Zvelebil, C Mathew, and E Solomon, *Genes Chromosomes Cancer*, 26, 286–294, 1999.

110. I Vorechovsky, L Luo, E Ortmann, D Steinmann, and T Dork, *Lancet*, 353, 1276, 1999.

111. D Shen, Y Wu, R Chillar, and JV Vadgama, *Anticancer Res.*, 20, 1129–1132, 2000.

112. JP Kraus, M Janosik, V Kozich, R Mandell, V Shih, MP Sperandeo, G Sebastio, R de Franchis, G Andria, LA Kluijtmans, H Blom, GH Boers, RB Gordon, P Kamoun, MY Tsai, WD Kruger, HG Koch, T Ohura, and M Gaustadnes, *Hum. Mutat.*, 13, 362–375, 1999.

113. CN Huang, KC Lee, HP Wu, TY Tai, BJ Lin, and LM Chuang, *Pancreas*, 18, 151–155, 1999.

114. SP Miller, GR Anand, EJ Karschnia, GI Bell, DC LaPorte, and AJ Lange, *Diabetes*, 48, 1645–1651, 1999.

115. WM Macfarlane, TM Frayling, S Ellard, JC Evans, LI Allen, MP Bulman, S Ayres, M Shepherd, P Clark, A Millward, A Demaine, T Wilkin, K Docherty, and AT Hattersley, *J. Clin. Invest.*, 104, R33–39, 1999.

116. DJ Hughes, AJ Hill, M Macek, Jr., AO Redmond, NC Nevin, and CA Graham, *Hum. Mutat.*, 8, 340–347, 1996.

117. R Besancon, A Lorenzi, M Cruts, S Radawiec, F Sturtz, E Broussolle, G Chazot, C van Broeckhoven, G Chamba, and A Vandenberghe, *Hum. Mutat.*, 11, 481, 1998.

118. O Murayama, M Murayama, T Honda, X Sun, N Nihonmatsu, and A Takashima, *Prog. Neuropsychopharmacol. Biol. Psychiatry*, 23, 905–913, 1999.

119. D Campion, C Dumanchin, D Hannequin, B Dubois, S Belliard, M Puel, C Thomas-Anterion, A Michon, C Martin, F Charbonnier, G Raux, A Camuzat, C Penet, V Mesnage, M Martinez, F Clerget-Darpoux, A Brice, and T Frebourg, *Am. J. Hum. Genet.*, 65, 664–670, 1999.

120. DN Cooper, EV Ball, and M Krawczak, *Nucleic Acids Res.*, 26, 285–287, 1998.

121. JG Hacia and FS Collins, *J. Med. Genet.*, 36, 730–736, 1999.

122. M Cargill, D Altshuler, J Ireland, P Sklar, K Ardlie, N Patil, N Shaw, CR Lane, EP Lim, N Kalyanaraman, J Nemesh, L Ziaugra, L Friedland, A Rolfe, J Warrington, R Lipshutz, GQ Daley, and ES Lander, *Nat. Genet.*, 22, 231–238, 1999.

123. MK Halushka, JB Fan, K Bentley, L Hsie, N Shen, A Weder, R Cooper, R Lipshutz, and A Chakravarti, *Nat. Genet.*, 22, 239–247, 1999.

124. SK Burley and JB Bonanno, *Annu. Rev. Genomics Hum. Genet.*, 3, 243–262, 2002.

125. T Zhou, M Daugherty, NV Grishin, AL Osterman, and H Zhang, *Structure Fold Des.*, 8, 1247–1257, 2000.

126. Z Wang and J Moult, *Hum. Mutat.*, 17, 263–270, 2001.

127. D Chasman and RM Adams, *J. Mol. Biol.*, 307, 683–706, 2001.

128. V Ramensky, P Bork, and S Sunyaev, *Nucleic Acids Res.*, 30, 3894–3900, 2002.

129. S Sunyaev, V Ramensky, I Koch, W Lathe III, AS Kondrashov, and P Bork, *Hum. Mol. Genet.*, 10, 591–597, 2001.

130. S Sunyaev, V Ramensky, and P Bork, *Trends Genet.*, 16, 198–200, 2000.

131. NO Stitziel, YY Tseng, D Pervouchine, D Goddeau, S Kasif, and J Liang, *J. Mol. Biol.*, 327, 1021–1030, 2003.

132. CT Saunders and D Baker, *J. Mol. Biol.*, 322, 891–901, 2002.

133. EL Sonnhammer, G von Heijne, and A Krogh, *Ismb*, 6, 175–182, 1998.

134. A Lupas, *Curr. Opinions Struct. Biol.*, 7, 368–393, 1997.

135. AK Dunker, E Garner, S Guilliot, P Romero, K Albrecht, J Hart, Z Obradovic, C Kissinger, and JE Villafranca, *Pac. Symp. Biocomput.*, 473–484, 1998.

136. DT Jones, *J. Molecular Biol.*, 292, 195–202, 1999.

137. A Bateman, E Birney, R Durbin, SR Eddy, RD Finn, and EL Sonnhammer, *Nucleic Acids Res.*, 27, 260–262, 1999.

138. F Corpet, J Gouzy, and D Kahn, *Nucleic Acids Res.*, 26, 323–326, 1998.

139. E Portugaly, I Kifer, and M Linial, *Bioinformatics*, 18, 899–907, 2002.

140. A Heger and L Holm, *Bioinformatics*, 17, 272–279, 2001.

141. D Frishman, K Albermann, J Hani, K Heumann, A Metanomski, A Zollner, and HW Mewes, *Bioinformatics*, 17, 44–57, 2001.

142. M Linial and G Yona, *Prog. Biophys. Mol. Biol.*, 73, 297–320, 2000.

143. JM Sauder, JW Arthur, and RL Dunbrack, Jr., *J. Molecular Biol.*, 300, 241–248, 2000.

144. L Hong, G Koelsch, X Lin, S Wu, S Terzyan, AK Ghosh, XC Zhang, and J Tang, *Science*, 290, 150–153, 2000.

145. M Farzan, CE Schnitzler, N Vasilieva, D Leung, and H Choe, *Proc. Natl. Acad. Sci. U.S.*, 97, 9712–9717, 2000.

146. B Rupp, *Acc. Chem. Res.*, 36, 173–181, 2003.

147. B Rupp, BW Segelke, HI Krupka, T Lekin, J Schafer, A Zemla, D Toppani, G Snell, and T Earnest, *Acta Crystallogr. D. Biol. Crystallogr.*, 58, 1514–1518, 2002.

148. AT Brunger, PD Adams, GM Clore, WL DeLano, P Gros, RW Grosse-Kunstleve, JS Jiang, J Kuszewski, M Nilges, NS Pannu, RJ Read, LM Rice, T Simonson, and GL Warren, *Acta Crystallogr. D. Biol. Crystallogr.*, 54, Part 5, 905–921, 1998.

149. CR Kissinger, DK Gelhaar, and D Fogel, *Acta Cryst. D.*, 55, 484–491, 1999.

150. A Perrakis, M Harkiolaki, KS Wilson, and VS Lamzin, *Acta Crystallogr. D. Biol. Crystallogr.*, 57, 1445–1450, 2001.

151. MW Adams, HA Dailey, LJ DeLucas, M Luo, JH Prestegard, JP Rose, and BC Wang, *Acc. Chem. Res.*, 191–198, 2003.

152. Y Kyogoku, Y Fujiyoshi, I Shimada, H Nakamura, T Tsukihara, H Akutsu, T Odahara, T Okada, and N Nomura, *Acc. Chem. Res.*, 36, 199–206, 2003.

153. GE Dale, C Oefner, and A D'Arcy, *J. Struct. Biol.*, 142, 88–97, 2003.

154. KL Longenecker, SM Garrard, PJ Sheffield, and ZS Derewenda, *Acta Crystallogr. D.*, 57, 679–688, 2001.

155. A Mateja, Y Devedjiev, D Krowarsch, K Longenecker, Z Dauter, J Otlewski, and ZS Derewenda, *Acta Crystallogr. D.*, 58, 1983–1991, 2002.

156. S Dasgupta, GH Iyer, SH Bryant, CE Lawrence, and JA Bell, *Proteins*, 28, 494–514, 1997.

157. O Carugo and P Argos, *Protein Sci.*, 6, 2261–2263, 1997.

158. J Janin and F Rodier, *Proteins*, 23, 580–587, 1995.

159. F Glaser, DM Steinberg, IA Vakser, and N Ben-Tal, *Proteins*, 43, 89–102, 2001.

160. MA McTigue, JA Wickersham, C Pinko, RE Showalter, CV Parast, A Tempczyk-Russell, MR Gehring, B Mroczkowski, CC Kan, JE Villafranca, and K Appelt, *Structure Fold Des.*, 7, 319–330, 1999.

8 *Ab initio* Modeling

Jeffrey Skolnick, Yang Zhang, and Andrzej Kolinski

CONTENTS

8.1 THE IMPORTANCE OF PROTEIN STRUCTURE IN THE POSTGENOMIC ERA

A paradigm shift in biology brought about by the sequencing of the genomes of many organisms [1–5] has happened over the last several years. There is increasing emphasis not on the study of individual molecules, but on the large-scale, high-throughput examination of all genes and gene products of an organism, with the aim of assigning their functions [1,2,6]. While DNA sequencing efforts have been incredibly successful at providing the gene sequences, this does not directly provide insight into what each molecule does in the cell. Indeed, function is multifaceted, ranging from molecular/biochemical to cellular or physiological to phenotypical [7–9]. Sequence-based methods (which detect evolutionary relationships) can provide insights into some aspects of the biological function of about 40 to 60% of the ORFs found in a given genome [10–11]. Representatives of such evolutionary approaches include iterative profile methods such as PSI-BLAST [10–11] and sequence motif (that is, local sequence descriptors) methods such as Blocks [12]; these define the standard against which all alternative approaches are assessed. However, evolutionary-based approaches increasingly fail as the protein families become more distant [13], and predicting the functions of the unassigned ORFs is an important challenge for molecular and cellular biology. Because the biochemical function of a protein is ultimately determined by its structure around the active site, protein structures are becoming essential tools in annotating genomes [14–18]. The recognition of the role that structure can play in elucidating function has spurred the rise of the structural genomics approach to structural biology that aims for high-throughput protein structure determination [19–24]; here, structure prediction can also play an important role in target selection [25] as well as in the more general problem of genome scale protein function prediction. Thus, protein structure prediction is both timely and important.

8.2 OVERVIEW OF PROTEIN STRUCTURE PREDICTION METHODS

There are three basic approaches to protein structure prediction: homology modeling, threading, and *ab initio* folding (although the differences among the approaches is becoming less pronounced). In homology modeling, the target sequence has a clear evolutionary relationship to another (template) protein whose structure has already been solved. As of CASP4 [26], the best way to proceed is to obtain the best sequence alignment to the template backbone, rebuild the side chains of the protein and fill in the gaps in the alignment, typically in the loops between secondary structural elements [27]. It remains to be established in CASP5 if progress has been made in improving the alignment on subsequent modeling.

The next, more difficult protein structure prediction method is threading [28–32]. Here, one attempts to find the closest matching structure in a library of already solved structures. The structures can be analogous, that is, the target and template proteins are not necessarily evolutionarily related, but could adopt very similar structures by convergent evolution [33]. Ideally, threading should extend sequence-based approaches.

Both threading and homology modeling have the disadvantage that an example of a related structure must already have been solved. To address this issue, the most difficult and most general approach is *ab initio* folding, where one attempts to fold a protein from a random conformation [34–37]. It has the advantage that an example of the fold need not have been seen before, but to date has been limited to relatively small proteins and in general, the various *ab initio* methods produce low-to-moderate resolution predicted structures. As detailed below, a number of approaches to *ab initio* folding use extensive information from threading [38–39]. Such information might include predicted local secondary structure, supersecondary structure and/or predicted tertiary contact information [38–39]. There are also very important issues associated with protein folding kinetics and thermodynamics that are related to the problem of protein structure prediction, but since the thrust of this review is protein structure prediction, we refer the interested reader to the literature [40–82].

8.3 *AB INITIO* STRUCTURE PREDICTION: HISTORICAL OVERVIEW

In a*b initio* protein structure prediction, there are two problems that must be simultaneously solved: the multiple minima problem and the development of an energy function having the global minimum in the native conformation. In what follows, we review the state of the art that attempts to address these issues.

A typical protein folds from the denatured to the native state on a time scale ranging from milliseconds to minutes [83]. At full atomic detail, with explicit water, only very small proteins can be simulated [84]. To access the requisite folding time scales for proteins on the order of several hundred residues, often the number of degrees of freedom of the protein are reduced (e.g., each residue is treated as an interaction center located on the protein backbone) and the solvent is treated implicitly [85–89].

8.3.1 CONTINUOUS SPACE REDUCED MODELS

The first reduced models were introduced more than 25 years ago. In their pioneering work, Levitt and Warshel [90–91] proposed a model with two interaction centers per residue, one on the backbone α-carbon and the other at the side group center of mass and a single degree of freedom involving the rotation around the Ca-Ca virtual bond [92]. A knowledge-based potential controlled the short-range interactions, while side group interactions were Lennard-Jones like. Sampling was done by classical molecular dynamics. Simulations of bovine pancreatic trypsin inhibitor sometimes produced nativelike structures, with the best structures having a coordinate root mean square deviation (RMSD) from native in the range of 6.5 Å. This was followed by similar studies of Kuntz et al. [93], Hagler and Honig [94], and Wilson and Doniach [95] with comparable results.

More recently, continuous-space models with greater structural detail were investigated. Sun [96] examined models having an all-atom representation of the main chain and single, united atom side groups. Knowledge-based statistical potentials described interactions between the side groups, and a genetic algorithm (GA) was used to search conformational space. For small peptides (mellitin, pancreatic polypeptide inhibitor,

PPI, and apamin), structures were predicted with an RMSD from native ranging from 1.66 Å to 4.5 Å. Wallqvist and Ullner [97] developed a similar model with slightly better results for PPI, probably due to the better side chain packing and related approaches that were used to predict folding dynamics [98–99].

Pedersen and Moult have developed an interesting protein structure prediction approach [100]. They assume an all-heavy atom protein representation and use knowledge-based potentials describing intra-protein interactions. Conformational space is searched by combining Monte Carlo (MC) and GA algorithms. MC yields structures for the GA starting population. Crossover points occur in the largest flexibility regions detected in the MC runs. MC simulations were also performed between crossover events in the GA scheme. The method successfully predicted low-to-moderate resolution protein fragments and the approximate folds of some small proteins.

8.3.2 LATTICE MODELS

One way of extending the size of proteins treated is to restrict the protein residues to a lattice. The first attempt at lattice-based *ab initio* protein structure prediction was due to Dashevskii [101], who used a diamond lattice chain to approximate the backbone and a chain growth algorithm to sample conformational space. Compact structures resembling native folds of small polypeptides were produced by a simple force field. Somewhat later, Covell used a simple cubic lattice model of real proteins [102] and knowledge-based tertiary interaction potentials to produce crude folds of comparable quality to early continuous space models. Covell and Jernigan [103] studied five small globular proteins by the exhaustive enumeration of all compact conformations on a body-centered cubic lattice. They found that the closest to native conformation could always be found within the top 2% of the lowest energy structures as assessed by a knowledge-based interaction scheme. In a related approach, Hinds and Levitt [104] developed a lattice model of proteins where a single diamond lattice vertex represents several (and a varying number of) residues. An elaborate statistical potential was employed and an exhaustive search of a compact space was done. Then, the actual identity of the residues was obtained from dynamic programming, and often correct, low resolution structures were found among the compact structures.

Over the years, Kolinski and Skolnick (46–55,105–109) have developed a series of high coordination lattice models of globular proteins. Lattices of various resolution were employed to mimic the conformation of the Cα-trace of real proteins, ranging from three-dimensional "chess-knight" type lattices to a high coordination lattice with 90 lattice vectors to represent possible subsequent locations of Cα-Cα virtual bonds. The models employed in the test predictions (107–112) had additional interaction centers to represent the side groups, described in a single sphere, multiple rotamer representation. The force field contained terms mimicking short-range interactions describing local conformational preferences for helices and beta strands, explicitly cooperative hydrogen bonds, one body, and pairwise and multibody long-range interactions with an implicit averaged effect of the water molecules. For several small globular proteins [107] and simple multimeric coiled coils [113–116], such models generated correct low-to-moderate resolution (high-resolution in the case of leucine zippers) structures from Monte Carlo simulated annealing simulations.

8.3.3 *AB INITIO* FOLDING USING SECONDARY STRUCTURE RESTRAINTS

A means of improving the quality of tertiary structure prediction is to use either known or predicted secondary structural information. Using an off-lattice model and exact secondary structure, Friesner et al. successfully folded two four-helix bundle proteins, cytochrome b562 and myohemerythrin, the large helical protein myoglobin and the relatively complicated fold of the L7/L12 ribosomal protein [117–120]. Furthermore, assuming known secondary structure and a genetic algorithm to search conformational space, Dandekar and Argos [1] reported encouraging results on a test set of 19 small proteins, where they succeeded in predicting a significant portion of these proteins with an RMSD from native of 5 Å. However, use of predicted (rather than known) secondary structure information substantially degraded the algorithm's performance. Mumenthaler and Braun [121] have developed a self-correcting distance geometry method that assumes known secondary structure and successfully identified the native topology for six of eight helical proteins. A very early example where predicted secondary structure information was successfully incorporated into a prediction algorithm was the folding of crambin that produced structures with a Cα RMSD below 4 Å from native [122]. Early work of Baker and coworkers (see below for additional discussion of their seminal studies) reported the successful assembly of the tertiary structures of helical proteins using fragments extracted on the basis of their local sequence homology to the sequence of interest (35,123-124). This study pointed out the promise of local sequence alignment-based approaches to tertiary structure prediction.

8.3.4 *AB INITIO* FOLDING USING SECONDARY AND TERTIARY RESTRAINTS

There have been a number of studies that incorporate correct secondary structure and a limited number of correct tertiary restraints to predict the global fold. Smith-Brown et al. [1] modeled a protein as a chain of glycines and find that a considerable number of restraints is required to assemble the native structure, thereby rendering the approach impractical. Another effort very much in the spirit of Mumenthaler and Braun is due to Aszodi et al. who predict interresidue distances from patterns of conserved hydrophobic amino acids extracted from multiple sequence alignments. In general, they find that to assemble structures below 5 Å RMSD, on average, typically more than N/4 restraints are required, where N is the number of residues. Even then, the Aszodi et al. method has problems selecting the correct fold from other alternatives. However, their approach is very rapid, with a typical calculation taking on the order of minutes. Again, a key problem with all these approaches is the relatively large number of exact tertiary restraints required for successful native topology assembly.

We have developed the MONSSTER algorithm [125] that is based on a high coordination lattice protein model [111–112] where an MC simulated annealing approach searches conformational space. The knowledge-based force field consisted of generic terms that generate proteinlike conformations, short-range interactions, one body burial, pairwise and multibody long-range interactions, and hydrogen bonds. Additionally, a weak bias towards predicted secondary structure (obtained from multiple sequence alignments using PHD [126–128]) and weak predicted long-range contact restraints obtained from correlated mutation analysis of multiple sequence alignments [129–133] are incorporated into the potential. Encouraging initial results were reported on a series of 19 nonhomologous proteins ranging in length from 28 to 100 residues that represented all secondary structure classes. The RMSD from native ranged from 3.1 to 6.7 Å. The importance of this study is that it demonstrated that relatively inaccurate tertiary restraints could be used to help guide the fold assembly process. However, correlated mutation analysis proved to be very difficult to automate, and the accuracy of the predicted contacts was rather sensitive to the set of sequences used to generate the multiple sequence alignments. As such, it did not prove to be suitable for large scale tertiary structure prediction.

8.3.5 REPRESENTATIVE CONFORMATIONAL SEARCH METHODS

In general, the choice of the simulation/optimization algorithm depends on the aim of the research. The study of protein dynamics and folding pathways requires different and in general more complex procedures (and perhaps force fields) than those designed to identify the native conformation of a protein that, according to the thermodynamic hypothesis [134], is the global minimum of the conformational energy. A variety of strategies have been developed to find this global minimum. These include the diffusion equation method and its variants, which deform the energy surface until a single minimum remains, which when traced back to the original energy surface corresponds to the global energy minimum on the non-deformed surface [135]. For relatively simple, but nontrivial systems, this method works very well, but for more complex situations (i.e., realistic protein models), existing methods do not guarantee that the lowest energy conformation will be found.

The most straightforward of the MC algorithms is simulated annealing, where the system starts out at a relatively high temperature and then the temperature is gradually lowered until a temperature below the folding transition is reached. If on repeated runs starting from different initial states, the same conformation is recovered, this is circumstantial evidence that the global minimum has been located (of course, this is not a proof, just a suggestion). However, for difficult problems, simulated annealing runs often become trapped in local energy minima that could be significantly different from the global minimum. Unfortunately, there is no simple test of convergence. The efficiency of simulated annealing could be considerably improved by modification of the conformational transition acceptance criteria. For instance, one might perform local minimization before and after the transition and then apply the Metropolis criterion to the locally lowest energy pair of conformations [136]. This way, the sampling procedure can avoid visits to a large fraction of irrelevant local states.

In contrast to simulated annealing, sampling techniques that use a multicanonical ensemble have convergence tests [137]. In a variant called Entropy Sampling Monte Carlo (ESMC) [79-81,138]), the estimation of the system's entropy is constructed by a sampling process that is controlled by the density of states of the energy. When converged, all energy levels, including the lowest energy, should be sampled with the same frequency, and if not, one knows the energy regions that have not converged. The ESMC method is quasi-deterministic — data from the preceding simulations could be used to improve the accuracy in successive runs. In principle, ESMC should find the lowest energy state. In practice, the energy spectrum near the lowest energy state could have large entropic barriers, and the lowest energy state might not be detected. The rate of convergence could be accelerated by the artificial deformation of the entropy versus the energy curve in the less important, high-energy range [116]; nevertheless convergence remains very slow.

The Replica Exchange Monte Carlo (REMC) method and its variants [139–144] have a different philosophy. Here, a number of copies are simulated by means of a standard Metropolis scheme at various temperatures spanning the high to low temperature region. Occasionally (on the order of the high temperature simulation's longest relaxation time), replicas are randomly swapped according to a criterion that depends on their temperature and energy differences. Thus, low-energy conformations at a higher temperature could move to lower temperature. At high temperatures, the energy barriers could be surmounted easily, while at low temperatures, the energy "valleys" are efficiently sampled.

Simulated annealing, entropy sampling MC, ESMC, [79–80,145) GA (96,146–148), and the combination of GA with MC sampling have been successfully used in the past to find the near-native conformations of reduced models of small proteins [149]. A number of recent studies have compared the efficiency of various Monte Carlo strategies for finding the global minimum of protein models (142,144–150). Comparison of the computational expense of finding the lowest energy state for a simple protein-like copolymer model [150] shows that REMC is much more efficient than an MC based simulated annealing protocol despite the fact that multiple copies of the system have to be simulated. REMC method also finds the low-energy conformations many times faster than ESMC does. Thus, it appears that the REMC method (or recently developed variants, parallel hat REMC and parallel hyperbolic Monte Carlo, [151–152]) is the method of choice. Due to the very efficient sampling by REMC, the conformations at various temperatures could be used for the umbrella type estimation of the density of states as a function of energy from which all thermodynamic quantities can be obtained [153].

8.3.6 *AB INITIO* PREDICTION RESULTS IN CASP3 AND CASP4

Over the past decade, *ab initio* protein structure prediction methods were tested during the CASP1–5 exercises (Critical Assessment of Techniques for Protein Structure Prediction, a communitywide experiment to make blind predictions using proteins whose structure is about to be determined but not publicly available in the prediction season), with CASP3 and CASP4 concluding in December 1998 and 2000, respectively, in Asilomar, California [154].

CASP3 marked the first time that significant progress in *ab initio* prediction was evident [155]. The ROSETTA method developed by Baker and coworkers [156] was very innovative and consists of several steps. First, a multiple sequence alignment is prepared, and secondary structure prediction done [126–128]. Combined secondary structure predictions and sequence alignments provide the most plausible 3- to 9-residue structural fragments (25 fragments for each segment of the query sequence) extracted from a structural database, and an MC algorithm randomly inserts these 3- and 9-residue fragments. Conformations are scored by a function that contains a hydrophobic burial term, a potential mimicking some aspects of electrostatics, a disulfide bond bias, and a sequence-independent term that evaluates the packing of secondary structure elements. The top 25 (of 1200 generated) structures frequently contained the proper fold. The best five structures that exhibited a single hydrophobic core were selected by visual inspection. This could be considered a drawback of the method, as it would be difficult to do a manual evaluation of predictions on a massive scale. Nevertheless, for 18 targets, 4 predictions were globally correct (with an RMSD from native in the range of 4–6 Å). Furthermore, the majority of their predictions contained correct fragments. It should be noted that a somewhat similar idea of protein structure assembly using predefined fragments and a Monte Carlo method was previously tested by Jones during the CASP2 exercise [157]. The Baker group also performed extremely well in CASP4, with many outstanding predictions [158] (see below for a summary of CASP4 results).

In CASP3, other groups also made good predictions for a number of difficult *ab initio* protein targets. Our group [36] applied MONSSTER to a total of 11 proteins, of which 10 had available structures. Targets 65 and 74 (targets in CASP are sequentially numbered) were correct, targets 46, 57 and 77 were almost correct, and each of the two domains of target 79 was correct. Four targets (52, 56, 59 and 63) were wrong; this is the same level of success as reported for test predictions [131]. Using knowledge-based potentials, Osguthorpe [159] employed a continuous space model and sampled conformations by molecular dynamics-based simulated annealing. He correctly predicted substantial portions of the targets that he attempted, and his prediction was the best for one of the difficult targets. Samudrala et al. developed a very interesting hierarchical procedure [160]. As previously [161], using a diamond lattice model with multiple residues per lattice point, all compact conformations are enumerated. The best structures are selected, and all-atom models built by fitting the predicted secondary structure fragments to the lattice models. Using an all-atom force field, these structures were energy minimized, and spatial restraints from the lattice models were extracted. The optimized structures are scored by a combination of all-atom and residue-based knowledge-based potentials [162]. Then, using distance geometry [163], possible consensus models are generated; all-atom structures are rebuilt, optimized and then ranked by their energy. A number of qualitatively correct, significant size protein fragments were correctly predicted. Perhaps, the major weakness of this approach was the small fraction of good structures in the initial pool of lattice models. Scheraga and coworkers [164] have also developed a hierarchical approach. First, they globally optimize the potential energy of a united atom representation of the α-carbons, peptide bonds, and side groups [165]. Their force field is based on physical principles rather than evolutionary information, and this distinguishes their approach from that of many participants in both CASP3 and subsequently in CASP4. Conformational space was searched using their Conformational Space Annealing [165] algorithm that narrows the conformational search regions on successive interations and finally finds distinct families of low-energy conformations. Then, the lowest energy, reduced model conformations are converted to the all-atom models and optimized by electrostatically driven MC [166]. For some CASP3 targets, this method produced exceptionally good predictions. The method performed better on helical proteins than on β or α/β proteins.

In the novel fold (formerly *ab initio*) category of CASP4, three kinds of predictions were assessed: the quality of tertiary, secondary, and residue-residue contact predictions [158]. The Baker group's ROSETTA method performed the best in the area of protein tertiary structure prediction [167]. Relative to CASP3, three types of improvements were introduced: several different secondary structure prediction schemes (PSIPRED [168], SAMT99 [169–171], and PHD [172–175]) were used to bias the fragment selection scheme and improved sampling was introduced via the use of local moves [176] and by locally minimizing the backbone torsional angles to reduce distortion. The second class of improvements involved a set of filters that eliminated nonprotein-like conformations (e.g., poorly formed β-sheets), and the third set involved the simultaneous clustering of conformations generated by folding different homologous sequences. They found large, correctly predicted segments (>50 residues with an RMSD below 6.5 Å) for 16/21 protein domains below 300 residues. The method performed best on helical and α/β proteins, with very impressive predictions for target 91, 106, 110 and 116. Interestingly, for targets 102 and 116, the ROSETTA models outperformed threading, a trend seen previously in CASP3 for some *ab initio* structure predictions by the Baker [156] and Skolnick groups [36]. Similar results were seen subsequent to CASP4, when ROSETTA was applied to the families of Pfam proteins below 150 residues in length [177]. Of 131 such protein families of known structure, approximately 60% had one of the top 5 models correctly predicted for 50 or more residues, and for approximately 35%, the correct SCOP superfamily [178–181] was identified in a structure-based search of the Protein Data Bank (PDB) [182].

In a very similar spirit to ROSETTA [167], the FRAGFOLD [183] algorithm of Jones, which assembles supersecondary structural fragments, gave some good results for 46–61 residue fragments of five α and α/β proteins. The Friesner [38], (as well as the Skolnick [39]) group combined fold recognition with *ab initio* folding, presenting early examples of the convergence of structure prediction methods. For α-proteins, global optimization and a statistical potential are used to generate structures that are then screened against a fold library; any structures that match are subject to further refinement. For β-proteins, sequence and secondary structure alignments are used to select template proteins from which spatial constraints are extracted and employed in subsequent global sampling/optimization. A number of proteins had significant fragments approximately predicted, including target 98 and 102, with a very good global model for the helical target 102 reported. In a related approach, the Oakridge group used their threading algorithm PROSPECT, which uses pair interactions explicitly and which is guaranteed to find a global optimum. They predicted fragments of a number of new fold/fold recognition and new fold targets; most impressive is a 97-residue fragment prediction for the new fold target 124, with an RMSD from native of 3.8 Å.

The results of our efforts in CASP are summarized below. For two proteins, targets 102 (RMSD for model 1 [of 5 submitted] was 3.6 Å) and 110 (RMSD for model 1 of 5.1 Å), the method generated the best models in CASP4, models that were better than could be obtained from matching to a threading template structure. Similarly, for a number of proteins, significant fragments ranging from 30–77 residues in length, were predicted below 5 Å. Interestingly, in a number of cases, the mirror image topology (helices are right handed, but the chirality of the turns was reversed) was generated. The best example of this is target 96 where the C-terminal domain (in the new fold category [184]) is predicted to have the mirror image of the native topology. In principle, both topologies should have been generated, but given the size of this protein fragment (212) residues, the simulations had not converged. The importance of this prediction is that although our threading algorithm did not identify a good global template (nor could it, as one was not present in the PDB), nevertheless side chain contacts were predicted from proteins having a different global structure from the native one that allowed us (modulo the topological mirror image) to assemble this novel topology.

Turning to the performance of secondary structure algorithms in CASP4, the average accuracy was 76%, an increase of 4% over CASP3, with the best methods getting segments correct and performing well on strands [185–187]. These are very encouraging results. With regards to contact prediction [158], Valencia and coworkers have developed the CORNET server that is based on a neural network approach [188]. For the 29 proteins considered, the average contact prediction accuracy was 0.14, but the method did not perform well on helical proteins not in the training set. Marginal improvement was noted on β and α/β proteins, with an average contact prediction accuracy of 0.16. Unfortunately, this level of prediction accuracy is too low to be used in structure assembly. In contrast, our *TOUCHSTONE* method that uses consensus contacts seen in threading (even when the global fold of target and template structures are different) for 11 proteins (see Table II of Skolnick, et al. [39]) had an average accuracy of 0.17, comparable performance to the Valencia group [158], but the average accuracy of contact prediction within ±2 residues (that is, a predicted contact is within two residues of an actual contact) was 0.62; a partially acceptable result if a sufficient number of contacts had been predicted. Unfortunately, in CASP4 too few contacts were generated. Clearly, progress in contact prediction (which has been made in the period following CASP4, see section 8.4.3) was required.

8.4 THE *TOUCHSTONE AB INITIO* FOLDING ALGORITHM

8.4.1 Overview

Over the past several years, we have developed the *TOUCHSTONE* folding algorithm [189], an overview is depicted in Figure 8.1. Using our threading algorithm *PROSPECTOR* [190], one first threads against a representative set of PDB templates. If

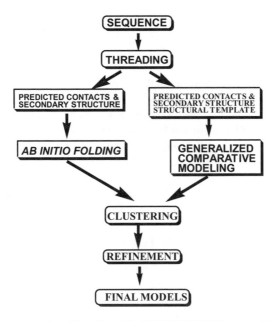

FIGURE 8.1 Flow chart describing the unified *TOUCHSTONE* approach to protein structure prediction. First, threading is done. If a significant hit to a template is found, then generalized comparative modeling in the vicinity of the template but supplemented by predicted secondary structure and contacts, possibly from other templates, is done. If no significant target sequence-template structure match is found, then consensus contacts and sets of local distances in the top 20 scoring structures are extracted and employed as restraints in our *ab initio* folding algorithm. Once a sufficient number of simulations (typically 100) are done, the structures are clustered, full atomic models are built in the refinement step and then, using a distance dependent atomic pair potential, the top 5 scoring structures are selected.

a template is significantly hit, generalized comparative modeling is undertaken. If not, then consensus contacts from weakly threading templates are pooled and incorporated into our *ab initio* folding algorithm. As a protein model, originally, we developed the SICHO lattice protein model (152,189,191,192) where the interaction center per residue is located at the center of mass of the side chain heavy atoms and the Cαs. To improve conformational sampling, we initially implemented REMC [189] and then developed parallel hat tempering [151] and a new scheme, termed local energy flattening, that improved the efficiency of conformational energy searches[152]. The structures are then clustered, atomic detail added, the models relaxed and the lowest energy families of structures selected.

We benchmarked the original *TOUCHSTONE* algorithm against a library of 65 proteins [189] and then applied it to fold all the small proteins in the *M. genitalium* genome [193], which was the first application of *ab initio* folding on a genomic scale and to a real (albeit small) genome. To improve the geometric fidelity of the protein representation to enable *TOUCHSTONE* to span from the regime from homology modeling to new fold prediction, we have developed, benchmarked and begun to optimize a Cα plus Cβ based lattice protein model, CABS, that addresses the inability of the SICHO model to treat target proteins that have closely related structures to known template structures. (In SICHO, the quality of the structures is too poor for homology modeling, with nativelike structures having a typical RMSD from native of 3 Å). We also applied the resulting modified version, *TOUCH-STONEX*, [194] to assemble proteins using a sparse amount of experimental data such as tertiary restraints as would be provided from NOEs [194], and in an earlier study with the SICHO protein model, we assembled a number of proteins using residual dipolar coupling [195–200] information. As part of an effort to develop better potentials, we have examined simplified models to determine the minimum requirements for an all-or-none, cooperative folding transition [201]. Finally, we have participated in CASP3-5 [39–125] as another means of independently validating and testing our *ab initio* prediction methodology.

8.4.2 BENCHMARKING ON A LARGE TEST SET

Successful protein structure prediction requires an efficient conformational search algorithm and an energy function with a global minimum in the native state. As a step towards addressing both issues, *TOUCHSTONE*, a novel algorithm that employs threading-based secondary and tertiary restraint predictions in *ab initio* folding, has been developed [189]. Such restraints are derived by extracting consensus contacts and local secondary structure from at least weakly scoring structures that can lack any global similarity to the sequence of interest (see section 8.4.3). In our original published work, the SICHO model [191] is employed with REMC [140–153] to explore conformational space. SICHO itself contains generic terms to mimic proteinlike local geometries, predicted secondary structure from PSIPRED [168], local distance restraints extracted from threading, orientation dependent pairwise interactions, a side chain burial term, and predicted tertiary restraints. This methodology was applied to a set of 65 nonhomologous proteins (pairwise sequence identity below 35%) whose lengths ranged from 39 to 146 residues, and which represented all secondary structural classes, α, α/β and β. For 47 (40), a cluster centroid whose RMSD from native is below 6.5 (5) Å is found. The number of correctly folded proteins increases to 50 when atomic detail is added and a knowledge-based atomic potential is combined with clustered and nonclustered structures for candidate selection [202]. The number of clusters generated by the folding algorithm and the ratio of the number of contacts to protein length, *F*, are found to be reliable indicators of the likelihood of successful fold prediction. When *F* is close to unity (this happens in >50% of the cases, thus, the problem of too few restraints seen in CASP4 has at least been partially addressed), the native state is likely to be found in one of the clusters even if seven clusters are generated, thereby opening the way for genome-scale *ab initio* protein structure prediction. Similarly if five or fewer clusters are generated, folding is likely to be successful. If three or fewer clusters are found, folding was always successful.

8.4.3 ACCURACY OF TERTIARY CONTACT AND SECONDARY STRUCTURE PREDICTION

By pooling the tertiary contacts generated by variants of *PROSPECTOR* [190] that use protein specific and orientation dependent pair potentials along with quasichemical-based pair potentials, contacts are extracted using a two-step procedure. If one of the threading variants identifies a significantly related protein, then the resulting aligned templates provide contacts. If no template is hit, then the consensus contacts are predicted from the protein specific and orientation dependent, pair potential-based variants of *PROSPECTOR*. For the latter case, in a 125-protein test set, the average fraction of correctly predicted contacts is 0.34, and within ±1 residue, the fraction correct is 0.58, with the average ratio of the number of predicted contacts to the number of residues of 0.93 [189]. These results are a considerable improvement over CASP4 of 0.16, 0.52, and 0.66 for these quantities, respectively [39]. Furthermore, contact prediction protocol is fully automated and robust, unlike our previously developed correlated mutation analysis where it is difficult to separate out functional from spatial correlations [36].

If a similar threading based protocol is applied to predict secondary structure by calculating average local Ca distances (up to 5 residues apart), then 79% of the local distances are correctly predicted within 1 Å. This offers the advantage over more traditional approaches in that a specific protein geometry is predicted rather than, e.g., just the statement that there is a turn in the protein. Such information is then fed directly into our *ab initio* folding algorithm as local distance restraints [189].

8.4.4 A NEW LATTICE MODEL FOR TERTIARY STRUCTURE PREDICTION

A new lattice model has been developed and tested [194]. In contrast to our previous models that had at most two interaction centers per residue, the CABS models have three (Cα, Cβ, and the center of mass of the side group) and are designed to enhance the specificity of the interactions as well as the geometric fidelity of the protein model. The CABS model employs 312 basis vectors for the Ca-Ca virtual bonds and the mesh size of the underlying lattice is 0.87 Å. CABS has been tested in various contexts and proven to be qualitatively better than the older SICHO model [189]. This has been achieved by a very careful design of the higher resolution geometry found in the model, redesign of the force field that reweights the interactions so that fluctuating secondary structure is present even in the high temperature, denatured states, and elaborate tuning of the model parameters. The CABS force field was optimized by maximizing the correlation of energy and RMSD of $30 \times 60,000$ decoy and native structures as well as the energy gap between native and the decoy ensemble. Then, we used both our original SICHO and optimized CABS models on a representative nonhomologous test set of 60 hard proteins whose length ranges from 36 to 174 residues. The fraction of foldable cases (defined as having one of the top five lowest average energy clusters with an RMSD from native below 6.5 Å) by SICHO and CABS are one third and one half respectively (i.e., 20/60 and 32/60).

8.4.5 SELECTION OF THE NATIVE CONFORMATION

Current *ab initio* structure-prediction methods are sometimes able to generate families of folds, one of which is native (167,177,189,194,203,204), but are unable to single out the native one due to imperfections in the folding potentials. To address this issue, we developed a method for the detection of statistically significant folds in a pool of predicted structures [205] that clusters and averages the structures into representative families. Using a metric derived from the root mean square deviation from native (RMSD) that is less sensitive to protein size, we determine whether the structures are clustered in relation to a group of random structures. The method searches for cluster centers and iteratively calculates the clusters and their respective centroids. The centroid inter-residue distances are adjusted by minimizing a potential constructed from the corresponding average distances of the cluster structures. Application to selected proteins shows that it can detect the fold family that is closest to native, along with other misfolded families. We also developed a method to obtain sub-structures. This is useful when the folding simulations fail to give a global topology prediction but produce common sub-elements among the structures. We have a web server that clusters user submitted structures that can be found at http://www.bioinformatics.buffalo.edu/services/structclust.

8.4.6 UNIVERSAL SIMILARITY MEASURE FOR COMPARING PROTEIN STRUCTURES

A key question in comparing protein structures is the statistical significance of the structural comparison. One common measure for comparing structures is their RMSD. We have developed a new dimensionless measure of the RMSD, termed the relative RMSD, RRMSD, which is zero between identical structures and unity if the two structures are as dissimilar as the RMSD of an average pair of randomly selected, same length protein structures, $RMSD^0$, viz. $RRMSD = RMSD/RMSD^0$ [206]. The RRMSD probability curve becomes independent of protein size as the length of the protein increases. Computation of the correlation coefficients between aligned randomly related protein structures as a function of size has two characteristic lengths of 3.7 and 37 residues, which mark the separation between phases of different inherent structural order between protein fragments.

8.4.7 HOW CAN ONE SELECT NATIVE-LIKE CLUSTERS?

In protein structure prediction, without knowing the native state, three key questions must be addressed. What is overall success rate? Can one tell if the algorithm generates low RMSD structures? If there are low RMSD structures, can one identify them in a single cluster or among the top five clusters? One way of addressing the first issue is to simply rank the structures by their average energy. For the 65-test proteins described in section 8.4.2 and using the SICHO lattice model, in 47/65 cases, at least one of the top five lowest energy clusters has an RMSD below 6.5 Å from native [189]. To address the second question, when the number of restraints is more than the number of residues in the sequence, in 32 out of 41 cases (78.0%), a cluster centroid closer than 6.5 Å RMSD to native is obtained. When the ratio of

TABLE 8.1
**Correlation of Successful Predictions with the Number
of Clusters and Tertiary Restraints**

		Number of Restraints/Protein Length	
		1.0 or more	**1.5 or more**
Number of Clusters	3 or less	13/13 (100%)	8/8 (100%)
	5 or less	23/26 (88.5%)	13/13 (100%)
	7 or less	30/36 (83.3%)	16/18 (88.9%)

restraints to residues is 150%, the success rate improves to 88.0% (22 out of 25 proteins). A proper fold is always obtained in two cases: (1) the number of obtained clusters is less than or equal to three, and (2) as shown in Table 8.1, when the number of clusters is less than or equal to five, and the ratio of predicted restraints to sequence length is 150% or more.

It is important to note that in contrast to other methods (25–38,177), both the accuracy of contact prediction and the success rate when the number of predicted contacts is sufficiently large is completely independent of the type of secondary structure class of the protein.

8.4.8 Relationship Between Cluster Population and Likelihood of Prediction Success

We next explored the issues of fold success and fold selection for the more powerful CABS model. Using the CABS model, we examined a representative set of 125 test proteins (65 of these proteins are the same as in the original *TOUCHSTONE* paper [189] plus the additional, harder 60 protein test). Our overall success rate is 66% (= 83/125), where at least one of the five lowest energy clusters has an RMSD from native below 6.5 Å.

To address the relationship between cluster population and the presence of native-like structures, we expanded on ideas due to the Baker group [156–167] who suggested that the most populated cluster is native. (In practice, we did not find this true in general, but the modified variant defined below often holds). In particular, we examined the following ratio:

$$R = \frac{N}{N_t}, \tag{8.1}$$

where N is the number of structures in the most populated cluster and N_t is the total number of structures clustered. Using a threshold of $R > 0.55$, 61 proteins satisfy this threshold, and in 58/61 (95%) cases, the native fold is in the top five clusters. This is significantly greater than our overall success rate of 83/125 (66%) proteins. Furthermore, the lowest energy cluster is below 6.5 Å in 50/61 cases.

The selection of the lowest energy cluster neglects information about the conformational entropy, since the lowest energy structure is not necessarily the lowest free energy structure. Thus, we considered a combination of the energy and the microscopic free energy (proportional to kT log R)

$$Y = -E - kT \log R \qquad (8.2)$$

where E is the average energy of the structures in the given cluster, and R is defined in equation 8.1 and is the thermodynamic probability of occupying a given cluster. In practice, Y is more discriminative than either E or kT log R alone. For example, by selecting the lowest energy cluster, we identify 58/125 cases that have an RMSD below 6.5 Å to the native structure; if we choose the structure of lowest Y, we identify 65/125 successful cases. If we further restrict our analysis to those proteins where $R>0.55$, then the lowest energy criterion selects 50/61 good clusters in the top position, whereas Y selects 52. If we consider $R<0.55$, there are 64 proteins, of which 25 have at least one cluster (at any rank) with an RMSD below 6.5 Å. Selection of the top cluster by lowest energy finds 8/25 of these good folds, whereas use of the max Y identifies 13/25 good folds. This is a significant enhancement, but for 12 proteins the native state is not in the top cluster. Thus, we can tell where the majority of successfully folded proteins are, but for those with low cluster density, the identification of the native fold stills need improvement.

8.4.9 Prediction of Tertiary Structure Using Residual Dipolar Coupling Information

NMR residual dipolar couplings (RDC) in the form of the projection angles between the respective internuclear bond vectors have been used as structural restraints in the *ab initio* structure prediction of a six protein test set (196,198,200). The restraints are applied using the *SICHO* lattice protein model [192] that employs a replica exchange Monte Carlo algorithm (139,140,142,153) to search conformational space. Using a small number of *RDC* restraints improves the quality of the predicted structures relative to that when the *RDCs* are absent. This is reflected by the lower RMSD/dRMSD (distance root mean square deviation) values and the higher correlation of the most cooperative mode of motion of each predicted structure with the native structure. The latter has implications for the structure-based functional analysis of predicted structures. In contrast to the work of Baker and collaborators, [207–208], much fewer RDCS need to be employed to assemble comparable quality structures. This is partly due to the presence of predicted tertiary restraints and the use of a more sophisticated force field.

8.4.10 Benchmark Prediction of Tertiary Structure Using a Small Number of Exact Long Range Restraints (Specific Aim 7)

A new method for folding proteins using a small number of exact long-range contact restraints derived from NMR experimental nuclear Overhauser enhancement (NOE) data, *TOUCHSTONEX*, has been developed [194]. The method employs the CABS protein model and a force field that consists of terms to produce protein-like behavior, hydrogen binding, one-body, pairwise and multibody long-range interactions. Exact contact restraints were incorporated into the force field as a NOE-specific pairwise potential. The algorithm was evaluated using the same 125-protein test set (as in Section 8.4.8) of various secondary structure types and lengths up to 174 residues. Using N/8 exact long-range side chain contact restraints, with N the number of residues, 108 proteins were folded to an RMSD from native below 6.5 ≈ in one of the top 5 lowest energy clusters. The average RMSD of the lowest RMSD structures for all 125 proteins (both folded and unfolded) is 4.39 ≈. The algorithm was also applied to three proteins with limited experimental NOE data. Using sparse experimental side chain restraints and a small number of side chain–main chain and main chain–main chain contact restraints, all three proteins were folded to low-to-medium resolution structures. The algorithm can be applied to the NMR structure determination process or to other experimental methods that can provide tertiary restraint information. It should be useful in the early stage of structure determination when only limited experimental data are resolved.

8.4.11 PREDICTION OF THE TERTIARY STRUCTURE OF ALL THE SMALL PROTEINS IN THE *M. GENITALIUM* GENOME

Our *ab initio* protein structure prediction procedure, *TOUCHSTONE,* was applied to all the 85 small proteins (less than 150 residues) [193] of the *Mycoplasma genitalium* genome [209]. For threading-based fold assignment, *PROSPECTOR* [190] was used with a Z-score threshold of 10.0, with the more positive Z-score being the more significant. Using these criteria, *PROSPECTOR* obtained folds for 34 proteins. FASTA [210] and PSI-BLAST [211–215] combined identify a total of 27 proteins as having homologous PDB structures. All 27 are a subset of the *PROSPECTOR* predictions. Although none of the native structures of the 85 proteins has been experimentally determined, it is possible to examine the performance of the prediction procedure for the 34 cases identified from threading. In all but two cases (MG087 and MG175), when a protein has a threading template hit, at least one of the cluster centroids obtained from the simulations has the same fold as the predicted threaded structure with at least 60% of the structure aligned.

One of the two cases, the C-terminal 60 residues of 1fjfM, which is the predicted template structure of MG175, lack a stable structure [216]. The best RMSD of the N-terminal domain (first 66 residues) is 6.1 Å. For another, 1fjfL, the template hit of MG087 has a 26-residue-long dangling N-terminal tail and a C-terminal 6 residue-long tail that is similarly impossible to reproduce. Excluding both tails, the best coverage of the structurally well-defined regions is 78.3%. Note that 1fjfM and 1fjfL, together with several other proteins, are part of the 30S ribosomal subunit. Besides these three proteins, the 1div(MG093), 1fjfI(MG417), 1fjfJ(MG150), 1fjfK(MG176), 1fjfN(MG164), 1fjfS(MG155)), 1a91(MG404), 1eg0K(081), 1hueA(MG353), 1aonO(MG393), 1a32(MG424) and 1bxeA(MG156) templates also have dangling structures that were not reproduced well (the corresponding *Mycoplasma genitalium* genes are shown in parentheses). Furthermore, 1div, the template hit of MG093, is a two-domain structure.

Since our *ab initio* potential contained a compactness term that forces the protein to adopt a single domain structure, it was impossible to predict a two-domain structure using the then current methodology. But the structures of both two domains are independently reproduced quite well. (The RMSD of the N-terminal 50 residue domain is 4.0 Å and the RMSD of the C-terminal 85 residue domain is 4.5 Å from the respective templates).

Of the remaining 51 proteins that do not have a significant threading hit, 29 converged to 5 or fewer clusters, indicating that one may have the correct fold. Selection of the native fold was done using the convergence from two different conformational search schemes, and the lowest energy structure evaluated using a knowledge-based atomic-detailed potential. Based on these results, *ab initio* protein structure prediction is becoming practical. The list of predicted structures can be found on our website: http://www.bioinformatics.buffalo.edu/services/mgen.

8.5 SUMMARY

The past several years have witnessed significant improvement in *ab initio* folding algorithms, especially for helical and mixed motif proteins, as evidenced by CASP3 and CASP4, but β proteins remained problematic [158]. Many of the most successful algorithms used a reduced protein model to assemble representative topologies, followed by addition of atomic detail (38,165–167). In addition, there have been the first attempts at unifying approaches to protein structure prediction to span the range from threading/homology modeling to *ab initio* folding [38–39]. While additional improvements in methodologies are clearly necessary, the results have been quite encouraging.

In the past three years, we have developed *TOUCHSTONE* (189,193,194), a unified approach to predict protein structure that spans the range from homology modeling to threading to *ab initio* folding. Initially, we used the SICHO model, but because it lacked the requisite resolution, we then developed the CABS model that describes each protein residue by its Cα, Cβ, and side chain center of mass. With respect to *ab initio* folding, the method has been applied to a representative 125-protein test set, 83 of which have a native-like cluster in the top five lowest energy clusters. If one of the clusters comprises more than 55% of the population, we can select the native-like cluster as the best single cluster in 50/61 cases, and when no cluster is significantly populated, in 13/25 cases. Using the SICHO model [191], we undertook the folding of all (85) small proteins <150 residues in the *M. genitalium* genome [209]. In addition to recovering the native topology of all 34 proteins in that thread, it is likely that 29/51 remaining proteins have a significant fold prediction. We also considered the use of experimental restraints (RDC and NOE like information) to enhance the yield and range of applicability of protein structure prediction methods. If N/8 randomly selected, long range restraints are used, then 108 proteins of the same 125-protein test set can be folded. The method is fast and should assist in the structural genomics efforts.

Key outstanding unsolved issues are the following: The potentials used to fold proteins need to be further improved, as does the conformational sampling scheme. While the native state is often among the clusters generated, often, it is not the lowest energy cluster. Improved force fields will also enable the folding of proteins larger than 150 residues or so. To assess the range of validity of the methods and to point out directions for future improvement, large scale benchmarking is clearly necessary. To address all these issues, we are currently folding a representative set of all proteins in the PDB below 200 residues. This will establish the strengths and weaknesses of our algorithm in a statistically significant way. The resulting set of nativelike structures and decoys will enable more specific potentials to be derived. Thus, while additional progress on *ab initio* structure prediction is clearly necessary, there is significant reason for optimism that such advances will be made.

ACKNOWLEDGMENT

The research described in this chapter was supported in part by NIH grant GM-37408 of the Division of General Medical Sciences of the National Institutes of Health.

REFERENCES

1. Clark, M.S., *Bioessays*, 21, 121–130, 1999.
2. DellaPenna, D., *Science*, 285, 375–379, 1999.
3. Moxon, E.R., D.W. Hood, N.J. Saunders, E.K. Schweda, and J.C. Richards, *Philos. Trans. R. Soc. Lond. B. Biol. Sci.*, 357, 109–116, 2002.
4. Waring, J.F. and D.N. Halbert, *Curr. Opinions Mol. Ther.*, 4, 229–235, 2002.
5. Anzick, S.L. and J.M. Trent, *Oncology*, 16, 7–13, 2002.
6. Wiley, S.R., *Curr. Pharm. Des.*, 4, 417–422, 1998.
7. Skolnick, J. and J.S. Fetrow, *Trends Biotechnol.*, 18, 34–39, 2000.
8. Skolnick, J., J.S. Fetrow, and A. Kolinski, *Nat. Biotechnol.*, 18, 283–287, 2000.
9. Fetrow, J.S., N. Siew, J.A. Di Gennaro, M. Martinez-Yamout, H.J. Dyson, and J. Skolnick, *Protein Sci.*, 10, 1005–1014, 2001.
10. Altschul, S.F., W. Gish, W. Miller, E.W. Myers, and D.J. Lipman, *J. Mol. Biol.*, 215, 403–410, 1990.
11. Pearson, W.R., *Methods Enzymol.*, 266, 227–258, 1996.
12. Henikoff, S. and J.G. Henikoff, *Genomics*, 19, 97–107, 1994.
13. Fetrow, J.S. and J. Skolnick, *J. Mol. Biol.*, 281, 949–968, 1998.
14. McKusick, V.A., *Genomics*, 45, 244–249, 1997.
15. Wilson, C.A., J. Kreychman, and M. Gerstein, *J. Mol. Biol.*, 297, 233–249, 2000.
16. Frishman, D., *Protein Eng.*, 15, 169–183, 2002.
17. Pandit, S.B., D. Gosar, S. Abhiman, S. Sujatha, S.S. Dixit, N.S. Mhatre, R. Sowdhamini, and N. Srinivasan, *Nucleic Acids Res.*, 30, 289–293, 2002.
18. Gough, J. and C. Chothia, *Nucleic Acids Res.*, 30, 268–272, 2002.
19. Zarembinski, T.I., L.W. Hung, H.J. Mueller-Dieckmann, K.K. Kim, H. Yokota, R. Kim, and S.H. Kim, *Proc. Natl. Acad. Sci. U.S.*, 95, 15189–15193, 1998.
20. Gaasterland, T., *Nat. Biotechnol.*, 16, 625–627, 1998.
21. Gaasterland, T., *Trends Genet.*, 14, 135, 1998.
22. Cassman, M. and J.C. Norvell, *Science*, 286, 239–240, 1999.
23. Buchanan, S.G., *Curr. Opinions Drug Discovery Dev.*, 5, 367–381, 2002.
24. Burley, S.K. and J.B. Bonanno, *Annu. Rev. Genomics Hum. Genet.*, 3, 243–262, 2002.

25. Baker, D. and A. Sali, *Science,* 294, 93–96, 2001.
26. Moult, J., K. Fidelis, A. Zemla, and T. Hubbard, *Proteins,* Suppl. 5, 2–7, 2001.
27. Tramontano, A., R. Leplae, and V. Morea, *Proteins,* Suppl. 5, 22–38, 2001.
28. Finkelstein, A.V. and B.A. Reva, *Nature,* 351, 497–499, 1991.
29. Fischer, D. and D. Eisenberg, *Protein Sci.,* 5, 947–955, 1996.
30. Rice, D.W. and D. Eisenberg, *J. Mol. Biol.,* 267, 1026–1038, 1997.
31. Fischer, D., *Proteins,* Suppl. 3, 61–65, 1999.
32. Godzik, A. and J. Skolnick, *Proc. Natl. Acad. Sci. U.S.,* 89, 12098–12102, 1992.
33. Russell, R.B., M.A.S. Saqi, R.A. Sayle, P.A. Bates, and M.J.E. Sternberg, *J. Mol. Biol.,* 269, 423–439, 1997.
34. Levitt, M., M. Gerstein, E. Huang, S. Subbiah, and J. Tsai, *Annu. Rev. Biochem.,* 66, 549–579, 1997.
35. Bystroff, C. and D. Baker, *J. Mol. Biol.,* 281, 565–577, 1998.
36. Ortiz, A.R., A. Kolinski, P. Rotkiewicz, B. Ilkowski, and J. Skolnick, *Proteins,* 37, 177–185, 1999.
37. Sternberg, M.J., P.A. Bates, L.A. Kelley, and R.M. MacCallum, *Curr. Opinions Struct. Biol.,* 9, 368–373, 1999.
38. Standley, D.M., V.A. Eyrich, Y. An, D.L. Pincus, J.R. Gunn, and R.A. Friesner, *Proteins,* Suppl. 5, 133–139, 2001.
39. Skolnick, J., A. Kolinski, D. Kihara, M. Betancourt, P. Rotkiewicz, and M. Boniecki, *Proteins,* Suppl. 5, 149–156, 2001.
40. Go, N. and H. Taketomi, *Proc. Natl. Acad. Sci. U.S.,* 75, 559–563, 1978.
41. Krigbaum, W.R. and A. Komoriya, *Biochim. Biophys. Acta,* 576, 204–246, 1979.
42. Krigbaum, W.R. and S.F. Lin, *Macromolecules,* 15, 1135–1145, 1982.
43. Kolinski, A. and J. Skolnick, *Proc. Natl. Acad. Sci. U.S.,* 83, 7267–7271, 1986.
44. Kolinski, A., J. Skolnick, and R. Yaris, *J. Chem. Phys.,* 85, 3585, 1986.
45. Kolinski, A., J. Skolnick, and R. Yaris, *Biopolymers,* 26, 937–962, 1987.
46. Skolnick, J., A. Kolinski, and R. Yaris, *Proc. Natl. Acad. Sci. U.S.,* 85, 5057, 1988.
47. Skolnick, J. and A. Kolinski, *Annu. Rev. Phys. Chem.,* 40, 207–235, 1989.
48. Skolnick, J., A. Kolinski, and R. Yaris, *Biopolymers,* 28, 1059–1095, 1989.
49. Skolnick, J., A. Kolinski, and R. Yaris, *Proc. Natl. Acad. Sci. U.S.,* 86, 1229–1233, 1989.
50. Skolnick, J. and A. Kolinski, *J. Mol.Biol.,* 212, 787–817, 1990.
51. Skolnick, J. and A. Kolinski, *Science,* 250, 1121–1125, 1990.
52. Kolinski, A., M. Milik, and J. Skolnick, *J. Chem. Phys.,* 94, 3978–3985, 1991.
53. Kolinski, A. and J. Skolnick, *J. Phys. Chem.,* 97, 9412–9426, 1992.
54. Kolinski, A. and J. Skolnick, *Proteins,* 18, 338–352, 1994.
55. Kolinski, A., M. Milik, J. Rycombel, and J. Skolnick, *J. Chem. Phys.,* 103, 4312–4323, 1995.
56. Sikorski, A. and J. Skolnick, *Biopolymers,* 28, 1097–1113, 1989.
57. Sikorski, A. and J. Skolnick, *Proc. Natl. Acad. Sci. U.S.,* 86, 2668–2672, 1989.
58. Sikorski, A. and J. Skolnick, *J. Mol. Biol.,* 215, 183–198, 1990.
59. Sikorski, A. and J. Skolnick, *J. Mol.Biol.,* 212, 819–836, 1990.
60. Chan, H.S. and K.A. Dill, *J. Chem. Phys.,* 92, 492–509, 1989.
61. Chan, H.S. and K.A. Dill, *Macromolecules,* 22, 4559–4573, 1989.
62. Chan, H.S. and K.A. Dill, *Proc. Natl. Acad. Sci. U.S.,* 87, 6388–6392, 1990.
63. Chan, H.S. and K.A. Dill, *Annu. Rev. Biophys. Chem.,* 20, 447–490, 1991.
64. Dill, K.A., *Biochemistry,* 24, 1501–1509, 1985.
65. Dill, K.A., D.O.V. Alonso, and K. Hutchinson, *Biochemistry,* 28, 5439–5449, 1989.
66. Dill, K.A., *Curr. Biol.,* 3, 99–103, 1993.

67. Dill, K.A., S. Bromberg, K. Yue, K.M. Fiebig, D.P. Yee, P.D. Thomas and H.S. Chan, *Prot. Sci.*, 4, 561–602, 1995.
68. Abkevich, V.I., A.M. Gutin, and E.I. Shakhnovich, *J. Chem. Phys.*, 101, 6052–6062, 1994.
69. Sali, A., E. Shakhnovich, and M. Karplus, *Nature,* 369, 248–251, 1994.
70. Shakhnovich, E.I. and A.V. Finkelstein, *Biopolymers,* 28, 1667–1680, 1989.
71. Shakhnovich, E.I. and A.M. Gutin, *Biophys. Chem.,* 34, 187–199, 1989.
72. Shakhnovich, E., G. Farztdinov, and A.M. Gutin, *Phys. Rev. Lett.*, 67, 1665–1668, 1991.
73. Shakhnovich, E.I. and A.M. Gutin, *Proc. Natl. Acad. Sci. U.S.*, 90, 7195–7199, 1993.
74. Shakhnovich, E.I. and A.M. Gutin, *Protein Eng.*, 6, 793–800, 1993.
75. Shakhnovich, E.I., *Phys. Rev. Lett.*, 72, 3907–3910, 1994.
76. Abkevich, V.I., A.M. Gutin, and E.I. Shakhnovich, *Folding & Design*, 1, 221–230, 1996.
77. Dinner, A.R., A. Sali, M. Karplus, and E. Shakhnovich, *J. Chem. Phys.*, 101, 1444–1451, 1994.
78. Dinner, A.R., A. Sali, and M. Karplus, *Proc. Natl. Acad. Sci. U.S.,* 93, 8356–8361, 1996.
79. Hao, M.-H. and H.A. Scheraga, *J. Phys. Chem.*, 98, 4940–4948, 1994.
80. Hao, M.-H. and H.A. Scheraga, *J. Phys. Chem.*, 98, 9882–9893, 1994.
81. Hao, M.-H. and H.A. Scheraga, *J. Chem. Phys.*, 102, 1334–1348, 1995.
82. Kolinski, A. and P. Madziar, *Biopolymers,* 42, 1997.
83. Plaxco, K.W., K.T. Simons, and D. Baker, *J. Mol. Biol.,* 277, 985–994, 1998.
84. Duan, Y., L. Wan, and P. Kollman, *Proc. Natl. Acad. Sci. U.S.,* 95, 9897–9902, 1998.
85. Go, N. and H. Taketomi, *Int. J. Pept. Protein Res.*, 13, 447–461, 1979.
86. Go, N. and H. Taketomi, *Int. J. Pept. Protein Res.*, 13, 235–252, 1979.
87. Go, N. and H. Taketomi, *Proc. Natl. Acad. Sci. U.S.* 75, 559–563, 1978.
88. Taketomi, H., Y. Ueda, and N. Go, *Int. J. Pept. Protein Res.*, 7, 445–459, 1975.
89. Skolnick, J. and A. Kolinski, *Annu. Rev. Phys. Chem.*, 40, 207–235, 1989.
90. Levitt, M. and A. Warshel, *Nature,* 253, 694–698, 1975.
91. Levitt, M., *J. Mol. Biol.,* 104, 59–107, 1975.
92. Levitt, M., *J. Mol. Biol.,* 104, 59–107, 1976.
93. Kuntz, I.D., G.M. Crippen, P.A. Kollman, and D. Kimelman, *J. Mol. Biol.,* 106, 983–994, 1976.
94. Hagler, A.T. and B. Honig, *Proc. Natl. Acad. Sci. U.S.,* 75, 554–558, 1978.
95. Wilson, C. and S. Doniach, *Proteins,* 6, 193–209, 1989.
96. Sun, S., *Protein Sci.,* 2, 762–785, 1993.
97. Wallqvist, A. and M. Ullner, *Proteins,* 18, 267–289, 1994.
98. Hoffmann, D. and E.W. Knapp, *Eur. Biophys. J.*, 24, 387–403, 1996.
99. Hoffmann, D. and E.W. Knapp, *Phys. Rev. E*, 53, 4221–4224, 1996.
100. Moult, J., T. Hubbard, K. Fidelis, and J.T. Pedersen, *Proteins,* Suppl, 2–6, 1999.
101. Dashevskii, V.G., *Molekulyarnaya Biologiya* (Translation from) 14, 105–117, 1980.
102. Covell, D.G., *Proteins*, 14, 409–420, 1992.
103. Covell, D.G. and R.L. Jernigan, *Biochemistry*, 29, 3287–3294, 1990.
104. Hinds, D.A. and M. Levitt, *Proc. Natl. Acad. Sci. U.S.*, 89, 2536–2540, 1992.
105. Kolinski, A. and J. Skolnick, *Lattice Models of Protein Folding, Dynamics and Thermodynamics*, Austin, TX, 1996, R.G. Landes, 200.
106. Skolnick, J. and A. Kolinski, *J. Mol. Biol.,* 221, 499–531, 1991.
107. Godzik, A., A. Kolinski, and J. Skolnick, *J. Comp. Aided Mol. Design*, 7, 397–438, 1993.

108. Kolinski, A. and J. Skolnick, *Acta Biochimica Polonica*, 44, 389–422, 1998.
109. Kolinski, A., P. Rotkiewicz, and J. Skolnick, Application of high coordination lattice model in protein structure prediction, in *Monte Carlo Approaches to Biopolymers and Protein Folding*, P. Grassberger, G.T. Barkema, and W. Nadler, Eds., 1998, World Scientific, Singapore, 110–130.
110. Kolinski, A., A. Godzik, and J. Skolnick, *J. Chem. Phys.*, 98, 7420–7433, 1993.
111. Kolinski, A. and J. Skolnick, *Proteins,* 18, 353–366, 1994.
112. Kolinski, A. and J. Skolnick, *Proteins,* 18, 338–352, 1994.
113. Vieth, M., A. Kolinski, C.L. Brooks III, and J. Skolnick, *J. Mol. Biol.,* 1994, 361–367, 1994.
114. Vieth, M., A. Kolinski, I. Brooks, C. L., and J. Skolnick, *J. Mol. Biol.,* 251, 448–467, 1995.
115. Vieth, M., A. Kolinski, and J. Skolnick, *Biochemistry*, 35, 955–967, 1996.
116. Mohanty, D., A. Kolinski, and J. Skolnick, *Biophys. J.*, 77, 54–69, 1999.
117. Monge, A., R.A. Friesner, and B. Honig, *Proc. Natl. Acad. Sci. U.S.,* 91, 5027–5029, 1994.
118. Friesner, R. and J.R. Gunn, *Annu. Rev. Biophys. Biomol. Struct.*, 25, 315–342, 1996.
119. Eyrich, V.A., D.M. Standley, A.K. Felts, and R.A. Friesner, *Proteins,* 35, 41–57, 1999.
120. Standley, D.M., J.R. Gunn, R.A. Friesner, and A.E. McDermott, *Proteins,* 33, 240–252, 1998.
121. Mumenthaler, C. and W. Braun, *Prot. Sci.*, 4, 863–871, 1995.
122. Godzik, A., A. Kolinski, and J. Skolnick, MATCHMAKER, *TRIPOS* , St. Louis, 1994.
123. Simons, K.T., C. Klooperberg, E. Huang, and D. Baker, *J. Mol. Biol.,* 268, 209–225, 1997.
124. Shortle, D., K.T. Simons, and D. Baker, *Proc. Natl. Acad. Sci. U.S.*, 95, 11158–11162, 1998.
125. Ortiz, A., A. Kolinski, P. Rotkiewicz, B. Ilkowski, and J. Skolnick, *Proteins,* Suppl., 3, 177–185, 1999.
126. Rost, B. and C. Sander, *J. Mol. Biol.,* 232, 584–599, 1993.
127. Rost, B. and C. Sander, *Proteins,* 19, 55–72, 1994.
128. Rost, B. and C. Sander, *Proteins,* 23, 295–300, 1996.
129. Ortiz, A.R., W.-P. Hu, A. Kolinski, and J. Skolnick, *J. Mol. Graphics*, 1997.
130. Ortiz, A.R., W.-P. Hu, A. Kolinski, and J. Skolnick, Method for low resolution prediction of small protein tertiary structure, in *Pacific Symposium on Biocomputing '97*, R.B. Altman, et al., Eds., World Scientific, Singapore, 1997, 316–327.
131. Ortiz, A.R., A. Kolinski, and J. Skolnick, *Proc. Natl. Acad. Sci. U.S.,* 95, 1020–1025, 1998.
132. Ortiz, A.R., A. Kolinski, and J. Skolnick, *J. Mol. Biol.,* 277, 419–448, 1998.
133. Ortiz, A.R., A. Kolinski, and J. Skolnick, *Proteins,* 30, 287–294, 1998.
134. Anfinsen, C.B., *Science,* 181, 223–230, 1973.
135. Scheraga, H.A., *Biophys. Chem.,* 59, 329–339, 1996.
136. Abola, E.E., F.C. Bernstein, S.H. Bryant, T.F. Koetzle, and J. Weng, Protein Data Bank in crystallographic databases — Information content, software systems, scientific application, F.H. Allen, Bergerhoff, G., Sievers, R. Eds., Data Commission of the International Union of Crystallography, Bonn, 1987, 107–132.
137. Berg, B.A. and T. Neuhaus, *Phys. Rev. Lett.*, 68, 9–12, 1991.
138. Hukushima, K. and K. Nemoto, *J. Phys. Soc.*, (Japan) 65, 1604–1608, 1996.
139. Swendsen, R.H. and J.S. Wang, *Phys. Rev. Lett.*, 57, 2607–2609, 1986.
140. Sugita, Y. and Y. Okamoto, *Chem. Phys. Lett.*, 314, 141–151, 1999.
141. Hansmann, U.H.E., and Y. Okamoto, *J. Comput. Chem.*, 14, 1333–1338, 1993.

142. Hansmann, U.H.E., *Chem. Phys. Lett.*, 281, 140–150, 1997.
143. Hansmann, U.H.E. and Y. Okamoto, *J. Comput. Chem.*, 18, 920–933, 1997.
144. Hansmann, U.H.E. and Y. Okamoto, *Curr. Opinions Struct. Biol.*, 9, 177–181, 1999.
145. Scheraga, H.A. and M.-H. Hao, *Adv. Chem. Phys.*, 105, 243–272, 1999.
146. Rabow, A.A. and H.A. Scheraga, *Protein Sci.*, 5, 1800–1815, 1996.
147. Dandekar, T. and P. Argos, *J. Mol. Biol.*, 256, 645–660, 1996.
148. Sun, Z., X. Xia, Q. Guo, and D. Xu, *J. Protein Chem.*, 18, 39–46, 1999.
149. Pedersen, J.T. and J. Moult, *Proteins,* Suppl. 1, 179–184, 1997.
150. Gront, D., A. Kolinski, and J. Skolnick, *J. Chem. Phys.*, 113, 5065–5071, 2000.
151. Zhang, Y. and J. Skolnick, *J. Chem. Phys.*, 115, 11, 5027–5032, 2001.
152. Zhang, Y., D. Kihara, and J. Skolnick, *Proteins,* 48, 192–201, 2002.
153. Gront, D., A. Kolinski, and J. Skolnick, *J. Chem. Phys.*, 115, 1569–1574, 2001.
154. Moult, J., T. Hubbard, K. Fidelis, and J.T. Pedersen, *Proteins,* Suppl. 3, 2–6, 1999.
155. Venclovas, C., A. Zemla, K. Fidelis, and J. Moult, *Proteins,* Suppl. 3, 231–237, 1999.
156. Simons, K.T., R. Bonneau, I. Ruczinski, and D. Baker, *Proteins,* Suppl. 3, 171–176, 1999.
157. Jones, D.T., *Proteins,* Suppl. 1, 185–191, 1997.
158. Lesk, A.M., L. Lo Conte, and T.J. Hubbard, *Proteins,* Suppl. 5, 98–118, 2001.
159. Osguthorpe, D.J., *Proteins,* Suppl. 3, 186–193, 1999.
160. Samudrala, R., H. Xia, E. Huang, and M. Levitt, *Proteins,* Suppl. 3, 194–198, 1999.
161. Hinds, D. and M. Levitt, *J. Mol. Biol.*, 243, 668–682, 1994.
162. Park, B., E. Huang, and M. Levitt, *J. Mol. Biol.*, 266, 831–846, 1997.
163. Huang, E., R. Samudrala, and J. Ponder, *Protein Sci.*, 7, 1998–2003, 1998.
164. Lee, J., A. Liwo, D.R. Ripoll, J. Pilardy, and H.A. Scheraga, *Proteins,* Suppl. 3, 204–208, 1999.
165. Lee, J., A. Liwo, and H.A. Scheraga, *Proc. Natl. Acad. Sci. U.S.*, 96, 2025–2030, 1999.
166. Ripoll, D.R., A. Liwo, and H.A. Scheraga, *Biopolymers,* 46, 117–126, 1988.
167. Bonneau, R., J. Tsai, I. Ruczinski, D. Chivian, C. Rohl, C.E. Strauss, and D. Baker, *Proteins,* Suppl. 5, 119–126, 2001.
168. McGuffin, L.J., K. Bryson, and D.T. Jones, *Bioinformatics*, 16, 404–405, 2000.
169. Karplus, K., R. Karchin, C. Barrett, S. Tu, M. Cline, M. Diekhans, L. Grate, J. Casper, and R. Hughey, *Proteins,* Suppl. 5, 86–91, 2001.
170. Karplus, K. and B. Hu, *Bioinformatics*, 17, 713–720, 2001.
171. Madera, M. and J. Gough, *Nucleic Acids Res.*, 30, 4321–4328, 2002.
172. Orengo, C.A., J.E. Bray, T. Hubbard, L. LoConte, and I.I. Sillitoe, *Proteins,* 37, 149–170, 1999.
173. Przybylski, D. and B. Rost, *Proteins,* 46, 197–205, 2002.
174. Rost, B., *Methods Enzymol.*, 266, 525–539, 1996.
175. Rost, B., *Proteins,* Suppl. 1, 192–197, 1997.
176. Gunn, J.R., *Proc. Int. Conf. Intell. Syst. Mol. Biol.*, 6, 78–84, 1998.
177. Bonneau, R., C. Strauss, C. Rohl, D. Chivian, P. Bradley, L. Malmstrom, T. Robertson, and D. Baker, *J. Mol. Biol.*, 322, 65, 2002.
178. Lo Conte, L., S.E. Brenner, T.J. Hubbard, C. Chothia, and A.G. Murzin, *Nucleic Acids Res.*, 30, 264–267, 2002.
179. Lo Conte, L., B. Ailey, T.J. Hubbard, S.E. Brenner, A.G. Murzin, and C. Chothia, *Nucleic Acids Res.*, 28, 257–259, 2000.
180. Hubbard, T.J., B. Ailey, S.E. Brenner, A.G. Murzin, and C. Chothia, *Acta Crystallogr. D. Biol. Crystallogr.*, 54, 1147–1154, 1998.
181. Hubbard, T.J., B. Ailey, S.E. Brenner, A.G. Murzin, and C. Chothia, *Nucleic Acids Res.*, 27, 254–256, 1999.

182. Berman, H.M., T. Battistuz, T.N. Bhat, W.F. Bluhm, P.E. Bourne, K. Burkhardt, Z. Feng, G.L. Gilliland, L. Iype, S. Jain, P. Fagan, J. Marvin, D. Padilla, V. Ravichandran, B. Schneider, N. Thanki, H. Weissig, J.D. Westbrook, and C. Zardecki, *Acta Crystallogr. D. Biol. Crystallogr.*, 58, 899–907, 2002.

183. Jones, D.T., *Proteins,* Suppl. 5, 127–132, 2001.

184. Murzin, A.G. and T. Hubbard, *Proteins,* Suppl. 5, 8–12, 2001.

185. Fischer, D., A. Elofsson, L. Rychlewski, F. Pazos, A. Valencia, B. Rost, A.R. Ortiz, and R.L. Dunbrack, Jr., *Proteins,* Suppl. 5, 171–183, 2001.

186. Rost, B. and V.A. Eyrich, *Proteins,* Suppl. 5, 192–199, 2001.

187. Eyrich, V.A., M.A. Marti-Renom, D. Przybylski, M.S. Madhusudhan, A. Fiser, F. Pazos, A. Valencia, A. Sali, and B. Rost, *Bioinformatics*, 17, 1242–1243, 2001.

188. Fariselli, P., O. Olmea, A. Valencia, and R. Casadio, *Proteins,* Suppl. 5, 157–162, 2001.

189. Kihara, D., H. Lu, A. Kolinski, and J. Skolnick, *Proc. Natl. Acad. Sci. U.S.,* 98, 10125–10130, 2001.

190. Skolnick, J. and D. Kihara, *Proteins,* 42, 319–331, 2001.

191. Kolinski, A. and J. Skolnick, *Proteins,* 32, 475–494, 1998.

192. Kolinski, A., M.R. Betancourt, D. Kihara, P. Rotkiewicz, and J. Skolnick, *Proteins,* 44, 133–149, 2001.

193. Kihara, D., Y. Zhang, H. Lu, A. Kolinski, and J. Skolnick, *Proc. Natl. Acad. Sci. U.S.,* 99, 5993–5998, 2002.

194. Li, W., Y. Zhang, D. Kihara, Y. Huang, D. Zheng, G. Montelione, A. Kolmski, and J. Skolnick, *Proteins*, 53, 290–306, 2003.

195. Tjandra, N. and A. Bax, *Science,* 278, 1111–1114, 1997.

196. Tjandra, N., J.G. Omichinski, A.M. Gronenborn, G.M. Clore, and A. Bax, *Nat. Struct. Biol.*, 4, 732–738, 1997.

197. Haliloglu, T., A. Kolinski, and J.S., *J. Biomol. Biopolymers*, 70, 548–562, 2003.

198. Clore, G.M. and A.M. Gronenborn, *Proc. Natl. Acad. Sci. U.S.,* 95, 5891–5898, 1998.

199. Cai, M., H. Wang, E.T. Olejniczak, R.P. Meadows, A.H. Gunasekera, N. Xu, and S.W. Fesik, *J. Magn. Reson.*, 139, 451–453, 1999.

200. Graf, R., D.E. Demco, S. Hafner, and H.W. Spiess, *Solid State Nucl. Magn. Reson.*, 12, 139–152, 1998.

201. Pokarowski, P., A. Kolinski, and J. Skolnick, *Biophys. J.*, 84, 1518–1526, 2003.

202. Lu, H. and J. Skolnick, *Proteins,* 44, 223–232, 2001.

203. Bonneau, R., I. Ruczinski, J. Tsai, and D. Baker, *Protein Sci.*, 11, 1937–1944, 2002.

204. Simons, K.T., C. Strauss, and D. Baker, *J. Mol. Biol.,* 306, 1191–1199, 2001.

205. Betancourt, M.R. and J.Skolnick., *J. Comput. Chem.*, 22, 339–353, 2001.

206. Betancourt, M.R. and J. Skolnick, *Biopolymers,* 59, 305–309, 2001.

207. Wedemeyer, W.J., C.A. Rohl, and H.A. Scherag, *J. Biomolecular NMR*, 22, 137–151, 2002.

208. Rohl, C.A. and D. Baker, *J. Am. Chem. Soc.*, 124, 2723–2729, 2002.

209. Fraser, C.M., J.D. Gocayne, O. White, M.D. Adams, R.A. Clayton, R.D. Fleischmann, C.J. Bult, A.R. Kerlavage, G. Sutton, J.M. Kelley, and et al., *Science,* 270, 397–403, 1995.

210. Pearson, W.R., *Methods Mol. Biol.*, 24, 307–331, 1994.

211. Altschul, S.F., T.L. Madden, A.A. Schaffer, J. Zhang, Z. Shang, W. Miller, and D.J. Lipman, *Nucleic Acids Res.,* 25, 3389–3402, 1997.

212. Schaffer, A.A., L. Aravind, T.L. Madden, S. Shavirin, J.L. Spouge, Y.I. Wolf, E.V. Koonin, and S.F. Altschul, *Nucleic Acids Res.,* 29, 2994–3005, 2001.

213. Schaffer, A.A., Y.I. Wolf, C.P. Ponting, E.V. Koonin, L. Aravind, and S.F. Altschul, *Bioinformatics*, 15, 1000–1011, 1999.

214. Pearl, F.M., N. Martin, J.E. Bray, D.W. Buchan, A.P. Harrison, D. Lee, G.A. Reeves, A.J. Shepherd, I. Sillitoe, A.E. Todd, J.M. Thornton, and C.A. Orengo, *Nucleic Acids Res.,* 29, 223–227, 2001.

215. Altschul, S.F. and E.V. Koonin, *Trends Biochem. Sci.*, 23, 444–447, 1998.

216. Kanehisa, M., S. Goto, S. Kawashima, and A. Nakaya, *Nucleic Acids Res.,* 30, 42–46, 2002.

9 Determining Function from Structure

Roman A. Laskowski

CONTENTS

9.1 INTRODUCTION

Traditionally, one of the principal aims of solving a protein's three-dimensional structure has been to help determine how the protein achieves its biological function, that is, to gain a better understanding of how it does what it does in its specific organism and cellular location. To date (May 2005), the structures of over 28,000 proteins have been deposited in the Protein Data Bank (PDB) (although, due to redundancy, this number corresponds to only around 8000 unique proteins). Virtually all of these structures have contributed in some way to our biological understanding of how proteins function. They have revealed how proteins recognize other molecules, how they bind their natural substrates, where in their structures the active region(s) are located, how they perform their specific biochemical reactions, and whether the proteins act alone or in concert with other molecules such as nucleotides, sugars, or other proteins.

Structural Proteomics differs from the traditional approach in that the elucidation of functional mechanisms is not one of the primary goals of solving structures. Indeed, in many cases, the actual function of the protein being solved may be completely unknown beforehand; function is often not one of the factors governing which targets are selected for structure determination. Of more interest may be proteins that are perhaps representatives of large, uncharacterized protein families, or proteins that are predicted from their sequence as having novel folds. Consequently, many of the 3-D structures obtained at the end of the high-throughput structure determination process might turn out to be of hypothetical proteins — i.e., proteins of unknown function.

As of May 2005, the PDB held 1758 entries identified as relating to protein structures solved by the various structural genomics consortia around the world. Of these, 699 entries (40%) were annotated as of "unknown function." On the face of it, such entries appear to be somewhat pointless, being the 3-D structures of proteins about which nothing else is known. Of what possible use might such entries be, other than as mere curiosities?

Firstly, as already mentioned, a protein's function is intimately related to its 3-D structure, and the PDB contains thousands of examples of how this intimate relationship manifests itself. Surely it must be possible to determine what a given protein does from its 3-D coordinates and so to work out what these hypothetical proteins do? The answer to this question forms the basis of this chapter. Here are described some of the computational methods that have been devised to determine function from structure, how much they are able to say about function, and how reliable they are. The final verification of a predicted function can only come from experiment, so the purpose of these computational methods is merely to suggest potential functions and guide the experimentalists in what to test for. The chapter gives several examples of how some of these methods have been successfully used by the structural genomics consortia on their hypothetical protein structures.

A second important point to make about the 699 hypothetical proteins in the PDB is that there is a strong pressure for the different structural genomics groups to submit structures pretty much as soon as they are solved to guard against being beaten to a specific structure by another group. Any subsequent analyses, including questions about the protein's function, have to wait, and might be published quite some time after the PDB deposition. Thus many of the hypothetical proteins do eventually reveal information about their function although this may not be apparent from their PDB entries.

The approaches covered in this chapter include analyses of: the protein's fold, its surface properties such as overall shape, electrostatic potential and residue conservation, its clefts and potential binding sites, its structural motifs and functionally important residues, and local features such as significant clusters of side chain conformations.

9.2 BIOCHEMICAL VS. BIOLOGICAL FUNCTION

Protein function has several definitions. A protein's biological function describes its role in the organism; for example, in digestion, blood clotting, fertilization, and so on. Its biochemical function, on the other hand, describes the atomic level interactions it makes with other molecules and any chemical changes it effects or undergoes. One can further characterize a protein in terms of its molecular activity (e.g., as a proteolytic enzyme), and cellular function (e.g., in protein degradation) [1]. To complicate matters, some proteins are known to perform more than one function, depending on their location and context, these being the so-called moonlighting proteins [2]. Many proteins are merely components of larger, multi-protein or protein-RNA complexes such as the ribosome, so determining function from the structure of what is merely a cog in a much larger machine is likely to be impossible.

Thus, the most one can hope for, given just the 3-D coordinates of a protein, and no other information about it, is an idea of what the biochemical function might be and how this function may be achieved. The biological role requires information not available from the structure alone such as the protein's location in the cell or organism, factors affecting its transcription, metabolic pathways, and so on.

How, then, can the 3-D coordinates of a protein tell us anything about its biochemical function? There are a number of different approaches, involving both global and local features of the structure, which can yield snippets of information about the protein. Where several methods agree or complement one another, so confidence in the assignment is increased.

Not all methods provide information for every structure; some will work for some structures and others will work for other structures. And there will be some structures for which no information at all is provided, and in these cases, the proteins must remain annotated as hypothetical.

9.3 PROTEIN FOLD COMPARISON

The first aspect of any newly solved protein that is usually checked is its overall fold or, in the case of multi-domain proteins, the fold of each domain. It is known that proteins having the same fold are often related — that is, are descended from a common ancestor protein — as protein structure tends to be better conserved over evolutionary time than protein sequence [3]. Often, the relationship between two proteins is so distant that it is only apparent in the similarity of their folds. In such cases, whereas the structure has remained largely unchanged over time, the sequences have diverged so far apart that they are no longer detectable by sequence comparison methods; their similarity is below what would be expected by chance between two unrelated sequences.

Preservation of structure sometimes goes with preservation of function, so finding a structural match to a protein of known function can sometimes imply that the hypothetical protein has the same, or similar, function to the known protein, as we shall see in the examples described below.

Furthermore, certain folds are known to be associated with certain types of function. For example, proteins having TIM-barrel or Rossmann folds tend to be enzymes, so a hypothetical protein having such a fold is almost certainly also an enzyme.

9.3.1 FOLD DATABASES

There are a number of databases on the web that provide classifications of protein folds and methods of identifying the fold of a given structure. The best-known classifications are those given by the SCOP, CATH and FSSP databases (see this volume, Chapter 10 by Kleywegt). Table 9.1(a) lists the servers that can identify the fold of a submitted structure. The most commonly used is the Dali server at the EMBL which compares the structure against the FSSP database and ranks any matched folds in order of Z-score to provide a measure of the significance of each match [4]. Another well-known server is the NCBI's Vector Alignment Search Tool (VAST) [5]. A relatively new server is the Secondary Structure Matching (SSM) program at the EBI, which computes the secondary structure elements (SSEs) of the two structures and compares them using a fast graph-matching algorithm before superposing the 3-D structures themselves. A similar method is employed by GRATH [6], which performs a graph-match against the structural domains classified by CATH. The highest scoring matches can then be more precisely aligned using the slower, but more robust, SSAP algorithm [7].

9.3.2 FOLD AND FUNCTION

A case where a fold-search was able to strongly suggest the function of a hypothetical protein is that of MJ0882, a protein from *Methanococcus jannaschii* [8]. The top five hits returned by Dali for this structure were all methyltransferases, with Z-scores ranging from 8.8 to 13.8. None of the matched proteins had any significant sequence identity to MJ0882. However, biochemical experiments confirmed this protein to indeed have methyltransferase activity.

Not all results are so clear-cut. Similarity of fold does not always imply similarity of function. This is particularly true of the ten superfolds, folds found to occur far more commonly than others [9], each of which encompasses a wide range of functions. The most striking example is the TIM-barrel fold. As already mentioned, TIM-barrel proteins tend to be enzymes. What is more, their active site tends to be located at the same end of the barrel. However, the enzymatic roles they perform and the substrates they act on differ hugely across the members of this fold family [10–13].

A useful rule of thumb, derived for enzymes, is that, if the sequence identity between two aligned protein sequences is greater than 40%, the proteins probably perform the same function. Where the identity drops to around 25 to 30%, there is

TABLE 9.1
Programs and Web Servers of Use in Predicting Function from Structure

Method	Programs/Servers	URL	Notes
a. Fold Match	DALI	http://www.ebi.ac.uk/dali	Similarity of fold sometimes, but not always, provides clues about function. For a sequence identity >40% the function is probably the same; at 25–30% it is likely to be similar; below 25% it is likely to differ.
	VAST	http://www.ncbi.nlm.nih.gov:80/Structure/VAST/vastsearch.html	
	SSM	http://www.ebi.ac.uk/msd-srv/ssm/ssmstart.html	
	Grath + SSAP	http://www.biochem.ucl.ac.uk/cgi-bin/cath/Grath.pl	
		http://www.biochem.ucl.ac.uk/cgi-bin/cath/GetSsapRasmol.pl	
b. Residue Conservation	Evolutionary Trace	http://imgen.bcm.tmc.edu/molgen/labs/lichtarge/ETServerHome.html	Residue conservation is a powerful method of reliably locating active sites and protein–protein interaction regions, but often says little more about function than that. Requires that the multiple alignment contains a sufficiently diverse set of relatives.
		http://www-cryst.bioc.cam.ac.uk/~jiye/evoltrace/evoltrace.html	
	Consurf	http://consurf.tau.ac.il	
	JevTrace	http://www.cmpharm.ucsf.edu/~marcinj/JEVTrace/index.html	
c. Surface Clefts	Surfnet	http://www.biochem.ucl.ac.uk/~roman/surfnet/surfnet.html	A protein's most likely binding site, particularly in the case of enzymes, can be located by analyzing its surface clefts, especially in conjunction with residue conservation.
	VOIDOO	http://xray.bmc.uu.se/usf/voidoo.html	
	CAST	http://cast.engr.uic.edu/cast	
d. Binding Site Comparison	eF-site	http://ef-site.protein.osaka-u.ac.jp/eF-site	Binding site comparison, in terms of surface shape and chemical properties, can detect sites binding of similar molecules even in nonhomologues.
	CASTp	http://cast.engr.uic.edu	
e. 3-D Templates	Rigor	http://xray.bmc.uu.se/usf/rigor_man.html	3-D templates are precompiled 3-dimensional motifs, possibly associated with function, that can be scanned for in the query structure.
	PINTS	http://www.russell.embl.de/pints	
	CSA	http://www.ebi.ac.uk/thornton-srv/databases/CSA	
f. Macromolecular Interactions	Protein-protein	http://www.biochem.ucl.ac.uk/bsm/PP/server	Search for likely interaction sites between the protein and other macromolecules.
	Protein-DNA	http://www.ebi.ac.uk/thornton-srv/databases/DNA-motifs	
g. Biological Unit	PQS	http://pqs.ebi.ac.uk	Before analyzing any protein structure it is important to ensure it represents the true biological unit.
	Pita	http://www.ebi.ac.uk/thornton-srv/databases/pita	
h. Function Prediction	ProFunc	http://www.ebi.ac.uk/thornton-srv/databases/ProFunc	As no single method is likely to work in all circumstances, the best strategy is to try many.

a likelihood of a broad conservation of functional class, whereas below 25% functions tend to diverge [14]. Of course, one can find exceptions to this rule. The enzymes chicken lysozyme (PDB entry 3lzt) and goat α-lactalbumin (1hfyA) have a sequence identity of 41%, yet they belong to completely different enzyme classes — the first is a hydrolase while the second is a transferase — indicating very different functions [15], as illustrated in Figure 9.1. Furthermore, there are examples where an enzyme is clearly homologous to a protein which has no enzymatic activity, presumably because, as the proteins have diverged over evolutionary time, either the

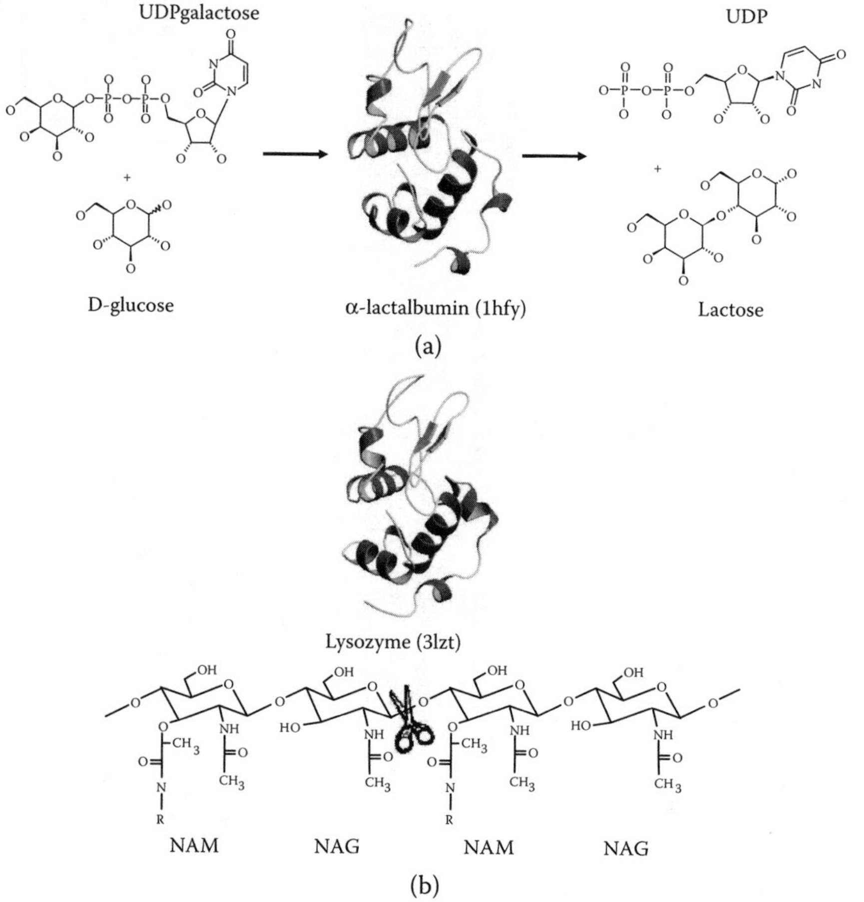

FIGURE 9.1 An example of two proteins having the same fold, and a high sequence identity (41%), yet performing completely different functions. Both are enzymes, but the first is a transferase, while the second a hydrolase. a) Goat α-lactalbumin (PDB entry 1hfyA) is involved in lactose synthesis in mammary glands. It removes a sugar from UDPgalactose, combining it with D-glucose to form lactose. b) Chicken lysozyme (3lzt) helps fight against bacterial infection. It breaks up bacterial cell-wall polysaccharides, consisting of alternating N-acetylmuramic acid (NAM) and N-acetylglucosamine (NAG) residues, by cutting the bond linking the NAG and NAM residues.

first has gained a catalytic ability or, conversely, the second has lost it. One such case is PDB entry 1ppf, the enzyme neutrophil elastase, which is 44% identical to the nonenzyme 1ae5, a human heparin binding protein. Similarly, 2vhm, the enzyme hevamine, is 41% identical to the nonenzyme 1cnv, concanavalin B. And there are many other similar examples [16].

Thus, while a match to a known fold may be suggestive, it does not directly imply a function, and can sometimes be misleading. Further complementary evidence, ideally experimental, is required to verify the protein's activity. Of course, in the case of proteins with novel folds, no match to any known fold may be obtained in the first place.

9.4 STRUCTURAL MOTIFS

Below the level of the fold come structural motifs: regions of the structure that are characteristic of specific functions. Perhaps the best known are the various DNA-binding motifs such as the helix-turn-helix (HTH) motif, zinc fingers and leucine zippers. Several groups have developed structural templates capable of detecting HTH motifs with a reasonable degree of accuracy [17–19]. Two example motifs are shown in Figure 9.2. In the first of the methods, the templates consist of the C-alpha coordinates of known HTH motifs involved in DNA binding. The motifs often generate false positives, but the majority of these can be filtered out by taking the solvent accessibility of the detected structural segment into consideration. A web server allows submission of protein structures for a HTH search – see Table 9.1(*f*).

A subtler structural motif is the "nest." This is an anion or cation binding site formed by three or more amino acids in the sequence whose main chain ψ–ϕ dihedral angles alternate between the left-handed α or α regions of the Ramachandran plot and the corresponding right-handed regions. Such an alternating conformation of the main chain results in an atom-sized concavity, flanked by the main-chain NH groups of the nest residues and the residue that follows. Such motifs are easy to locate and are found to be frequently associated with the functional sites of proteins [20,21].

9.5 COMMON RESIDUE GROUPINGS

Similar functions are frequently achieved by similar 3-D arrangements of amino acid residues in the protein. A method called PINTS (Patterns In Nonhomologous Tertiary Structures) detects the largest common 3-D arrangement of residues between any two structures or between a query structure and all those in the PDB [22]. It works by first sorting the residues of each protein being compared into alphabetical order. Each residue in the target protein is then compared against each residue of the same type in the other protein. The procedure is iterated for each amino acid type, checking at each step whether the equivalenced groups of residues have the same relative separations, within specified distance criteria. So the matching pattern is built up as equivalent residues in the two proteins, matching the distance criteria, are added to it. Once the pattern can grow no larger, an RMSD

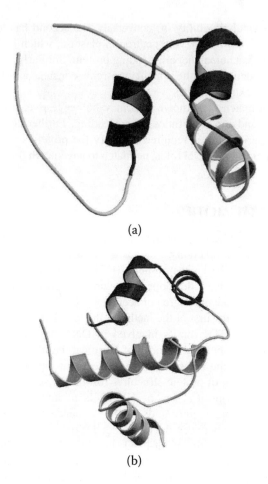

(a)

(b)

FIGURE 9.2 Two DNA-binding helix-turn-helix motifs, shown in black, with the remainder of the proteins in which they are found colored gray. a) Structure of the DNA-binding domain of HIN recombinase (taken from PDB entry 1hcr). b) Structure of the bacteriophage lambda repressor/operator (1lmb).

between the equivalenced residues is calculated and a significance score assigned to it [23].

9.6 DETECTION OF BINDING SITES

Where comparison against other proteins of known structure provides few clues as to function, it is important to at least be able to identify the parts of the structure that are likely to be functional. Perhaps the easiest regions to locate are the protein's likely binding sites. Proteins that interact with, or operate on, small molecules tend to contain a sizeable cleft in their surface into which the molecule(s) bind and where the active site is located. Often such a cleft is the largest one in the protein's surface.

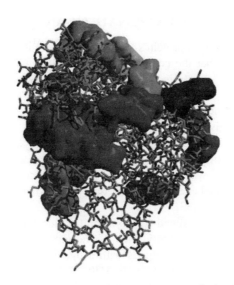

FIGURE 9.3 Clefts in the surface of a protein structure as calculated by the sphere-filling method of the SURFNET program. Each cleft is shown as the solid region, representing a "mould" of the void region between the protein atoms. The protein shown is the enzyme UDP-*N*-acetylenolpyruvoyglucosamine (PDB entry 1hsk). The bonds between its atoms are represented by sticks.

In enzymes, for example, it has been shown that the binding site corresponds to the largest cleft in over 80% of cases, and the second largest in a further 9% [24]. The exceptions tend to be cases where a protein can have an open and closed conformation, so the binding site is closed and difficult to detect unless something is bound there.

There are many methods for detecting and visualizing clefts in a protein's surface. A common algorithm involves filling the void regions between protein atoms with spheres. The resultant sphere-clusters pick out different clefts in the protein's surface (as well as any totally enclosed cavities that might be within it). The SURFNET program [25] is one that employs this method (Figure 9.3). Other methods include the atom-fattening procedure of VOIDOO [26], the alpha shapes used by CAST [27] and methods involving Voronoi polyhedra [28].

As the largest cleft is not absolutely guaranteed to be the site of interest, it is helpful to have additional information. Residue conservation, discussed later, provides very good confirmation of important regions of the protein's surface. Thus a large cleft, lined by highly conserved residues, is almost certainly the functionally important binding site.

Where residue conservation data are nonexistent, or poor, one can look at various properties of the residues lining the cleft. It has been shown that ligand-binding sites tend to be hydrophobic in nature [29], so a highly hydrophobic surface cleft is also worth investigating further.

9.7 IDENTIFICATION OF COGNATE LIGANDS

While the identification of a protein's binding site is useful, it rarely follows that it can say much about what the cognate ligand that binds there is. Although some progress has been made in trying to identify the characteristics of sites that bind specific molecules (e.g., adenine-binding moieties [30,31,32]), there is no single method that can currently say with certainty what molecule, or even what class of molecule, binds to a given binding site.

Perhaps an obvious-sounding approach would be to try docking different molecules such as a selection representing the organism's known metabolites, into the binding site and seeing which scores the highest. Unfortunately, current docking methods, and the potential energy functions they use, are not able to provide reliable results for such a strategy [33].

The only method guaranteed to work, and which has proved successful in several cases to date, is serendipity. This is where the experimentalists get lucky and the structure of the protein comes with a ligand molecule already bound in the active site! One such case was that of MJ0577 from *M. jannaschii*, one of the first structural genomics structures to be solved. In the crystal structure, the protein was found to have an ATP molecule bound [34]. The protein was deemed to be either an ATPase or an ATP-mediated molecular switch. Biochemical experiments suggested that it probably was the latter. Another example was that of MT150 from *Methanobacterium thermoautotrophicum*, which was found to have an NAD$^+$ molecule bound in the crystal structure. Its fold was similar to that of a number of nucleotidyltransferases, and biochemical studies showed it to have nicotinamide mononucleotide adenylyltransferase (NMNATase) activity [35]. In a further example, the structure of TM841 from *Thermatoga maritima* proved to have a fatty acid molecule, a palmitate, bound between the two domains, suggesting the protein to be a fatty acid binding protein possibly involved in fatty acid transport or metabolism [36].

Such fortuitous cases are not the norm, but they do occur. Often a structure can have either a metal bound or a molecule from the buffer solution used during protein extraction and purification. Even in the latter cases the presence of a bound molecule can occasionally help identify the protein's active site. For example, the crystal structure of the BioH protein from *E. coli* was found to have a buffer molecule covalently bound to a serine that later proved to be one of the protein's catalytic residues [37].

9.8 BINDING SITE COMPARISON

The protein structures in the PDB provide a wealth of examples of different ligands binding to different, and to the same, proteins. If one could compare, in some way, the binding site of the query protein against all the known binding sites in the PDB one might be able to infer the type of molecule that the protein binds. Several different approaches have been developed for comparing binding sites and have reported some successes.

9.8.1 Geometrical Matching

In a purely geometrical approach [38], Connolly's well-known MS program [39] is used to define a protein surface as a set of points and faces between them. Various critical locations are defined, being those points lying on exposed atoms and in the dents and seams between them. The faces defined at these critical locations are then represented by three properties that are stored: a surface point, its surface normal and the face area. Geometrical hashing is then used to compare surfaces defined in this way. The best superposition of the surfaces, giving the maximum number of matching critical points, is then reported.

9.8.2 Physicochemical Properties

It is unlikely that the comparison of binding site shapes alone can always be successful. The main problem with any purely geometrical approach is that it generates a large number of potential matches, most of which are wrong. The physicochemical properties of the residues and atoms underlying each surface point provide a means of filtering out the majority of these false positives.

Several methods "color" each surface point by its physicochemical properties at the start so that these become an integral part of the matching process, rather than being used as a postmatching filter. In one method, the property assigned to each surface point is the underlying atom's ability to form hydrogen bonds [40]; in this method the optimal matching is computed using a genetic algorithm rather than geometric hashing.

The Nakamura group defines points on the surface by a normal vector, a measure of the surface curvature and the electrostatic potential at that point. They have compiled a database of known binding sites, called eF-site (electrostatic surface of Functional site) [41] against which unknown binding sites can be searched using a clique-detection algorithm. The known binding sites in the database are of various types: active sites, phosphate binding sites, motifs from PROSITE, and the antigen binding sites of antibodies.

Despite some successes, it is unlikely that such binding site comparison methods will work in all cases. Proteins are not solid, static molecules with rigidly defined binding sites. Apart from the thermal motions of their atoms, many proteins exhibit conformational changes, either large or small, particularly on binding of substrate or cofactor. The problem of binding sites being rendered difficult to detect in cases of domain closure has already been mentioned. However, even if the motions are relatively small, their effect on the electrostatic properties of the surface can be significant [42]. Furthermore, very different protein surfaces can bind the very same molecule whereas similar surfaces may bind very different molecules [43].

9.8.3 Nonsurface Methods

To avoid the problem of a shifting, changing surface, one can ignore the surface altogether and consider only the residues that surround the binding site and their relative arrangement. Klebe's group represents each residue flanking a cleft by one of five types of "pseudosphere." The clefts are identified by the group's Ligsite program [44]. The pattern of pseudospheres surrounding each cleft is represented by a graph, and a clique detection algorithm is used to identify the similarity between any two binding sites. Chance hits are filtered out by taking into account the number of surface patches the two binding sites have in common. A database, called Cavbase, holds the sites of all the proteins in the PDB and can be rapidly scanned for matches to the binding site of the query protein. The method is able to pick up similar sites, such as phosphate-binding motifs and adenine binding pockets, in proteins showing no sequence or fold homology whatsoever [45].

Related methods, which aim to match more general, or, in the case of the template methods, more specific, groupings of residues will be described later.

9.8.4 SURFACE AND SEQUENCE

A very recent method [46], which also looks at the residues around clefts, represents each cleft in the protein surface by listing, as a short sequence fragment, the residues that line it. This sequence fragment preserves the order of the original sequence. Comparison of binding sites is then reduced to the commonplace problem of aligning protein sequences, albeit in some cases rather short ones. Any similarities are assessed for statistical significance, and the alignments obtained from any strong matches can be used to align the clefts in the 3-D structures and further compare the similarity of the binding sites themselves in terms of their 3-D shape.

In essence the method is identifying local regions that have changed less over time than the remaining regions, possibly due to the preservation of function, and so can pick up remote homologues that purely sequence-based methods will fail to find. Having said that, the authors present two matches between proteins of different folds and classes.

The sequence fragments defining all clefts in the proteins in the PDB are precomputed and stored in a database called CASTp (Computed Atlas of Surface Topography of Proteins) — see Table 9.1(*d*).

9.9 IDENTIFICATION OF FUNCTIONALLY IMPORTANT RESIDUES

There are several methods that aim to pinpoint which residues in a given protein structure are likely to be functionally important. As these various methods differ greatly, they can provide complementary and confirmatory information.

9.9.1 RESIDUE CONSERVATION

Probably the best guide is evolutionary conservation. The functionally important parts of the protein are more likely to have been preserved over evolutionary time simply as a result of evolutionary pressure on the protein to keep doing its job. Some modification to the protein's sequence and structure may have occurred in the functional regions to fine-tune the protein's performance or to alter it in some way, but the business end is likely to have remained largely unchanged since the days of the protein's distant ancestors.

Hence, given a multiple sequence alignment of the protein against all its related proteins, it is possible to calculate which residue positions show the highest conservation and which the highest variability (see this volume, Chapter 6 by Norin). The calculations take into account the relatedness of the sequences in the alignment, the number of changes in residue type at each position in the alignment and to what degree these changes are conservative (i.e., in relation to residues of similar character) [47]. The first of these factors, that of the relatedness of the sequences, is important as each sequence in the alignment may contribute information of varying quality about residue conservation. The simplest case is where all sequences in the alignment are identical to the sequence of interest; clearly, none of them provide any information about residue conservation. On the other hand, the more the sequences differ, the further they have diverged from the query protein and so the more they can say about which residue positions have stood the test of evolutionary time to remain constant. In essence one needs to determine the phylogenetic relationships (or evolutionary trace) between the sequences in an alignment to get a more accurate estimate of the conservation of each residue.

Having obtained a conservation score for each residue in the protein it is useful to map these scores onto the protein's 3-D structure and observe whether any regions show particularly strong conservation (Figure 9.4). It has been demonstrated that

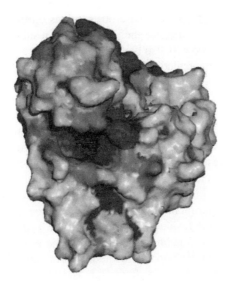

FIGURE 9.4 Residue conservation mapped onto the surface of a protein helps highlight the highly conserved regions (here shown as the darker shading) and suggest which might be functionally important sites on the surface. In this example, which is of the enzyme UDP-*N*-acetylenolpyruvoyglucosamine (PDB entry 1hsk), the very highly conserved regions are principally those lining a deep cleft in the protein's surface which, in the crystal structure, has the enzyme's cofactor, flavin-adenine dinucleotide (FAD), bound. Hence in this case the conserved regions have correctly picked out the active site.

such regions on the surface can reliably pinpoint ligand binding sites and pro-tein–protein interfaces [48,49,50]. Specifically, clusters of highly conserved *polar* residues are strongly associated with functional sites [51], while an enzyme's cata-lytic residues are virtually always among the protein's most conserved residues [52].

So, residue conservation is a strong predictor of functional sites. However, for some proteins — particularly hypothetical ones — it is not possible to calculate meaningful conservations scores, either because there are too few related sequences in the sequence databanks, or all the relatives are all either too similar or too dissimilar. The ideal alignment contains sequences that are distant from the target protein (say, down to about 35% sequence identity) as well as sequences that are close.

9.9.2 Identification of Residues with "Unusual" Properties

A number of experimental studies have shown that functionally important residues are frequently ones that actually destabilize a protein's structure in some way. By mutating catalytic residues in enzymes, these studies showed that the protein actually became more stable as a result [53,54,55,56]. Similarly, the replacement of residues involved in protein–protein interfaces, or protein–ligand interactions, has been shown to result in more stable protein molecules [57,58]. Computational analysis of protein structures has also suggested that functional residues are often in strained confor-mations in terms of their backbone dihedral angles [59].

Several methods try to exploit these observations on the basis that, if one can computationally identify which are the destabilizing residues in the structure, the chances are high that these will also be functionally important. In one such method, continuum electrostatics calculations are performed to identify the most destabilizing residues. These are defined as the residues that undergo a positive change in elec-trostatic free energy, G_{elec}, when their side chain is transferred from aqueous solution to the fully folded protein [60].

A different strategy is used in the THEMATICS program that specifically hunts for catalytic residues. The method involves calculating theoretical microscopic titra-tion curves for each residue in the structure [61]. A titration curve is a plot of the pH of a solution as a function of the volume of a standardized acid or base added; the shape of the curve reflects the gross properties of the molecule under study. Performing such an experiment on a protein in a test tube would not provide much information about specific residues. So, the aim of the THEMATICS program is to simulate the local effect of adding base or acid to the protein solution, and, using finite difference Poisson–Boltzmann methods, to compute a theoretical titration curve for each ionizable residue in the protein (i.e., Lys, Arg, Asp, Glu, His, Tyr, Cys, plus the N- and C-termini). It has been shown that the method works for several different types of enzyme, with the catalytic residues displaying titration curves differing markedly from the norm [61,62].

Two more recent approaches employ combinations of methods to identify cat-alytic residues. In the first, each protein residue is assessed according to its conser-vation score, stability profile, and geometrical location in the protein (i.e., buried, on the surface, in a hole or in a cleft) [63]. Each residue's stability profile is computed by estimating the effect on the protein's stability of replacing the residue with any of the 19 other amino acids. The effect on stability is computed using a knowledge-based potential that includes such factors as side chain packing, hydration, and local conformation.

The second method uses a neural network to judge which residues are likely to be catalytic on the basis of residue conservation, residue type, solvent accessibility, depth within the protein, secondary structure (helix, strand or coil), and the size of any neighboring cleft [64].

9.10 3-D TEMPLATE METHODS

The 3-D template methods are a different class of method that, rather than searching for residues that look like they are functional, search the protein structure for specific clusters or constellations of residues that are known to be associated with specific functions. For example, it has been shown that sequence motifs such as the PROSITE patterns [65] (see this volume, Chapter 6 by Norin), which are powerful indicators of certain functions or sequence families, have 3-D conformations that are strikingly well preserved [66]. Similarly, the relative spatial dispositions of the catalytic residues in enzymes are strongly maintained across different members of the same enzyme family. The best-known example is the Ser-His-Asp catalytic triad of the serine proteases [67]. This triad even occurs in almost identical conformations in proteins having totally different folds (see Figure 9.5).

Such templates can be defined either manually or in an automated fashion and stored in a database. Some fuzziness can be introduced into the search such that matches need not be exact but can, say, allow specific substitutions at any or all of the template positions.

9.10.1 ENZYME CATALYTIC RESIDUES

An example of a template database is the manually curated PROCAT database that aims to hold a 3-D template, consisting of the coordinates of the active site residues (or key atoms), for each type of enzyme of known structure in the PDB. Figure 9.6 shows two example templates. PROCAT's associated search program, TESS, uses a geometrical hashing technique to rapidly compare potential sites in the protein structure against the templates in the database [68]. The database and program were used for recognizing a Ser-His-Asp catalytic triad in BioH, a structural genomics structure of unknown function [37]. The catalytic nature of these residues, shown in Figure 9.7, was confirmed experimentally, and the protein was found to be a novel carboxylesterase acting on short acyl chain substrates.

The PROCAT database has recently been restyled as the Catalytic Site Atlas, or CSA (Table 9.1(e)), and currently contains around 400 enzyme active site templates [69].

9.10.2 AUTOMATICALLY GENERATED TEMPLATES

The creation of a manually compiled database such as PROCAT is very labor intensive, involving much literature searching and curation to keep up to date. More

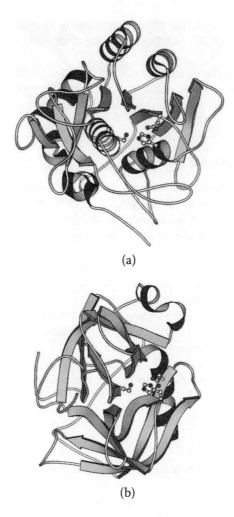

(a)

(b)

FIGURE 9.5 An example of two serine proteinases with completely different folds but having identical functions. The three crucial active-site residues (or catalytic triad) of His-Asp-Ser are shown as ball-and-stick sidechains in both structures and are approximately aligned. *a*) Subtilisin (PDB code 1cse) which has an alternating α/β fold, and *b*) trypsin (PDB code 8gch) which has a β-sandwich fold.

rapid strategies involve the automatic generation of templates. The RIGOR motif database is automatically generated on the basis of interesting arrangements of residues in the protein structures of the PDB. There are three rules distinguishing interesting arrangements from what are presumably boring ones: a) the protein contains *n* sequential residues of the same type (e.g., four consecutive arginine residues), b) a set of neighbouring residues are all hydrophobic, or all polar/charged, or a mixture of hydrophobic and polar/charged, and c) residues that all make contact to a single hetero compound [70]. The RIGOR database contains around 5000 motifs

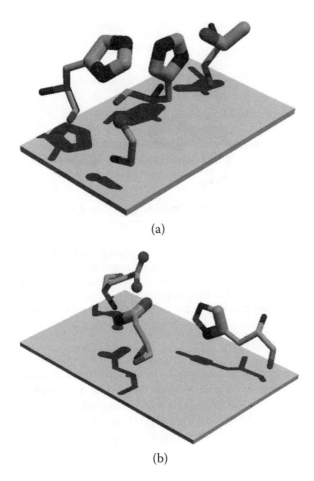

(a)

(b)

FIGURE 9.6 Two example enzyme active site templates. In each case the main chain and side chain atoms of the residue are shown, but the atoms, or groups of atoms, that are included as part of the searchable template, are represented by the larger spheres and thicker bonds. For example, in the first figure, both the histidine rings belong to the template whereas the main chain and Cβ atoms do not. a) The active site residues of IIAglc histidine kinase — Thr66, His68, His83 and Gly85 — (taken from PDB entry 1gpr). b) The active site residues of cholesterol oxidase — Glu361, His447, and Asn485 — (taken from PDB entry 1coy).

and the RIGOR motifs can be searched for in a query structure using the SPASM program (Spatial Arrangements of Side chains and Main chains).

A more general strategy for automatically generating templates is that of Oldfield [71]. This uses purely mathematical techniques to extract commonly recurring local structural motifs in known protein structures. In all, the method generated 1972 templates, ranging in size from 3 to 7 amino acids, from the proteins in the PDB as of March 2001. These included expected conformations such as the catalytic triads, metal binding sites, ligand-binding sites, 3-residue salt bridges, and N-linked glycosylation sites, as well as conformations not previously characterized.

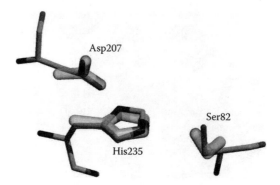

FIGURE 9.7   The close match between the Ser82, His235, and Asp207 residues in the crystal structure of BioH from **E. coli** and the TESS template for the Ser-His-Asp catalytic triad of the triacylglycerol lipase fold. The template side chains are depicted by the thicker, transparent bonds, whereas the BioH residues are represented by the thinner, solid bonds and include the main chain atoms. The r.m.s.d. between equivalent atoms in the template and matched side chains is 0.28Å. (From R Sanishvili, AF Yakunin, RA Laskowski, T Skarina, E Evdokimova, A Doherty-Kirby, GA Lajoie, JM Thornton, CH Arrowsmith, A Savchenko, A Joachimiak, and AM Edwards, *J. Biol. Chem.*, 278, 26039–26045, 2003. With permission.)

9.11 PROTEIN–PROTEIN INTERACTION SITES

Another useful piece of information that a protein structure can reveal is the location of any protein–protein interaction sites on its surface. Such sites tend to occur at predominantly flat regions of the protein surface and, not surprising, usually consist of highly conserved residues [49]. Other than relying purely on residue conservation, a method called *patch analysis* looks at various properties of the surface — including flatness, hydrophobicity, protrusion, charge, as well as residue conservation — and is able to identify protein–protein interaction sites with a reasonable success rate [72–75].

9.12 DETERMINING THE BIOLOGICALLY RELEVANT UNIT

In the discussion so far, no consideration has been given to whether the protein structure being analyzed is the biologically relevant unit. Often it is not. This is quite apart from the case already mentioned where the protein might actually be merely a component of a large multi-protein or protein-RNA complex, or where it is merely a single domain of a multi-domain protein.

The data from an X-ray crystallographic experiment provide information on the structure of the molecule(s) in the crystal's asymmetric unit. This may contain a single molecule of the protein, or two or three, naively suggesting it to be a monomer, dimer or trimer, respectively. However, whatever the arrangement in the asymmetric unit, it is not necessarily the biologically relevant one, but possibly an artefact of crystal packing. For example, many dimers, because of the symmetrical way the two molecules of the same protein bind to one another, result in an asymmetric unit containing a single copy of the molecule; to obtain the coordinates of the dimer requires applying the crystallographic symmetry operations to the x-, y-, z-coordinates of the monomer.

In the case of protein structures determined by NMR spectroscopy, the deposited coordinates are, again, not necessarily those of the true biological unit. The majority of NMR structures appear to be monomers — around 88% of the nearly 4000 NMR structures of proteins as of May 2005 were monomeric. This apparent bias is largely due to its being easier to determine the structures of small proteins or protein domains (<20kD) than larger ones. Consequently, naturally monomeric proteins or their constituent fragments (often domains) are preferentially selected for structure determination. Furthermore, as it is difficult to distinguish between a monomer and a symmetrical multimeric assembly of the same protein from the data alone, it is common practice to mutate native proteins or manipulate solution conditions to eliminate oligomerization and simplify the NMR structure determination process. Having said that, because it is difficult to determine an oligomeric structure by NMR, if one is present in an NMR PDB file, it is very likely to be the biological unit.

There exist experimental methods, such as analytical ultracentrifugation and chemical crosslinking in combination with gel electrophoresis, that can determine a protein's true biological unit, but in the case of X-ray crystal structures it is sometimes also possible to do so from the 3-D structure itself. By applying the crystal symmetry operations to the molecule(s) in the asymmetric unit, one can generate all the symmetry-related copies surrounding it. Then, from an examination of the resultant protein–protein interfaces, and the interactions across them, it is often possible to determine which are likely to be the true, biological interfaces and which are merely artefacts of crystal packing. Such analyses have been performed for all X-ray structures deposited in the PDB and can be accessed via the PQS (Protein Quaternary Structure) server [76] — see Table 9.1(g). For new structures, not in the PDB, the PITA program (Protein InTerfaces and Assemblies) performs a similar analysis [77]. It takes a standard PDB file and suggests the most likely biological unit for the macromolecules it contains. A PITA web server is available — see Table 9.1(g).

If the true biological unit can be obtained, then it is this that should be analyzed, otherwise crucial information relating to the protein's function might be lost. For example, in dimeric proteins, the substrate often binds at the dimer interface, so using just the coordinates of the monomer would miss out a very significant part of the binding site.

9.13 CONCLUDING REMARKS

This chapter has described a number of approaches for identifying the function of a protein solely from its structure in cases where sequence methods have failed. Different methods have different strengths and can provide different functional clues. The true test of such functional clues has to be the test tube, but the hope is that the structure can narrow down the possibilities and guide the experimentalists as to what they should test for. And, of course, experimental verification of any proposed function is crucial, otherwise erroneous assignments will propagate through the databases from the protein under study to its close relatives, and then, as the source of further false hypotheses, through to its more distant relatives [78].

It seems clear that no single method can hope to work in all cases, and there will be difficult proteins about which no method is able to say very much at all. So a sensible strategy is to combine a number of methods and see what they predict. Common predictions from several methods are worthier of further investigation than functional assignments that come from only a single method. A step in this direction is ProFunc [79], see Table 9.1(*h*), which aims to be a web server that applies many of the methods described above to a submitted target protein and compiles and summarizes the functional assignments obtained, highlighting any that are particularly strong or seem to be common to two or more methods.

A protein structure of known function is far more valuable and useful to the scientific community than the structure of a protein whose purpose is unknown. The structure should not be the end in itself, but a significant step towards that end. At the moment, function prediction is not a fully automated process, and many of the methods are still somewhat new and requiring of further development. But this will change and, very soon, more and more of the structures emerging from the structural genomics and proteomics initiatives will be functionally annotated and will help fill in the missing pieces of the functional, as well as the structural, jigsaw of the proteome.

ACKNOWLEDGEMENT

Thanks to Dr. Mark Williams for his valuable comments on the text of this chapter.

REFERENCES

1. S Rastan and LJ Beeley, *Curr. Opinions Genet. Dev.*, 7, 777–783, 1997.
2. CJ Jeffery, *Trends Biochem. Sci.*, 24, 8–11, 1999.
3. C Chothia and AM Lesk, *EMBO J.*, 5, 823–826, 1986.
4. L Holm and C Sander, *Trends Biochem. Sci.*, 20, 478–480, 1995.
5. JF Gibrat, T Madej, and SH Bryant, *Curr. Opinions Struct. Biol.*, 6, 377–385, 1996.
6. A Harrison, F Pearl, I Sillitoe, T Slidel, R Mott, J Thornton, and C Orengo, *Bioinformatics,* 19, 1748–1759, 2003.
7. WR Taylor and CA Orengo, *J. Mol. Biol.*, 208, 1–22, 1989.
8. L Huang, LW Hung, M Odell, H Yokota, R Kim, and S-H Kim, *J. Struct. Funct. Genomics,* 2, 121–127, 2002.
9. CA Orengo, DT Jones, and JM Thornton, *Nature,* 372, 631–634, 1994.
10. N Nagano, EG Hutchinson, and JM Thornton, *Prot. Sci.*, 8, 2072–2084, 1999.
11. H Hegyi and M Gerstein, *J. Mol. Biol.*, 288, 147–164, 1999.
12. N Nagano, CA Orengo, and JM Thornton, *J. Mol. Biol.*, 321, 741–765, 2002.
13. V Anantharaman, L Aravind, and EV Koonin, *Curr. Opinions Chem. Biol.*, 7, 12–20, 2003.
14. AE Todd, CA Orengo, and JM Thornton, *J. Mol. Biol.*, 307, 1113–1143, 2001.
15. D Devos and A Valencia, *Proteins: Struct. Funct. Genet.*, 41, 98–107, 2000.
16. AE Todd, CA Orengo, and JM Thornton, *Structure,* 10, 1435–1451, 2002.

17. S Jones, JA Barker, I Nobeli, and JM Thornton, *Nucleic Acids Res.*, 31, 2811–2823, 2003.
18. WA McLaughlin and HM Berman, *J. Mol. Biol.*, 330, 43–55, 2003.
19. EW Stawiski, LM Gregoret, and Y Mandel-Gutfreund, *J. Mol. Biol.*, 326, 1065–1079, 2003.
20. JD Watson and EJ Milner-White, *J. Mol. Biol.*, 315, 171–182, 2002.
21. JD Watson and EJ Milner-White, *J. Mol. Biol.*, 315, 183–191, 2002.
22. A Stark and RB Russell, *Nucleic Acids Res.*, 31, 3341–3344, 2003.
23. A Stark, S Sunyaev, and RB Russell, *J. Mol. Biol.*, 326, 1307–1316, 2003.
24. RA Laskowski, NM Luscombe, MB Swindells, and JM Thornton, *Prot. Sci.*, 5, 2438–2452, 1996.
25. RA Laskowski, *J. Mol. Graph.*, 13, 323–330, 1995.
26. GJ Kleywegt and TA Jones, *Acta Cryst.*, D50, 178–185, 1994.
27. J Liang, H Edelsbrunner, P Fu, PV Sudhakar, and S Subramaniam, *Proteins: Struct. Functional Genet.*, 33, 18–29, 1998.
28. S Chakravarty, A Bhinge, and R Varadarajan, *J. Biol. Chem.*, 277, 31345–31353, 2002.
29. L Young, RL Jernigan, and DG Covell, *Prot. Sci.*, 3, 717–729, 1994.
30. SL Moodie, JB Mitchell, and JM Thornton, *J. Mol. Biol.*, 263, 486–500, 1996.
31. I Nobeli, RA Laskowski, WSJ Valdar, and JM Thornton, *Nucleic Acids Res.*, 29, 4294–4309, 2001.
32. N Kobayashi and N Go, *Eur. Biophys. J.*, 26, 135–144, 1997.
33. I Halperin, B Ma, H Wolfson, and R Nussinov, *Proteins: Struct. Functional Genet.*, 47, 409–443, 2002.
34. TI Zarembinski, L-W Hung, H-J Mueller-Dieckmann, K-K Kim, H Yokota, R Kim, and S-H Kim, *Proc. Natl. Acad. Sci. U.S.*, 95, 15189–15193, 1998.
35. D Christendat, A Yee, A Dharamsi, Y Kluger, A Savchenko, JR Cort, V Booth, CD Mackereth, V Saridakis, I Ekiel, G Kozlov, KL Maxwell, N Wu, LP McIntosh, K Gehring, MA Kennedy, AR Davidson, EF Pai, M Gerstein, AM Edwards, and CH Arrowsmith, *Nat. Struct. Biol.*, 7, 903–909, 2000.
36. U Schulze-Gahmen, J Pelaschier, H Yokota, R Kim, and S-H Kim, *Proteins: Struct. Functional Genet.*, 50, 526–530, 2003.
37. R Sanishvili, AF Yakunin, RA Laskowski, T Skarina, E Evdokimova, A Doherty-Kirby, GA Lajoie, JM Thornton, CH Arrowsmith, A Savchenko, A Joachimiak, and AM Edwards, *J. Biol. Chem.*, 278, 26039–26045, 2003.
38. M Rosen, SL Lin, H Wolfson, and R Nussinov, *Prot. Eng.*, 11, 263–277, 1998.
39. ML Connolly, *J. Appl. Crystallogr.*, 16, 548–558, 1983.
40. AR Poirrette, PJ Artymiuk, DW Rice, and P Willett, *J. Computer-Aided Mol. Design*, 11, 557–569, 1997.
41. K Kinoshita, J Furui, and H Nakamura, *J. Struct. Functional Genomics*, 2, 9–22, 2001.
42. X Fradera, X de la Cruz, CHTP Silva, JL Gelpí, FJ Luque, and M Orozco, *Bioinformatics*, 18, 939–948, 2002.
43. A Via, F Ferrè, B Brannetti, and M Helmer-Citterich, *Cell. Mol. Life Sci.*, 57, 1970–1977, 2000.
44. M Hendlich, F Rippmann, and G Barnickel, *J. Mol.Graph. Model*, 15, 359–363, 1997.
45. S Schmitt, D Kuhn, and G Klebe, *J. Mol. Biol.*, 323, 387–406, 2002.
46. TA Binkowski, L Adamian, and J Liang, *J. Mol. Biol.*, 332, 505–526, 2003.
47. WS Valdar and JM Thornton, *J. Mol. Biol.*, 313, 399–416, 2001.
48. O Lichtarge, HR Bourne, FE Cohen. *J. Mol. Biol.*, 257, 342–358, 1996.
49. O Lichtarge and ME Sowa, *Curr. Opinions Struct. Biol.*, 12, 21–27, 2002.

50. S Madabushi, H Yao, M Marsh, DM Kristensen, A Philippi, ME Sowa, and O Lichtarge, *J. Mol. Biol.,* 316, 139–154, 2002.
51. P Aloy, E Querol, FX Aviles, and MJE Sternberg, *J. Mol. Biol.,* 311, 395–408, 2001.
52. GJ Bartlett, CT Porter, N Borkakoti, and JM Thornton, *J. Mol. Biol.,* 324, 105–121, 2002.
53. K Yutani, K Ogasawara, T Tsujita, and Y Sugino, *Proc. Natl. Acad. Sci., U.S.,* 84, 4441–4444, 1987.
54. EM Meiering, L Serrano, and AR Fersht, *J. Mol. Biol.,* 225, 585–589, 1992.
55. BK Shoichet, WA Baase, R Kuroki, and BW Matthews, *Proc. Natl. Acad. Sci. U.S.,* 92, 452–456, 1995.
56. S Kanaya, M Oobatake, and Y Liu, *J. Biol. Chem.,* 271, 32729–32736, 1996.
57. JH Zhang, Z-P Liu, TA Jones, LM Gierasch, and JF Sambrook, *Proteins: Struct. Functional Genet.,* 13, 87–99, 1992.
58. G Schreiber, AM Buckle, and AR Fersht, *Structure,* 2, 945–951, 1994.
59. O Herzberg and J Moult, *Proteins: Struct. Functional Genet.,* 11, 223–229, 1991.
60. AH Elcock, *J. Mol. Biol.,* 312, 885–896, 2001.
61. MJ Ondrechen, JG Clifton, and D Ringe, *Proc. Natl. Acad. Sci. U.S.,* 98, 12473–12478, 2001.
62. IA Shehadi, H Yang, and MJ Ondrechen, *Mol. Biol. Rep.,* 29, 329–335, 2002.
63. M Ota, K Kinoshita, and K Nishikawa, *J. Mol. Biol.,* 327, 1053–1064, 2003.
64. A Gutteridge, GJ Bartlett, and JM Thornton, *J. Mol. Biol.,* 330, 719–734, 2003.
65. L Falquet, M Pagni, P Bucher, N Hulo, CJ Sigrist, K Hofmann, and A Bairoch, *Nucleic Acids Res.,* 30, 235–238, 2002.
66. A Kasuya and J Thornton, *J. Mol. Biol.,* 286, 1673–1691, 1999.
67. AC Wallace, RA Laskowski, and JM Thornton, *Prot. Sci.,* 5, 1001–1013, 1996.
68. AC Wallace, N Borkakoti, and JM Thornton, *Prot. Sci.,* 6, 2308–2323, 1997.
69. CT Porter, GJ Barlett, and JM Thornton, *Nucleic Acids Res.,* 32, D129–D133, 2004.
70. GJ Kleywegt, *J. Mol. Biol.,* 265, 1887–1897, 1999.
71. T Oldfield, *Proteins: Struct. Functional Genet.,* 49, 510–528, 2002.
72. S Jones and JM Thornton, *Prog. Biophys. Mol. Biol.,* 63, 31–65, 1995.
73. S Jones and JM Thornton, *J. Mol. Biol.,* 272, 133–143, 1997.
74. HX Zhou and Y Shan, *Proteins: Struct. Functional Genet.,* 44, 336–343, 2001.
75. P Fariselli, F Pazos, A Valencia, and R Casadio, *Eur. J. Biochem.,* 269, 1356–1361, 2002.
76. K Henrick and JM Thornton, *Trends Biochem. Sci.,* 23, 358–361, 1998.
77. H Ponstingl, T Kabir, and JM Thornton, *J. Appl. Crystallogr.,* 36, 1116–1122, 2003.
78. SE Brenner, *Trends Genet.,* 15, 132–133, 1999.
79. RA Laskowski, JD Watson, JM Thornton. *J. Struct. Funct. Genomics,* 4, 167–177, 2003.

10 Retrieval and Validation of Structural Information

Gerard J. Kleywegt and Henrik Hansson

CONTENTS

10.1 INTRODUCTION

In order to understand the biological function of molecules at the atomic level, knowledge of their three-dimensional (3-D) structures is essential. Hence, structural biology and structural bioinformatics will play key roles in many areas of scientific endeavour in the post-genome-sequencing era. The number of 3-D structures of biomacromolecules that is available in public databases has been increasing exponentially since the late 1970s. Early in 2005, the 30,000th structure was released by the Protein Data Bank (PDB), which is the major primary resource for deposition and distribution of 3-D structures of biomacromolecules. This may seem modest compared to the number of known protein sequences, but thanks to sequence homology, the number of sequences for which structural (and functional) information can be inferred is orders of magnitude larger. Usually fewer than half of all genes in newly sequenced genomes are related to a protein of known 3-D structure. This situation will improve since the various structural genomics initiatives around the globe are expected to deliver thousands of new structures over the next decade.

To make the best use of this wealth of structural information, it must be easy to access. Although the PDB is the main source of 3-D information, at present it offers only rudimentary search mechanisms and limited annotation in terms of biological function, roles in disease states, phylogeny, etc. This makes it difficult for nonexpert users to access the information contained in the PDB. As a result, many secondary (derivative) databases have been developed that are organised around a certain class of molecules that are related in sequence, structure, or function. Typically, in addition to structures, such added-value databases contain sequences and information about function, tissue distribution, substrates, ligands, mutational studies, etc. Conversely, another set of databases has been developed that contain purely structural information (e.g., fold-classification, protein-ligand interactions, metal-binding sites).

In the first part of this chapter (sections 10.2 and 10.3), we discuss a number of (mostly) publicly available, web-based resources from which structures and derivative structural information can be retrieved. Obviously, an overview such as this is both incomplete and static. Every year in January, the journal *Nucleic Acids Research* (http://nar.oupjournals.org/) publishes a special issue devoted to new and updated biological databases [1] (http://www3.oup.co.uk/nar/database/a), and in July another special issue is devoted to web-based servers that perform useful calculations on sequences and structures of biomacromolecules. Furthermore, a number of scientists maintain annotated lists of links to structural and other databases on their web sites. For example, the links mentioned in this chapter are all listed at http://xray.bmc.uu.se/embo/structdb/links.html.

The second part of this chapter (section 10.4) deals with structure validation. Contrary to what many inexperienced users of the PDB believe, the fact that a structure is available from the PDB is by no means a measure of its quality. The PDB has traditionally been a repository of structural information that accepts whatever is deposited without being too judgmental. The few quality criteria that have to be satisfied are either of a clerical nature or deal with issues of precision rather than accuracy. As a result, the PDB contains a number of structures that are so seriously flawed as to be of no value except for teaching students how to recognise poor structures. This means that, prior to using or analysing a structure from the PDB, the user must ensure that the structure is of sufficient quality, a process known as *validation*. In the last part of this chapter we discuss the kind of errors that may occur and how they can be detected. We also discuss a number of web-based resources that can be used to obtain information about the quality of structures. This includes databases of quality-related information about structures that are available from the PDB and servers that can be used to assess the quality of new structures.

10.2 PRIMARY STRUCTURAL DATABASES

In this section, we discuss the primary structural databases, i.e., the databases that store and distribute the experimentally determined, 3-D structures of biomolecules, as well as the experimental data. Links to the web sites of these resources are also listed in Table 10.1.

10.2.1 PROTEIN DATA BANK

Whereas a number of independent sequence database centres exist that exchange their often overlapping holdings on a regular basis, essentially all macromolecular structures are stored in a single repository, the Protein Data Bank [2,3,4] (http://www.pdb.org/). The PDB was first established at Brookhaven National Laboratory on Long Island around 1971 [5], at the initiative of Walter Hamilton. Since 1998, management of the PDB is in the hands of the Research Collaboratory for Structural Bioinformatics (RCSB), comprising members from Rutgers University, the San Diego Supercomputer Center, and the Center for Advanced Research in Biotechnology. This responsibility is shared with the European Bioinformatics Institute (EBI) and the Protein Data Bank of Japan (PDBj).

At its inception, the databank contained only a handful of structures [6]. By 1977, their number had grown to 77 [2] and continued to grow slowly but steadily.

TABLE 10.1
URLs of Primary Structural Databases and Web-Based Front-Ends

Resource	URL
PDB	http://www.pdb.org
E-MSD	http://msd.ebi.ac.uk
NDB	http://ndbserver.rutgers.edu
EDS	http://eds.bmc.uu.se
BMRB	http://www.bmrb.wisc.edu
PDBobs	http://pdbobs.sdsc.edu/index.cgi
CSD	http://www.ccdc.cam.ac.uk
OCA	http://www.pdblite.org
PDBsum	http://www.ebi.ac.uk/thornton-srv/databases/pdbsum
MMDB	http://www.ncbi.nlm.nih.gov:80/Structure/MMDB/mmdb.shtml
Jena Image Library	http://www.imb-jena.de/IMAGE_PDB_SEARCH.html
PDBFINDER	http://www.cmbi.kun.nl/swift/pdbfinder
SRS6	http://srs6.ebi.ac.uk
SRS5	http://srs.embl-heidelberg.de:8000/srs5/
PDB at a Glance	http://cmm.info.nih.gov/modeling/pdb_at_a_glance.html

In the early 1990s, the number of structures was still below 1,000, but at the end of that decade it exceeded 10,000 and early in 2005 the 30,000th structure entry was released. Up-to-date information about the current holdings of the PDB can be found on the PDB web site at http://www.rcsb.org/pdb/holdings.html.

The original scope of the PDB was to store and disseminate atomic coordinates, structure factors and phases from X-ray and neutron diffraction studies of macro-molecules [2]. Over the years, this goal has changed to reflect developments in the field. Nowadays the stored information includes, for instance, details of the structure-determination process. In addition, structures obtained by other techniques such as Nuclear Magnetic Resonance (NMR) spectroscopy (accounting for ~15% of all PDB entries as of May 2005) and electron microscopy, electron diffraction, fibre diffrac-tion and neutron diffraction (together accounting for less than 1% of the PDB entries at present) are included. And although the first entries in the PDB were all proteins, the databank nowadays contains structures of peptides, proteins, nucleic acids, car-bohydrates, and their complexes (including viruses and ribosomes). In recent years, the PDB has also come to include more meta data such as stereochemical dictionaries for ligands and information related to structural genomics projects, and this trend is expected to continue.

One aspect of the original PDB that has not changed much is the fact that "it is essentially a depository of data" [2]. In practice, this means that the PDB accepts whatever the authors wish to deposit (with just a minimum of conditions). Deposition of experimental data, for instance, is not obligatory, and, now as in the 1970s, "no exhaustive verification is attempted" [2] of structures that are deposited. These two factors make that acceptance of a structure in the PDB cannot be taken as a sign of quality of the structure. In practice, then, validation of structures must be done by their users.

The PDB has traditionally been a collection of text files, which made advanced searches on it difficult. The lack of advanced search capabilities has lead to the development of a large number of external programs, databases and web-based servers. Ideally, such efforts should eventually be incorporated into the services of one or more of the primary data centres. This would ensure their continued devel-opment and support, as well as their synchronisation with the PDB. Another problem with the PDB is that there exist many inconsistencies, omissions, and errors in the annotations of, in particular, older entries [7]. Fortunately, both the RCSB and the EBI Macromolecular Structure Database (E-MSD) have made considerable progress in recent years in cleaning up the annotations [8,9].

At present, the PDB (at the RCSB and its mirror sites) can be searched in two ways (see section 10.2.3): on the textual information stored in the annotation of the files and on the sequence of the macromolecules. Purely structural queries (e.g., searching for all structures that contain a certain set of residues in a specific spatial orientation) are not possible and neither are more complex queries (e.g., "list all human enzymes involved in the Krebs cycle for which the 3-D structure is known"). The latter is because the annotation of PDB entries mainly focuses on the technical aspects of the structure determination and much less on the biology of the molecules involved (e.g., cellular localisation, function, phylogeny of parent organism) nor, somewhat surprisingly, on the description of the structures themselves (e.g., domain organisation, fold class, structural motifs).

10.2.2 OTHER PRIMARY DATABASES

- The EBI maintains the E-MSD [10] (http://msd.ebi.ac.uk), which is the European project for the collection, management, and distribution of data about macromolecular structures. Macromolecular structures can be deposited at this centre and are passed on to the RCSB after processing has been completed. The centre also hosts the 3-D electron-microscopy database (EMD). E-MSD offers various advanced query capabilities for retrieval of structural information.
- The Nucleic Acid Data Bank (NDB) [11] (http://ndbserver.rutgers.edu) is a resource that collects and distributes structural information about nucleic acids and was established in 1991. From 1992 to 1998, the NDB was responsible for the processing of all nucleic acid crystal structures deposited in the PDB. Since the NDB was designed from the outset as a relational database, its query and reporting facilities are much richer than those of the PDB.
- The PDB distributes the experimental crystallographic data that has been deposited for roughly half of all crystal structures. However, this in itself is not of much use to nonexpert users since software and know-how are required to calculate electron density maps from the combination of the structure and the experimental structure factor data. For this reason, the Uppsala Electron-Density Server (EDS) [13] (http://eds.bmc.uu.se/) provides access to precalculated electron density maps for most of the crystal structures in the PDB for which structure factors are available (see also section 10.4).
- The BioMagResBank (BMRB) [12] (http://www.bmrb.wisc.edu/), in collaboration with PDB and NDB, aims to become the collection site for structural NMR data on proteins and nucleic acids, including NMR-specific data such as assigned chemical shifts, J-couplings, relaxation rates, and chemical information (e.g., hydrogen exchange rates and pKa values).
- For a variety of reasons, PDB entries are sometimes retired by their depositors. For instance, a structure may be determined to considerably higher resolution than previously, or correction of errors in a structure may render it redundant. These superseded entries are removed from the PDB proper. However, for comparative and historical studies as well as for methodological development purposes, access to such entries can be very useful (e.g., to test new validation software) and therefore the RCSB maintains a separate database, PDBobs [14] (http://pdbobs.sdsc.edu/index.cgi), in which these obsolete entries are saved for posterity.
- The Cambridge Structural Database (CSD) (http://www.ccdc.cam.ac.uk/) is the central repository for bibliographic, chemical and structural information on more than 250,000 small molecules whose structures have been determined by X-ray or neutron diffraction [15,16]. Although macro-

molecular structures are beyond the scope of the CSD, many small molecules play key biological roles. To model such entities properly, the availability of their crystal structures is of major importance [17]. The CSD is maintained by the Cambridge Crystallographic Data Centre (CCDC) that was established in 1965 and also figured in the genesis of the PDB [5,18].

10.2.3 Search Mechanisms and Front-Ends

A number of laboratories around the world provide an official *mirror* (carbon copy) of the PDB web site. In addition, several institutes provide various search mechanisms (front-ends) to access the information in the PDB.

- The PDB offers two search facilities [4], namely a simple keyword search (SearchLite, useful to quickly locate entries by author or molecule name) and a more advanced one called SearchFields. The latter allows searches by one or more criteria involving authors, molecule types, experimental technique and details, etc. It is also possible to carry out sequence-based searches using either a FASTA [19] comparison of an entire sequence, or a search against a short sequence pattern. For every PDB entry, some basic information is listed and various methods for visualising and downloading entries are provided. A list of links to other web-based resources that carry information concerning each entry is also provided.
- OCA (http://www.pdblite.org/) allows searches on information found in the PDB entries, as well as kingdom, gene, related disease and function. For every PDB entry, key information is listed, but this is augmented with information about subunit structure, catalytic activity, posttranslational modifications, tissue distribution, subcellular localisation, function and any diseases in which the molecule has been implicated. OCA is at present also the search mechanism provided by E-MSD, although this will change in the future.
- PDBsum [20] (http://www.ebi.ac.uk/thornton-srv/databases/pdbsum) provides summary information and derivative data for all PDB entries. The summary information gives an at-a-glance overview of the contents of each PDB entry in terms of numbers of macromolecular chains, ligands, metal ions, etc. The derivative data include secondary structure analyses, validation statistics, schematic drawings of the interactions with ligands, etc. In addition, links to structure-classification, validation, and many other databases are provided.
- The Molecular Modelling Database (MMDB) [21,22] (http://www.ncbi. nlm.nih.gov:80/Structure/MMDB/mmdb.shtml) is part of NCBI's Entrez system and contains 3-D biomacromolecular structures. MMDB is linked to and integrated with the other NCBI databases (sequences, bibliographic citations, taxonomic classifications, and sequence and structure neighbours).
- The IMB Jena Image Library of Biological Macromolecules [23] (http://www.imb-jena.de/IMAGE_PDB_SEARCH.html) provides access

to all PDB and NDB entries. Basic information on the architecture of biopolymer structures is available. Searches can be carried out using various database codes, organism names, gene names, etc.

- PDBFINDER [24] (http://www.cmbi.kun.nl/swift/pdbfinder/) is a database that is constructed from the PDB and other databases. Many of the fields contained in the PDBFINDER database are difficult to access from the original databases. Moreover, some information is retrieved from the original literature. The database can be queried conveniently with the SRS system [25] (http://srs6.ebi.ac.uk/ or http://srs.embl-heidelberg.de:8000/srs5/).

- PDB at a Glance (http://cmm.info.nih.gov/modeling/pdb_at_a_glance. html) provides a different kind of access to the structural database. A number of chemically or biologically meaningful contexts have been defined, and entries in the PDB are automatically assigned to the applicable contexts. Useful contexts include molecular classes, secondary/tertiary structural classes, functional classes, species of origin, and structure-determination method.

10.3 SECONDARY STRUCTURAL DATABASES

In this section, we discuss a small selection of the enormous number of web-based resources that provide derivative structural information. Links to the web sites of these resources are also listed in Table 10.2.

10.3.1 STRUCTURAL CLASSIFICATION DATABASES

One of the first questions that is asked when a new structure has been determined is whether the fold of the protein (or each of its domains) is unique. Structural classification databases (as well as programs and servers to compare the fold of new structures to structures already in the PDB) address this issue. They also serve in the study of sequence/structure/function relationships, may help in discovering remote evolutionary relationships between proteins that have low levels of sequence identity, assist in the creation of multiple sequence alignments, and aid in the definition of structural cores for sequence-structure alignment.

The way in which the major classification databases are curated ranges from being completely automatic to being largely manual. FSSP (Families of Structurally Similar Proteins) [26] (http://www.bioinfo.biocenter.helsinki.fi:8080/dali/) is an example of an automatically generated database (and, hence, based solely on structural information), whereas SCOP (Structural Classification of Proteins database) [27] (http://scop.mrc-lmb.cam.ac.uk/scop/) is carefully curated manually, a process in which information from the literature is taken into account. CATH (Class, Architecture, Topology, Homologous superfamily) [28] (http://www.biochem.ucl.ac.uk/bsm/cath_new/) is intermediate between FSSP and SCOP, although still based only on structural information. The structure neighbours facility in MMDB is based on an FSSP-like automated structure-comparison method [21,22]. SCOP, CATH and FSSP have been subjected to a systematic comparison [29] that concluded that the three systems agree on the majority of their classifications.

TABLE 10.2
URLs of Secondary (Derivative) Structural Databases

Resource	URL
1. Structural Classification Databases	
FSSP	http://www.bioinfo.biocenter.helsinki.fi:8080/dali
SCOP	http://scop.mrc-lmb.cam.ac.uk/scop
CATH	http://www.biochem.ucl.ac.uk/bsm/cath_new
MMDB	http://www.ncbi.nlm.nih.gov:80/Structure/MMDB/mmdb.shtml
2. Structurally Specialised Structural Databases	
PROCAT	http://www.biochem.ucl.ac.uk/bsm/PROCAT/PROCAT.html
SPASM	http://portray.bmc.uu.se/spasm
PINTS	http://www.russell.embl.de/pints
ReLiBase	http://relibase.ccdc.cam.ac.uk
HIC-Up	http://xray.bmc.uu.se/hicup
ChemPDB	http://www.ebi.ac.uk/msd-srv/chempdb
3Dee	http://www.compbio.dundee.ac.uk/3Dee
MolMovDB	http://molmovdb.mbb.yale.edu/molmovdb
PROMISE	http://metallo.scripps.edu/PROMISE/MAIN.html
SPIN-PP	http://trantor.bioc.columbia.edu/cgi-bin/SPIN
PQS	http://pqs.ebi.ac.uk
SLoop	http://www-cryst.bioc.cam.ac.uk/~sloop
3. Biologically Specialised Structural Databases	
Enzyme Structures Database	http://www.biochem.ucl.ac.uk/bsm/enzymes
PKR	http://pkr.sdsc.edu/html/index.shtml
Antibodies - Structure and Sequence	http://www.bioinf.org.uk/abs
SACS	http://www.bioinf.org.uk/abs/sacs
HIVdb	http://xpdb.nist.gov/hivsdb/hivsdb.html
RNABase	http://www.rnabase.org
SDAP	http://fermi.utmb.edu/SDAP
Prolysis	http://prolysis.phys.univ-tours.fr/Prolysis
CAZy	http://afmb.cnrs-mrs.fr/CAZY
ESTHER	http://bioweb.ensam.inra.fr/ESTHER/general?what=index
NucleaRDB	http://receptors.ucsf.edu/NR
GPCRDB	http://www.gpcr.org/7tm
EF-hand Calcium-Binding Proteins	http://structbio.vanderbilt.edu/cabp_database/cabp.html
Serine Proteases	http://biochem.wustl.edu/~protease
4. Miscellaneous Databases	
HAD	http://www.bmm.icnet.uk/had/heavyatom.html
BMCD	http://wwwbmcd.nist.gov:8080/bmcd/bmcd.html
RESID	http://pir.georgetown.edu/pirwww/dbinfo/resid.html
PISCES	http://dunbrack.fccc.edu/PISCES.php/
ModBase	http://alto.compbio.ucsf.edu/modbase-cgi/index.cgi
LIGBASE	http://alto.compbio.ucsf.edu/ligbase
FAMSBASE	http://famsbase.bio.nagoya-u.ac.jp/famsbase/
Gene3D	http://www.biochem.ucl.ac.uk/bsm/cath_new/Gene3D
SUPERFAMILY	http://supfam.mrc-lmb.cam.ac.uk/SUPERFAMILY
Parts List	http://bioinfo.mbb.yale.edu/partslist
TargetDB	http://targetdb.pdb.org
SeqAlert	http://bioportal.weizmann.ac.il/salertb/main

However, to make use of these databases, one needs to understand the principles underlying them — each database offers certain advantages, depending on the biological requirements and knowledge of the user.

For newly determined protein structures, a large number of web-based servers is available with which their folds can be classified through comparison to a database of known protein structures. For a discussion of this area, references and links, see the review by Novotny et al. [30].

10.3.2 STRUCTURALLY SPECIALISED STRUCTURAL DATABASES

The PDB has engendered a large number of derivative databases that focus on one particular aspect of macromolecular structure. The structure classification databases discussed in the previous section are examples of this phenomenon. Other examples are discussed here.

- PROCAT [31] (http://www.biochem.ucl.ac.uk/bsm/PROCAT/PROCAT.html) provides facilities for interrogating a database of 3-D enzyme active-site templates. PROCAT can be thought of as the 3-D equivalent of the sequence motifs found in databases such as PROSITE. The PROCAT database allows one to search for the occurrence of 3-D enzyme active-site motifs in a protein structure.

- SPASM [32,33] (http://portray.bmc.uu.se/spasm) provides functionality that is in a sense the opposite of PROCAT. With SPASM, a database of structures can be queried to find if a certain 3-D motif of residues (e.g., an unusual loop, a putative active site, or a set of ligand-binding residues) occurs in any other known structures and, if so, in which ones and to what degree of structural similarity. Similar functionality is provided by the PINTS server [34] (http://www.russell.embl.de/pints).

- The Receptor-Ligand Interaction Database (ReLiBase) [35] (http:// relibase. ccdc.cam.ac.uk) is an object-oriented database system for storing and analysing structures of protein-ligand complexes currently available from the PDB. It has its own tools to analyse protein-ligand interaction patterns such as keyword searches, substructure searches, ligand similarity searches based on topological fingerprints, and 3-D query tools specified by 3-D constraints.

- HIC-Up (Hetero-compound Information Centre – Uppsala [36]) (http://xray. bmc.uu.se/hicup/) is a resource that contains information, including coordinates, refinement dictionaries, images, and links to related sites, about all hetero-entities (e.g., organic molecules, ions, metal centres) that occur in the PDB.

- ChemPDB [10] (http://www.ebi.ac.uk/msd-srv/chempdb) provides web access to the ligand and small molecule component of the E-MSD database, which is a consistent and enriched library of all the small molecules and monomers that are referred to in any macromolecular structure of a PDB entry, such as bound molecules and standard and modified amino and nucleic acids. For each such compound, the database contains descriptions of its connectivity and stereochemistry, 3-D coordinates, trivial and systematic names, etc.

- The 3Dee database (database of protein-domain definitions [37]; http://www.compbio.dundee.ac.uk/3Dee/) is a repository of protein structural domains that stores alternative domain definitions for the same protein. The domains are organised into both sequence and structural hierarchies and the database contains nonredundant sets of sequences and structures.

- The Molecular Movements Database (MolMovDB) [38] (http://molmovdb. mbb.yale.edu/molmovdb/) lists motions in proteins and other macromolecules. It is arranged around a multilevel classification scheme and includes motions of loops, domains, and subunits.

- Many proteins contain a metal ion in the active site. The Prosthetic groups and Metal Ions in Protein Active Sites Database (PROMISE) [39] (http://metallo.scripps.edu/PROMISE/MAIN.html) is a database containing information on bioinorganic motifs in proteins. It contains 10 groups of metal-containing proteins: chlorophyll proteins, cobalt proteins, copper proteins, diiron-carboxylate proteins, flavoproteins, haem proteins, iron-sulphur proteins, molybdopterin-containing proteins, mononuclear iron proteins, and nickel proteins. The database focuses on the active-site structure and combines sequential, 3-D structural, and physico-chemical information.

- Surface Properties of INterfaces – Protein Protein Interfaces (SPIN-PP) (http://trantor.bioc.columbia.edu/cgi-bin/SPIN/) is a database of all protein-protein interfaces found in the PDB. Physico-chemical properties have been mapped to the molecular surfaces of the protein–protein interfaces and used to classify the interfaces by a surface property taxonomy. These properties include surface curvature, electrostatic potential, hydrophobicity, and sequence variability from HSSP [40] alignments.

- PDB files of crystal structures tend to contain only the asymmetric unit of the crystal, which is not necessarily the same as the biologically active unit (which could be a dimer generated by crystallographic symmetry, for example). The Protein Quaternary Structure server (PQS) [41] (http://pqs.ebi.ac.uk/) provides access to probable quaternary structures for macromolecules solved by X-ray crystallography.

- Loops connecting secondary structure elements can sometimes contain binding sites for ligands. At present it is difficult to model the 3-D structure of loops correctly. SLoop [42] (http://www-cryst.bioc.cam.ac.uk/~sloop/) is a database of loop conformations connecting elements of defined protein secondary structure. The loops are classified according to length, type of secondary structure elements they are connected to, and conformation of the main chain.

10.3.3 BIOLOGICALLY SPECIALISED STRUCTURAL DATABASES

Databases that focus on a particular set of proteins that share one or more properties (e.g., function, substrate, mechanism, parent organism, or presence of metal ions) are in a sense orthogonal to the structurally specialised databases. There are quite a few such databases on the web, many of them containing substantial added value in the form of sequences, functional information, structural analyses, correlations, illustrations, etc.

- The Enzyme Structures Database [31] (http://www.biochem.ucl.ac.uk/bsm/enzymes/) lists, for every category in the Enzyme Classification (EC) system, the PDB entries associated with them. In essence, this is an inverted index to the PDB listing all known enzyme structures.
- The Protein Kinase Resource (PKR) (http://pkr.sdsc.edu/html/index.shtml) is a web-based compendium of information on the protein kinase family of enzymes. Besides structural data, it contains information about classifications and a list of diseases in which protein kinases play a role.
- The "Antibodies – Structure and Sequence" site (ABS) (http://www.bioinf.org.uk/abs/), as its name implies, summarises information on antibody structure and sequence. It provides a query interface to the Kabat antibody sequence data, general information on antibodies and crystal structures and links to other antibody-related information. As part of ABS, the SACS site (Summary of Antibody Crystal Structures) [43] (http://www.bioinf.org.uk/abs/sacs) provides up-to-date information about all antibody crystal structures available from the PDB.
- The HIV Protease Structural Database (HIVdb) [44] (http://xpdb.nist.gov/hivsdb/hivsdb.html) is an archive of experimentally determined 3-D structures of Human Immunodeficiency Virus 1 (HIV-1), Human Immunodeficiency Virus 2 (HIV-2), and Simian Immunodeficiency Virus (SIV) proteases and their complexes with inhibitors or products of substrate cleavage. Interestingly, the structures contained in HIVdb are not all taken from the PDB — quite a few were actually obtained directly from depositors for inclusion in HIVdb alone. The structures unique to HIVdb were contributed mostly by pharmaceutical companies and are not all completely refined.
- RNABase [45] (http://www.rnabase.org/) is a database of all structures that contain RNA and that have been deposited in either the PDB or the NDB. For every entry, brief summary information is provided as well as conformational information and identification of possible model errors.
- The Structural Database of Allergenic Proteins (SDAP) [46] (http://fermi.utmb.edu/SDAP/) gives access to integrated sequence, structure, and epitope data of allergenic proteins. The site also provides software tools to screen protein sequences for the presence of IgE epitopes that they may share with known allergens.
- Prolysis (http://prolysis.phys.univ-tours.fr/Prolysis/) is a web resource on proteases and their natural or synthetic inhibitors. It stores general information about proteases and inhibitors as well as links to other resources with related information.
- The CAZy site (Carbohydrate-Active enZymes) [47] (http://afmb.cnrs-mrs.fr/CAZY/) describes the families of structurally related catalytic and carbohydrate-binding modules (or functional domains) of enzymes that

degrade, modify, or create glycosidic bonds, including glycosidases and trans-glycosidases, glycosyl-transferases, polysaccharide lyases, and carbohydrate esterases.

- ESTHER (ESTerases and alpha/beta Hydrolase Enzymes and Relatives) [48] (http://bioweb.ensam.inra.fr/ESTHER/general?what=index) is a server dedicated to the analysis of protein and nucleic acid sequences belonging to the superfamily of alpha/beta hydrolases homologous to cholinesterases.

- The NucleaRDB site [49] (http://receptors.ucsf.edu/NR/) contains a wealth of primary and secondary data on nuclear receptors, including sequences, alignments, phylogenetic trees, 2-D and 3-D structures, genetic and mutational data, and information on binding partners.

- GPCRDB [50] (http://www.gpcr.org/7tm/) is related to NucleaRDB, but instead focuses on G-protein-coupled receptors. It provides similar kinds of information as NucleaRDB.

- The EF-hand Calcium-Binding Proteins web site (http://structbio.vanderbilt.edu/cabp_database/cabp.html) provides general information (e.g., literature references), sequence information (e.g., alignments), structural information (e.g., PDB files, inter-helical angles, inter-residue contacts, distance difference matrices, dihedral angles, solvent accessible surface area, and NMR assignments) as well as computational tools and web-based resources.

- The Serine Proteases site (http://biochem.wustl.edu/~protease/) contains information about classification, sequences, evolution, mutations, structures, ligands, substrates, inhibitors, and functions of these enzymes.

10.3.4 MISCELLANEOUS DATABASES

- The Heavy-Atom Databank (HAD) [51] (http://www.bmm.icnet.uk/had/heavyatom.html) contains information on the preparation and characterisation of heavy-atom derivatives of protein crystals, including experimental conditions for crystallisation, chemical details of the heavy-atom compounds used, bibliographic references, atomic coordinates of heavy-atoms, and details of the binding sites of the heavy-atoms in the protein crystal environment.

- The Biological Macromolecule Crystallization Database (BMCD) [52] (http://wwwbmcd.nist.gov:8080/bmcd/bmcd.html) contains information compiled from the literature about the crystallisation conditions of biomacromolecules (proteins, protein–protein complexes, nucleic acid, nucleic acid–nucleic acid complexes, protein–nucleic acid complexes and viruses) for which diffraction-quality crystals have been obtained. In addition to crystallisation data reported in the literature, the BMCD contains the NASA Protein Crystal Growth Archive, which includes the crystallisation data generated from studies carried out in a microgravity environment.

- The RESID database [53] (http://pir.georgetown.edu/pirwww/dbinfo/resid.html) is a comprehensive collection of annotations and structures of protein modifications, including amino-terminal, carboxyl-terminal and peptide chain cross-link, and pre-, co- and post-translational modifications. For each modification, chemical formulas, literature references, images, and a model in PDB format are provided.

- If one wants to carry out some statistical study on protein structures, it is often desirable to use a subset of structures that satisfy certain minimum quality criteria, and that are representative in terms of their sequences. PISCES [54] (http://dunbrack. fccc.edu/PISCES.php/) is a server that can be used to obtain such subsets. A number of such subsets are generated on a weekly basis. In addition, users can specify their own criteria and generate subsets on the fly.

- One way to obtain 3-D information for a protein whose structure has not been determined experimentally is by comparative (or homology) modelling using a similar protein whose 3-D structure is known. MODBASE [55] (http://alto.compbio.ucsf.edu/modbase-cgi/index.cgi) is a database of comparative protein structure models for sequences from several genomes. The database includes the fold assignments and alignments from which the models were derived. LIGBASE [56] (http://alto.compbio. ucsf.edu/ligbase/) is a related structural database containing 3-D models of ligand-binding proteins that are aligned to structural templates from the PDB. The 3-D models of the aligned sequences are from MODBASE, and the site provides schematic plots of the predicted protein-ligand interactions. The database is continuously growing as more sequences and structures become available.

  - FAMSBASE [57] (http://famsbase.bio.nagoya-u.ac.jp/famsbase/) is another database of comparative protein structure models generated with a different program (Full Automatic Modeling System, FAMS). It contains homology models for proteins derived from sequences of the whole genomes of dozens of species.

  - The Gene3D database [58] (http://www.biochem.ucl.ac.uk/bsm/ cath_new /Gene3D/) is somewhat different in that it provides structural assignments for genes within complete genomes. It links the assignments to their corresponding entries in PDBsum and CATH, so that structural information can be obtained, for instance, to annotate a possible function for an otherwise unknown protein sequence.

  - Whereas Gene3D is based on PSI-BLAST alignments and the CATH classification, the SUPERFAMILY database [59] (http://supfam.mrc-lmb.cam.ac.uk/SUPERFAMILY/) is based on the hierarchical SCOP classification of protein domains and uses a library of hidden Markov models to represent all proteins of known structure. At the SCOP level of "superfamily," distantly related proteins or protein domains that have a common evolutionary ancestor, as evidenced by sequence, function, and structure, are grouped together. The server allows for amino acid or nucleotide sequence queries and produces alignments of the query sequence to sequences in the PDB. Structural assignments for all completed genomes are available from the server.

- Yet another approach to structural classification of sequences is provided by Parts List [60] (http://bioinfo.mbb.yale.edu/partslist/). The creators of the Parts List server assume that there is only a limited number of protein folds. The database thus contains only one entry for each of the known folds. A structural classification, though, can be facilitated by attributes other than the fold itself such as frequency of occurrence of a fold in all completely sequenced genomes or in a certain genome, the occurrence of a fold in the structural databanks, expression level information relating to the fold, etc. The ranking within the database is based on 180 such attributes.

- Structural genomics is a worldwide initiative aimed at determining thousands of protein structures in a high-throughput mode. Targets include all Open Reading Frames (ORFs) that encode genes present in the genomes of a variety of organisms, proteins in specific biochemical pathways, and proteins associated with specific disease states. There are many sites devoted to Structural Genomics on the web, but the PDB maintains an overview of all current efforts at http://www.rcsb.org/pdb/strucgen.html. Data about targets from world-wide structural genomics and proteomics projects are collected in the target-registration database, TargetDB [61] (http://targetdb.pdb.org/). A search of this database in May 2005 revealed that 87,000 proteins had been selected by the various structural genomics consortia (up from 67,000 in September 2004 and 39,000 in December 2003.) Of these proteins, 10,300 (12%; up from 7400 in September 2004 and 4650 in December 2003) had been cloned, expressed, and purfied. More than 1700 structures (2%; up from 1200 in September 2004 and just over 500 in December 2003) had at that time been deposited in the PBB.

If one is interested in the structure of one or more specific proteins, their sequences can be submitted to an alerting service such as SeqAlert (http://bioportal.weizmann. ac.il/salertb/main). This service will periodically compare the sequences of interest to the sequences of all proteins in the PDB and in TargetDB, and report any hits by e-mail.

10.4 VALIDATION

10.4.1 NEED FOR VALIDATION

The structural models that are obtained from either X-ray crystallography or NMR spectroscopy are not calculated directly from the experimental data. Rather, both techniques involve a certain amount of subjective interpretation of the data and of subjective choices with respect to the parameterisation and refinement of the model [62,63,64]. Thus, the quality of the resulting model is limited by the amount and quality of the experimental data as well as by the experience, prejudices, expectations, and practices of the scientist who constructs and refines the model. In fact, even when two X-ray crystallographers or NMR spectroscopists are given the exact same set of experimental data, the models they produce will be different. In favourable cases, when the data are good and the scientist knows what he or she is doing, the models will differ only with respect to minor details. With high resolution data and good phases, a crystal structure should be to 95% a consequence of the data [63]. However, even at very high resolution, subjective choices must be made, for example with respect to the treatment and inclusion of alternative conformations, explicit hydrogen atoms, anisotropic temperature factors, and solvent molecules. At lower resolution, where the interpretation of electron density becomes more difficult and ambiguous, the subjectivity also involves the placement of the mainchain and sidechain atoms, the inclusion of parts of the model for which there is little or no density, the way in which temperature factors should be modelled and if and how any noncrystallographic symmetry should be exploited [65].

The unavoidable subjective aspects of the model-construction process need not be detrimental in and of themselves, provided that the scientist involved is experienced, is using the most appropriate techniques (e.g., for model refinement), and is aware of the limitations that the data impose on the model. Even less experienced people can do a good job, provided that they are properly supervised and use the tools that are available for producing the best possible model and for tracking down possible errors in intermediate models [62,64,66–70].

For users of structural data, however, all this has three important implications:

1. Almost all models contain some (hopefully small) errors, and some models can even be completely wrong.
2. Not all models (even of the same protein and at the same resolution) have the same quality.
3. Even within a given model, not all residues and atoms have the same level of quality or reliability.

As was pointed out earlier, the PDB is merely a repository of structural information, with essentially no minimum requirements on its quality. Therefore, point 1 above means that users of structural data themselves should ascertain that any model they want to use does not contain any serious errors. When more than one structure is available, point 2 implies that the user should assess the quality of each of these before deciding which one(s) to use (for instance, for homology modelling purposes or in docking studies). Finally, when a model has been selected for further study, point 3 implies that the user should ensure that residues of interest (e.g., the active site, substrate-binding, or ligand-binding residues) have been modelled reliably.

In the following discussion, it is assumed that the reader is familiar with the basic concepts of structure determination by X-ray crystallography (this volume, Chapter 5 by Rupp) and NMR spectroscopy (this volume, Chapter 4 by Lee et al.).

10.4.2 Validation of Crystallographic Models

In this section, we discuss model validation from the point of view of the users of structural data. Our emphasis will be on protein structures, because most of the work in this area has traditionally dealt with these molecules, but this is not to imply that they are inherently less (or more) reliable than the structures of nucleic acids (see, e.g., [71,72]) or hetero-compounds (see, e.g., [17,36,73,74]).

10.4.2.1 Errors

Errors in structural models vary in degrees of seriousness and frequency, the most serious being errors in the tracing of the protein chain through the electron density. In the worst case, a subunit, domain or entire structure may have been mistraced (see [62] and [63] for some examples). Alternatively, the crystallographer may have identified the secondary structure elements correctly, but, probably due to poor density in the loops between them, connected them erroneously (see [63] for an example). Finally, ambiguities in the density may have lead to register errors [70], where the sequence is frame-shifted with respect to the density (see [75] for an example). Such errors may involve as few as two residues, but they can also extend through much of a protein. All tracing errors have in common that a substantial number of residues in the model are nowhere near the position in 3-D space where they ought to be situated. This renders such models more or less useless for purposes of ligand design, structure analysis, and structural interpretation of functional, mutational, binding, and other data.

A second class of errors are local modelling errors that affect only one, two, or three residues. The residues involved are roughly in the correct part of 3-D space, but the identity, position, orientation, or conformation of the mainchain or sidechain atoms are incorrect. Errors of this kind need to be identified prior to use or analysis of the model. Although they do not necessarily invalidate the entire model, the erroneous residues are unreliable for modelling and design purposes. However, if they are far removed from the functionally important sites (e.g., in a flexible loop on the surface), the remainder of the model may still be useful.

A third class of errors is caused by over-fitting of the data. This happens whenever the crystallographer refines more parameters than is warranted by the information content of the experimental data. For instance, the crystallographer may have chosen to ignore experimental data below a certain resolution or to omit weak reflections. Alternatively, the model may have been parameterised inappropriately, for example, by selecting an inappropriate temperature-factor model, by ignoring constraints (e.g., those provided by any noncrystallographic symmetry), or by under-weighting restraints. Finally, the model may contain aspects that are not convincingly supported by the data (e.g., spurious water molecules, alternative conformations of protein or ligand atoms, or refined atomic occupancies).

In almost every structure determination, errors will be made at some stage. It is the task of the crystallographer to identify and correct such mistakes during the iterative process of model building and refinement. The model that is deposited should be the best possible representation of the data gathered in the crystallographic experiment. Nevertheless, the user of structural information is advised to always assess the quality of any models before embarking on any in-depth studies based on these models (such as ligand design, mutational studies, homology modelling, etc.).

10.4.2.2 Quality Indicators

Validation of protein models was not really an issue until the late 1980s, when several high-profile protein structures (often published in equally high-profile journals) turned out to be seriously flawed [62]. Since then, a lot of effort has been spent on the development of methods that identify problems in protein models. These methods can be classified in different ways. For instance, some methods only use the model itself (coordinates and, possibly, temperature factors and occupancies), whereas other methods assess the correspondence between the model and the experimental data of which it is an interpretation. Alternatively, one can distinguish between methods that assess the overall quality of a model (global methods) and those that focus on the details (individual atoms and residues; local methods). Finally, some weak methods assess model aspects that were explicitly or implicitly used during the construction of the model and are therefore more a test of the methodology (e.g., the refinement program's ability to impose proper bond lengths and angles) than of the validity or accuracy of the model. If a model performs poorly in such a test, it is clearly reason for concern, but the opposite is not true. Any refinement program worth its money will impose the restraints it is told to impose, and the model will conform to those restraints. However, this proves nothing about the inherent correctness of the model aspects that have not been restrained. For example, an intentionally backwards-traced protein model can be refined to reasonable R-values with very reasonable bond lengths and angles, while obviously being grossly incorrect as an interpretation of the experimental electron density [64]. More useful for validation purposes are strong methods that assess aspects of the model that are independent of (or orthogonal to) the information used in the creation of the model.

Users of structural information will initially want to know if a model is globally reliable or not, for which purpose strong, global criteria are the most useful. Once a model has been selected, the reliability of the aspects of it that will be studied needs to be assessed critically. For instance, prior to initiating docking studies in an enzyme's active site, it is worthwhile to assess the reliability of the residues that line the active site and its entrance(s). In general, however, a good model is one that fares well in all the validation tests it is subjected to, especially the strong ones. Conversely, a model that scores poorly on several tests should be treated with caution. The same is true with respect to the quality and reliability of individual residues. Here, however, it is important to realise that validation methods based on coordinates alone can only detect *outliers* (i.e., residues with unusual properties). These outliers are not necessarily errors, but can also be structurally or functionally significant features [76]. To assess whether an outlier is an *error in the model* (that needs to be corrected by the crystallographer) or a genuine *feature of the structure* (that may well merit discussion in the paper describing it), one requires access to the experimental electron density.

In the following sections, some statistics that can be used to evaluate the global and local quality of crystal structures of biomacromolecules are discussed. The selected validation tests and statistics are useful for, and reasonably easily accessible to, nonexpert users of structural data. For more details about these and other methods, the reader is referred to a compendium of quality indicators [63] and references therein.

10.4.2.3 Global Model Quality

Assessment of the global quality or reliability of a crystal structure has traditionally involved the resolution and the crystallographic R-value. Resolution is not an entirely objective statistic, but still useful as an initial filter. When faced with the choice between a 3.0 Å native and a 1.5 Å complex structure, for example, the latter will almost always be the safer bet. The crystallographic R-value, however, can be deceptive because it can be made almost arbitrarily small (especially at low resolution) by increasing the number of parameters in the model [62,64,77]. The free R-value measures the discrepancy between observed and calculated structure-factor amplitudes for a subset of the reflections that is not used for refinement of the model. Thus, any changes to the model that are not supported by the data (e.g., introduction of inappropriate parameters) are expected to lead to a constant or increasing free R-value, even if the R-value decreases. High values of the free R-value are indicative of problems in the fit of the model to the data and such models should be treated with caution. Large discrepancies between the free and conventional R-value suggest that the model may overfit the data (e.g., too many parameters) or, conversely, underfit it (e.g., a domain is absent in the model due to mobility that rendered it invisible in the electron-density maps).

Some validation methods assess the global quality of a model based only on the coordinates. These methods usually compare one or more properties calculated from the model (but not related directly to information used in the creation of the models) to expected values or distributions based on analysis of a large number of well-determined, high-resolution structures. Examples are the 3D–1D profile method [78], which assesses the compatibility of the sequence and the structure, and related methods that use threading and other potentials. The overall quality of the Ramachandran plot (e.g., expressed as a percentage of outliers) can also be used as a quick indication of the overall quality of the model [79], Figure 10.1. Even though an individual residue may be an "outlier-for-a-reason" in the Ramachandran plot, the model as a whole should have only a small fraction of outliers. Similarly, average or root-mean-square (RMS) values of many of the local quality statistics (discussed in the next section) can also serve as global quality indicators. In analogy to crystallographic R-values, geometric G-factors have been formulated [80] that combine the results of a number of quality checks in a single number.

In addition to the conventional R-value, similar scepticism should be extended to average temperature factors and RMS deviations from ideal geometry parameters as measures of the quality of the model [64].

10.4.2.4 Local Model Quality

The free R-value provides an indication of the overall fit of a model to the experimental data it is meant to explain. This fit can also be assessed on a local scale (e.g., for a sidechain, entire residue, ligand, secondary-structure element, or domain). It is then called the real-space fit and entails comparing the experimental electron density with the distribution of electrons calculated from the model [66]. The fit can be expressed either as an R-value (which is then called the real-space R-value) or

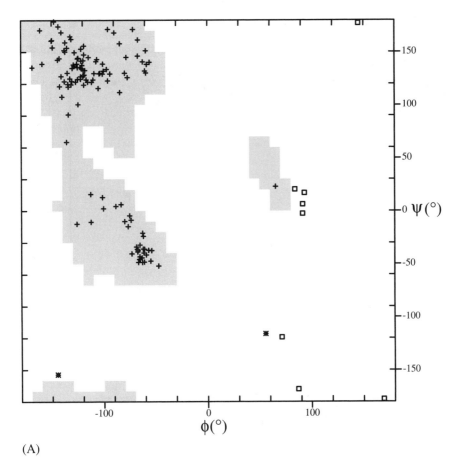

(A)

FIGURE 10.1 Ramachandran plot of (A) an essentially correct and (B) a completely incorrect protein model. The plot in (A) is of the 1.8 Å structure of cellular retinoic-acid-binding protein type II [121], and that in (B) of a deliberately backwards-traced model of the same protein refined to only 3 Å [64]. The core regions of the Ramachandran plot (in the definition of Kleywegt and Jones [79] are shaded in grey. In the plots, glycine residues are represented by squares, and all other residues by plus signs if they fall within the core regions, or asterisks if they are outliers. The model in (A) has 2 outliers (out of 127 residues), whereas the one in (B) has no fewer than 46. In general, the Ramachandran plots of good models will not only have few outliers (<5%) but also a strong concentration of residues within the core regions of the plot.

as a correlation coefficient. Inspection of a plot of the real-space fit as a function of residue number immediately reveals any parts of the model that fit the density poorly. Such residues should be treated with caution and their density should be inspected to decide if they are sufficiently reliable for the purposes of the study.

One of the simplest ways to assess the overall quality of a model rapidly is to inspect its Ramachandran plot [79,81], Figure 10.1. This is a scatter plot of ϕ, ψ-

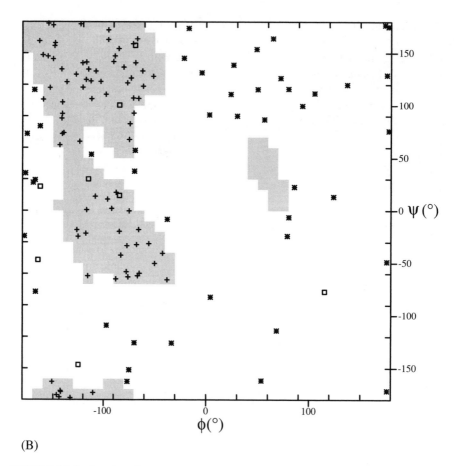

(B)

FIGURE 10.1 (continued)

value pairs for all nonterminal residues in a protein, where ϕ is the torsion angle C_{i-1}-N_i-$C\alpha_i$-C_i and ψ is N_i-$C\alpha_i$-C_i-N_{i+1}. (The third torsion angle along the backbone, $C\alpha_i$-C_i-N_{i+1}-$C\alpha_{i+1}$, is called ω and is restricted to values near 180° for *trans*-peptides and near 0° for *cis*-peptides.) The ϕ and ψ torsion angles are not normally restrained during the refinement of the model and since they have preferred combinations of values, this makes them useful for validation purposes. Residues that are outliers in the Ramachandran plot should be checked in the electron-density map to ascertain that they are features of the structure rather than errors in the model.

The side chains of many amino-acid residues also contain one or more torsion angles ($\chi 1$, $\chi 2$, etc.) whose values are not usually restrained, but that nevertheless have clearly preferred values [82] and combinations of values (rotamers) [83]. Residues with unusual sidechain conformations should be scrutinised in the density map.

Proteins contain preferred packing arrangements in which the hydrophobic residues tend to interact with each other, the charged residues tend to be involved in salt links, and hydrophilic residues prefer to interact with each other or to point out into the bulk solvent. If a model contains serious mistakes (i.e., some or many residues have incorrect positions, orientations, or conformations), these general rules of thumb will often be violated. Severe errors will introduce nonphysical interactions into the model (e.g., a charged arginine residue located inside a hydrophobic pocket). Conversely, such unusual interactions can be used as indicators of possible model errors. Directional atomic contact analysis (DACA) [84] is a method in which these empirical notions have been formalised. For every group of atoms in a protein, it yields a score that, in essence, expresses how comfortable that group is in its environment in the model under scrutiny (compared to the expectations derived from a database of well-refined, high-resolution structures). If a region in a model (or the entire model) has consistently low scores, this is a very strong indication of model errors.

Residues with other unusual features should of course be examined critically in the electron-density map as well. These features may pertain to unsatisfied hydrogen bonds, geometry (unusual bond lengths and angles), deviations from planarity, unusual chirality, very high temperature factors compared to the rest of the model or compared to related atoms or residues, or unusual conformational differences between corresponding residues in molecules related by noncrystallographic symmetry.

10.4.3 Validation of NMR Models

Many of the observations made with respect to the validation of crystal structures in the previous section also apply to models based on NMR spectroscopic data, and many of the purely structural quality indicators (e.g., Ramachandran and sidechain rotamer analysis) can also be applied to NMR models or ensembles. One should, however, be aware of the following regarding validation of NMR structure models using conformational torsion angles and Ramachandran plots. First, the Ramachandran plot statistics are generally better for X-ray structures than for NMR structures. Second, if the refinement protocol included use of a "conformational database potential" [85], one should expect that the final models have better Ramachandran statistics, etc. However, since information about the torsion-angle distributions was used in the creation of the model, they can no longer be used for validation purposes.

The remainder of this section will deal only with the aspects of validation that are specific to NMR-based models.

10.4.3.1 Precision and Accuracy

The best representation of NMR data is normally as an ensemble of structure models rather than a single structure [86,87,88], Figure 10.2. The overall precision of such an NMR ensemble has long been used by the NMR community as a quality indicator and as such has been evaluated in many studies (see, e.g., [88,89]). Heterogeneity in an NMR ensemble leads to lower precision, Figure 10.3. Such heterogeneity can be caused by thermal motion, presence of multiple conformations, and under-determination due to low-quality data or an insufficient number of structural restraints [90,91]. However, the underlying causes cannot be assessed without knowledge of experimental details such as spectral line widths and relaxation parameters, both of which can be probes of motions within a protein.

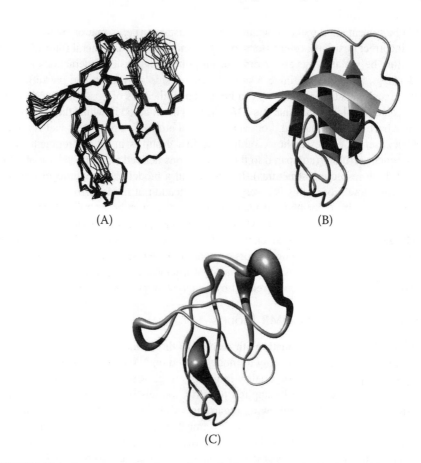

FIGURE 10.2 Structure model of a small protein domain determined using traditional NMR methods. (A) Backbone trace display of all models in the ensemble (superimposed by the backbone atoms). In regions with good precision, the backbone atoms will be on top of each other. (B) Traditional ribbon model of the first model in the ensemble's PDB file. (C) "Sausage" representation of the protein backbone of all ensemble models. The worse the precision is among the ensemble models, the thicker the sausage will be. The figures were prepared using the program MOLMOL [122].

Usually, the precision is expressed as the RMS deviation of the coordinates of an NMR ensemble compared to the mean coordinates, or as the RMS deviation across the entire ensemble. The lower the values are, the better the precision. What precision actually measures is the reproducibility of the calculations for a given set of restraints and constraints. Thus, the way in which the representative models of an ensemble are selected can obviously be a major determinant of the calculated value of the precision. The most common way to select models for a representative ensemble is to choose the N structures with the lowest energy or the N structures that do not violate the structural restraints too severely. The latter is preferred since it involves the quality of the fit of the models to the experimental data.

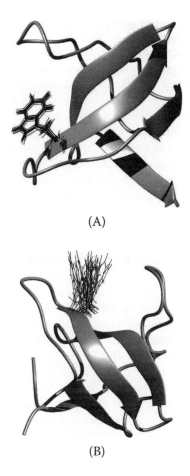

(A)

(B)

FIGURE 10.3 An illustration of locally good and poor precision for a functionally important sidechain in two different NMR structure models. In (A), the conformation of the sidechain is very similar among the various ensemble models leading to good local precision. In (B), the corresponding sidechain has a different conformation in each ensemble model, leading to poor local precision.

Nevertheless, the choice of both the selection method and the size of the ensemble is subjective.

Accuracy, on the other hand, is a measure of how close the calculated model is to the true structure, and thus it is a quantity that cannot normally be measured. In a comparison of 16 structures determined both by NMR and X-ray crystallography, it was found that a higher precision of a model ensemble was correlated to a higher accuracy, when the accuracy was measured as the global RMS deviation of the NMR ensemble coordinates from the coordinates of the X-ray structure [92]. This does, however, not necessarily imply that higher precision always corresponds to higher accuracy. In fact, under certain circumstances some types of errors that lower the accuracy do not influence the precision (see the discussion in the next section).

10.4.3.2 Data Quality and Quantity

A major determinant of the quality of NMR models is the total number of restraints per residue, and a useful quality indicator is the average number of restraints per residue. The most important experimental restraints for structure calculations are the NMR-measurable distances between proton–proton pairs that are derived from nuclear Overhauser enhancement (NOE) data. Other types of experimental restraints are commonly used as well, such as scalar couplings and dihedral angles derived from them, chemical shifts, and residual dipolar couplings [93,94].

The use of structural restraints other than NOEs for the calculation of NMR models has probably contributed to higher model quality in recent years. In general, it has been observed that the more restraints are used, the better the model will be in terms of precision, with respect to both the backbone and the sidechain conformations [89,95,96]. It is of course important that there is no redundancy among the restraints and that the restraints used are not violated in the models. For example, the common practice of converting the NOE intensities into loose distance restraints means that some of them, especially intra-residue ones, become redundant and should not be included in the calculations. Violations of restraints, if present in the final models, are usually quoted for the entire ensemble and not on a per-model basis. This can be difficult to interpret for a nonexpert, though. Both the redundancy of the restraints and the occurrence of violations can be assessed using programs such as AQUA and PROCHECK-NMR [97,98], provided that one has access to the restraint set (e.g., obtained from the BMRB [12] or the authors of the study).

The quality of a model will of course be affected by any errors in the restraints. The two main sources for errors in the distance restraints are incorrect estimates of distances derived from the NOE intensities and incorrect assignments. Normally, these types of errors will show up as violations in the usual controls during the iterative refinement process and can be eliminated after validation by the spectroscopist (e.g., by careful inspection of the spectra). Under certain circumstances, though, such errors may persist in the final model, as happened in the structure determination of the p53 oligomerisation domain [99]. This may occur in particular when there are few experimental restraints, so that the impact of an incorrect restraint will be considerably higher than in cases with many restraints. Therefore, the probability for such errors to occur can be reduced by collecting many restraints. Locally, for instance, in regions that are underdetermined by the data, and in interface regions of complexes with few intermolecular restraints, such problems may nevertheless persist and remain undetected.

A more informative indicator than the number of restraints per residue is the "completeness of NOEs" [98], which is calculated as the percentage of observed NOEs relative to the number of NOEs expected on the basis of the model. This quantity can be expressed both on a per-residue basis (local) and for the entire structure (global). In principle, all inter-proton distances shorter than 3.5 Å should be observed in NOE spectra [93]. In a subset of 97 NMR ensembles deposited in the PDB before 1998, the average completeness of NOEs was ~70% for distances up to 3 Å, ~50% for distances up to 4 Å and ~25% for distances up to 5 Å [98]. A higher degree of completeness of the NOEs was found to be correlated with a better Ramachandran score.

10.4.3.3 Validation Using Experimental Data

In recent years, quality factors based on residual dipolar couplings [87,100,101,102] have been shown to be useful for validation purposes, but their use is, as yet, not widespread. Residual dipolar couplings contain information about average bond orientations relative to a molecular alignment frame and can be measured (only) if the protein is weakly aligned.

The Q factor [87,100,101] is based on a comparison between experimentally derived and calculated residual dipolar couplings. A fraction of the measured couplings are omitted in the structure calculations and used to validate the models by comparing them to the couplings predicted by the models. The Q factor is expressed as a percentage and a lower value indicates better agreement between the measured and the predicted residual dipolar couplings. A Q factor of 36% was reported for a structure of the protein apo-S100B(ßß), whereas an older model of the same protein scored 62% for the same set of residual dipolar couplings [103]. The Q factor can be used to validate any structure model, provided that the residual dipolar couplings can be measured. High-resolution crystal structures typically score ~20-25%, whereas a 2.5 Å model scores ~40%. The backbone generally yields much lower Q factors than the side chains [87].

Related to the Q factor are the dipolar R-factor and free R-factor [102]. These R-factors are similar to their crystallographic counterparts only in name, because a single reflection in a crystallographic data set contains information about the entire structure, whereas a single NMR restraint usually contains information only about the local structure. Note that the dipolar R factors and the Q factor differ only by a factor of 2 [87].

The reliability of these quality factors is greatest when complete cross-validation is used, i.e., the structure calculation is repeated several times, omitting a different small subset of the residual dipolar couplings at each calculation, that is then used to validate the calculated model [102].

10.4.4 VALIDATION DATABASES

In this section, we describe a few databases that contain validation-related information for many or all structures in the PDB. A tutorial on the use of these databases is available on the web at URL http://xray.bmc.uu.se/embo2001/modval. The URLs for the resources discussed in this and the next section are also listed in Table 10.3.

- As mentioned earlier, the PDB [3] (http://www.pdb.org/) is essentially a repository for structural data and not a judge of the structural quality of its contents. The amount of validation-related data that is available for each entry is concomitantly small. It includes the resolution of the study as well as the conventional and free R-value, although more information can often be found in the header of the more recent PDB

TABLE 10.3
URLs of Structure-Validation Resources

Resource	URL
1. Validation Databases	
PDB	http://www.pdb.org
PDBsum	http://www.ebi.ac.uk/thornton-srv/databases/pdbsum
PDBReport	http://www.cmbi.kun.nl/gv/pdbreport
EDS	http://eds.bmc.uu.se
2. Validation Servers	
Biotech Validation Suite	http://biotech.ebi.ac.uk:8400
ADIT	http://pdb.rutgers.edu/validate
AQUA	http://www.nmr.chem.uu.nl/users/jurgen/Aqua/server/
STAN	http://xray.bmc.uu.se/cgi-bin/gerard/rama_server.pl
RP	http://dicsoft1.physics.iisc.ernet.in/rp
ERRAT	http://www.doe-mbi.ucla.edu/Services/ERRAT
VERIFY3D	http://www.doe-mbi.ucla.edu/Services/Verify_3D
MolProbity	http://kinemage.biochem.duke.edu/validation/valid.html
RamPage	http://www-cryst.bioc.cam.ac.uk/rampage
ANOLEA	http://www.sciences.fundp.ac.be/biologie/bms/CGI/test.htm

files. The PDB further provides some assessment of the quality of the geometry (adherence of bond lengths and angles as well as some torsion angles to standard values) of each entry [14]. The availability of structure factor data is also reported. If structure factors have been deposited, there is a good chance that the experimental electron density will be available at EDS (see below). Finally, the PDB provides links to a number of other sites that carry validation-related information, such as PDBsum, PDBReport, and EDS.

- PDBsum [20] (http://www.ebi.ac.uk/thornton-srv/databases/pdbsum) is a popular front-end to the PDB, since it provides a rich annotation of every entry as well as a comprehensive set of links to other resources. In terms of validation-related information, it lists the resolution and R-values and it shows a diagram of the secondary structure. Usually, around 60% of all residues in a protein are part of a regular secondary-structure element, although there are exceptions. PDBsum also includes a LIGPLOT diagram [104] showing the interactions between the protein and any ligands. If the fold of the protein model is correct, this should be reflected in the residues that are normally expected to interact with ligands, substrates, ions, etc. Finally, PDBsum provides access to a summary of the validation report created by PROCHECK [105]. This summary consists of a Ramachandran plot, statistics pertaining to the Ramachandran plot (where a good model has >90% residues in the core allowed areas and no more than 1 to 2% in the disallowed ones) and a list of G-factors that provide information as to how unusual various aspects of the model are. In the case of NMR models, only the statistics for the first model are provided.

- The PDBReport database [7] (http://www.cmbi.kun.nl/gv/pdbreport/) contains quality analyses carried out with the program WHAT IF [106] (or, rather, a subset called WHATCHECK [7]). At the top of each PDBReport page is a link to the full report that includes a huge number of checks and tests. The diagnostics come in three classes of stringency: note, warning and error. The most useful checks are the Ramachandran plot and several of the 3-D-database-related checks (including the packing scores, quality value plot and rotamers). At the bottom of each page are two summaries, one for users of a model (comparing the quality of the model to a set of high-resolution, reliable models) and one for the person who deposits the model (comparing the quality of this model to a set of structures solved at similar resolution). In particular the list of structure Z-scores in the first summary provides a quick overview of the overall quality of the model. For NMR ensembles, average per-model values are reported for many of the checks, which may help in comparing the quality of individual models.

- The Uppsala Electron-Density Server [13] (http://eds.bmc.uu.se/) provides information about the model and its fit to the experimental data. For ~85% of all crystal structures for which structure factors are available from the PDB, an electron-density map has been calculated and can be downloaded (together with the model) for inspection using programs such as O [66], WHAT IF, or DeepView [107]. In addition, real-space fit values, as well as several other quantities, have been calculated for all residues and these can be viewed as plots. The residues in these plots are clickable and doing so will start up a Java-based model and electron-density viewer. In this fashion, experts and nonexperts alike have access to the electron density for many of the crystal structures in the PDB and can thus obtain important information about the quality and reliability of a model and its details, Figure 10.4.

10.4.5 VALIDATION SERVERS

The resources discussed in the previous section can be used to assess the quality of models that are available from the PDB. However, when models are obtained from other sources (e.g., proprietary models from an in-house crystallography department or acquired from an external source), the user of the models will have to subject them to quality checks. This can be done either by acquiring software packages such as WHAT IF, WHATCHECK, PROCHECK or PROCHECK-NMR [97], SFCHECK [108], AQUA [97] and O, or by using web-based servers that run validation checks (which, obviously, involves submitting possibly proprietary information over probably insecure lines). Some of these servers are described in this section. The URLs for these resources are also listed in Table 10.3.

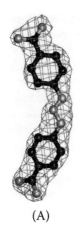 

(A)   (B)

FIGURE 10.4 Illustration of good and poor density for the same ligand in complex with very similar DNA molecules, both at ~2 Å resolution (carbon atoms are shown in black, nitrogen atoms in grey). The electron density (represented by the chicken-wire contours) covers the entire ligand in (A), leading to a real-space R-value of 0.11. In (B), however, there is much less density to support the presence and orientation of the ligand, and consequently its fit to the density is poor (real-space R-value 0.44).

- The Biotech Validation Suite server [109] (http://biotech.ebi.ac.uk:8400/) is the result of a collaboration involving half a dozen laboratories. It enables users to upload a structure and to run any or all of the quality checks implemented in PROCHECK and WHATCHECK. Results can be obtained for each check separately, or in the form of a table with one row for every residue, for instance. This server produces by far the most detailed quality control analyses, is easy to use, and well-documented.
- The PDB provides access to the validation software that is used when a structure is submitted to it. This ADIT server [4] (http://pdb.rutgers. edu/validate/) will run PROCHECK and some of the PDB's own software. If a structure-factor file is provided, SFCHECK is also run.
- AQUA and PROCHECK-NMR are two linked programs that have been specifically developed for the validation of NMR data and structure ensembles. Both information on redundancy and violations of a restraint list compared to the model ensemble, and analyses and comparisons of the geometry of the models comprising the ensemble, can be obtained with these programs. Some functions of the program AQUA can be accessed using the AQUA server (http://www.nmr.chem.uu.nl/users/jurgen/ Aqua/server/), namely redundancy checks of restraint lists, checks of restraint violations by the model, and calculation of the completeness of the NOEs. For users of structural information to be able to use this server, the restraint list has to be available (e.g., from BMRB) along with the structure model(s), which is, unfortunately, not always the case.

- The STAN (STructure ANalysis) server (http://xray.bmc.uu.se/cgi-bin/ gerard/rama_server.pl) runs the program MOLEMAN2 and will produce a Ramachandran plot for each protein chain in a submitted structure [79]. It will also produce a "Cα-Ramachandran" plot [110] for each protein chain and can thus even be used for validation of models for which only Cα coordinates are available. Further, for each nucleic acid chain a Duarte-Pyle plot [111] will be generated. The server also runs two other useful programs, namely WASP and CISPEP. WASP [112] scrutinises water molecules and reports those that are more likely to be small cations (sodium, lithium, magnesium, or calcium). CISPEP [113] can be used to detect any nonproline cis-peptide bonds that have accidentally been modelled as trans-peptides.
- A web-based server called RP [114] (http://dicsoft1.physics.iisc.ernet.in/rp/) can be used to create and analyse PROCHECK-style Ramachandran plots. This server can also be used with existing PDB entries.
- ERRAT [115] (http://www.doe-mbi.ucla.edu/Services/ERRAT/) is a program that analyses the relative frequencies of nonbonded interactions between various combinations of atoms (CC, CN, CO, NN, NO and OO). By comparing these to a normal distribution derived from a set of high-resolution structures, the likelihood that a certain region of the protein is in error can be estimated. The result is a plot of this likelihood as a function of residue number, which aids in identifying possibly incorrect residues or regions. In addition, an overall score is calculated that correlates well with the Ramachandran score obtained from PROCHECK [116].
- VERIFY3D [117] (http://www.doe-mbi.ucla.edu/Services/Verify_3D/) calculates a measure of the compatibility of a protein structure and its own sequence. The environment of each residue in the structure is characterised by three parameters: buried surface area, fraction of sidechain area covered by polar atoms, and secondary structure type. The sequence is then used to look up tabulated values that express how favourable the environment is for each residue. These numbers are averaged in a window of 21 residues and plotted. Models that contain substantial tracing errors can be detected by low-scoring regions in the plot.
- The MolProbity server [118] (http://kinemage.biochem.duke.edu/validation/ valid.html) provides geometrical validation tools based on atomic coordinates alone. A structure can either be uploaded or retrieved from the PDB. The Cβ validation tool measures deviations of observed Cβ atomic positions from calculated ideal positions. Large deviations are indicative of problems in the model. The validation process also includes an updated Ramachandran plot with new definitions of favoured, allowed, and outlier regions based on conformational statistics from 500 high-resolution structures. The analysis of atomic overlaps, such as clashes and hydrogen bonds, is another feature of the server. The analysis requires that hydrogen atoms are present in the model, but they can also be added by the server. All the results can be viewed in 3-D (using Java applets), which helps in identifying problematic regions in a model, or obtained as plots. RamPage (http://www-cryst.bioc.cam.ac.uk/

rampage) is a separate server that provides the Ramachandran plot service alone. The plot is produced with the same definitions for favoured, allowed, and outlier regions as the MolProbity server.

- The Atomic NOn-Local Environment Assessment (ANOLEA [119]; http://www.sciences.fundp.ac.be/biologie/bms/CGI/test.htm) server uses an atomic mean force potential to assess protein structures. The input for the server is a structure and the output is an energy profile calculated for nonlocal interactions between the heavy atoms of the model. The profile contains an energy value for each amino acid residue within the model. Residues or regions showing high energy values may contain model errors.

10.5 CONCLUDING REMARKS

It would be interesting to get a quick look at the world of structure databases ten years from now. However, making firm predictions is a hazardous affair from which we shall refrain. Even though organic growth can be expected and predicted, science and technology often progress by quantum leaps, which by their very nature defy prediction. For instance, in 1990 few could have predicted the enormous impact of the World Wide Web, since it didn't even exist at the time. We therefore conclude with some of our hopes and expectations with respect to the further development of structural databases over the next decade.

The worlds of biomolecular sequence and structure databases are very different. In the sequence world, a number of centres exist that collect and distribute the primary sequence data and many of the derivative servers and databases are also maintained by these sites. In the structure world, on the other hand, only a single major repository for primary information exists, but 99% of the derivative servers and databases are provided and maintained by external laboratories. At the moment, there are probably already at least a hundred such servers and databases, each covering its own niche. In science, diversity is to be encouraged, even when it comes to different approaches to solving the same problem (e.g., the various approaches to classifying protein structures are all different and all have their specific strengths and shortcomings). However, we are reaching the stage where even for experts it is almost impossible to maintain an overview of what is available and where. For casual, nonexpert users the situation is even worse. We are, indeed, facing an "embarrassment of riches" [120].

If developments in other areas of science and technology are any guide (e.g., videotape formats, molecular modelling software companies, computer hardware manufacturers), the field will eventually mature and the most useful facilities will survive or be copied by others. Hopefully, the major database-distribution centres (RCSB, EBI) will absorb much of these useful facilities. This will ensure that these facilities are up-to-date and synchronised with the PDB and that they will be maintained permanently (and not cease to work upon the departure of a postdoc, or be off-line for a year while a group gets settled in a new laboratory). An even greater benefit would be that the information would be integrated with all other structural, functional, and other information in the database. It would then be accessible through a uniform interface and could be included in advanced database searches. This would ultimately enable complex queries that combine issues of structure, quality, function, etc. Equally importantly, nontrivial actions on the results of the query would become possible. One could then issue queries such as "retrieve all enzymes that use ATP and are not TIM-barrels; for all these enzymes, extract the active-site residues unless any of these residues fit the density poorly; display the phi, psi angles of these residues in a Ramachandran plot," or "list the various fold classes of all human proteins that have been implicated in Alzheimer's disease, that have less than 30% mutual sequence identity, and whose structures have been determined by NMR spectroscopy." At present, finding answers to such complex queries involves a substantial amount of expertise, knowledge of the various resources, and time and jockeying between various databases. Yet, for the community that wishes to use structural information (molecular biologists, medicinal chemists, etc.), such queries are the kinds of questions they want to find answers to. The kind of query that can be answered with present-day technology, on the other hand, is of the type: "what is the largest protein crystal structure determined by T. A. Jones after 1990 with resolution better than 3 Å and R-value less than 0.25?"

Enriching the annotation of the available structures is a process that should proceed in two directions. Inclusion of descriptive information at the level of the molecule, complex or domain, be it about structure, function, genetic variation, ligand binding, roles in disease states, etc., is crucial to answer ever more complex queries and discover new knowledge. On the other hand, the annotation also needs to be extended downwards to the level of residues. Whereas the fold of a protein is important as the structural framework, the actual work (binding or catalysis) is done by atoms in residues. Other residues may play a crucial role in maintaining the structural integrity of the fold, or in orienting an active-site residue, or in binding a cofactor or metal ion. Mutations in yet others may be associated with loss of function or give rise to a disease. Annotation at this level will be of use in two settings. First, when the structure of a protein of unknown function is solved and found to be similar to that of a protein whose structure is known, annotation of residues conserved in both structures may provide clues as to the function of the new protein (this volume, Chapter 9 by Laskowski). Second, such annotation can be used to improve the alignments of sequences to known structures by providing clues as to which residues play crucial roles and whose mutation or deletion is therefore less likely.

Besides making it easier for nonexpert users to obtain answers to nontrivial queries, these users also need to be made aware that even experimentally determined structures are models. They represent the experimenter's best attempt to explain the experimental data, but that does not mean they are to be believed to the third decimal place. Hopefully, quality-related information (e.g., in the form of statistics or expert annotation, or both) about all structure entries will soon be available directly from the main database centres. This will facilitate the task of the nonexpert user to evaluate the fitness of a given structure for the intended purposes and to find the most suitable structure when several alternatives are available. Global structural quality criteria (resolution, R-values, Ramachandran plot quality, packing scores, profile scores, etc.) can be used in the latter process. However, even within a model the quality varies from residue to residue and therefore local quality statistics will also be required for a proper evaluation. These should include both purely structural criteria and those that measure how well a set of atoms or residues fit the experimental data (e.g., real-space fit values for crystal structures and restraint violations for NMR structures).

ACKNOWLEDGEMENTS

The authors wish to thank Prof. T. Alwyn Jones and Marian Novotny (Uppsala University) for a critical reading of the manuscript, Dr. Kim Henrick and Dr. Roman Laskowski (EBI) for insightful comments, and Terese Bergfors (Uppsala University) for indispensable linguistic comments and corrections. Henrik Hansson was supported by the Swedish Structural Biology Network (SBNet). Gerard J. Kleywegt is a Royal Swedish Academy of Sciences (KVA) Research Fellow, supported through a grant from the Knut and Alice Wallenberg Foundation. This work was supported by KVA, SBNet, Uppsala University and its Linnaeus Centre for Bioinformatics.

REFERENCES

1. AD Baxevanis, The Molecular Biology Database Collection: 2003 Update, *Nucleic Acids Res.*, 31, 1–12, 20 03.
2. FC Bernstein, TF Koetzle, GJB Williams, EF Meyer Jr, MD Brice, JR Rodgers, O Kennard, T Shimanouchi, and M Tasumi, The Protein Data Bank: a computer-based archival file for macromolecular structures, *J. Mol. Biol.*, 112, 535–542, 1977.
3. HM Berman, J Westbrook, Z Feng, G Gilliland, TN Bhat, H Weissig, IN Shindyalov, and PE Bourne, The Protein Data Bank, *Nucleic Acids Res.*, 28, 235–242, 2000.
4. HM Berman, T Battistuz, TN Bhat, WF Bluhm, PE Bourne, K Burkhardt, Z Feng, GL Gilliland, L Iype, S Jain, P Fagan, J Marvin, D Padilla, V Ravichandran, B Schneider, N Thanki, H Weissig, JD Westbrook, and C Zardecki, The Protein Data Bank, *Acta Crystallogr.,* D 58, 899–907, 2002.
5. EF Meyer, The first years of the Protein Data Bank, *Prot. Sci.,* 6, 1591–1597, 1997.
6. Protein Data Bank, *Acta Crystallogr.,* B29, 1746, 1973.
7. RWW Hooft, G Vriend, C Sander, and EE Abola, Errors in protein structures, *Nature,* 381, 272–272, 1996.
8. TN Bhat, P Bourne, Z Feng, G Gilliland, S Jain, V Ravichandran, B Schneider, K Schneider, N Thanki, H Weissig, J Westbrook, and HM Berman. The PDB data uniformity project, *Nucleic Acids Res.*, 29, 214–218, 2001.
9. J Westbrook, Z Feng, S Jain, TN Bhat, N Thanki, V Ravichandran, GL Gilliland, WF Bluhm, H Weissig, DS Greer, PE Bourne, and HM Berman, The Protein Data Bank: unifying the archive, *Nucleic Acids Res.*, 30, 245–248, 2002.
10. H Boutselakis, D Dimitropoulos, J Fillon, A Golovin, K Henrick, A Hussain, J Ionides, M John, PA Keller, E Krissinel, P McNeil, A Naim, R Newman, T Oldfield, J Pineda, A Rachedi, J Copeland, A Sitnov, S Sobhany, A Suarez-Uruena, J Swaminathan, M Tagari, J Tate, S Tromm, S Velankar, and W Vranken, E-MSD: the European Bioinformatics Institute Macromolecular Structure Database, *Nucleic Acids Res.*, 31, 458–462, 2003.
11. HM Berman, J Westbrook, Z Feng, L Iype, B Schneider, and C Zardecki, The Nucleic Acid Database, *Acta Crystallogr.,* D 58, 889–898, 2002.
12. BR Seavey, EA Farr, WM Westler, and JL Markley, A relational database for sequence-specific protein NMR data, *J. Biomol. NMR*, 1, 217–236, 1991.

13. GJ Kleywegt, MR Harris, JY Zou, TC Taylor, A Wählby, and TA Jones, The Uppsala Electron Density Server, *Acta Crystallogr.*, D60: 2240–2249, 2004.

14. H Weissig and PE Bourne. An analysis of the Protein Data Bank in search of temporal and global trends, *Bioinformatics*, 15, 807–831, 1999.

15. FH Allen, S Bellard, MD Brice, BA Cartwright, A Doubleday, H Higgs, T Hummelink, BG Hummelink-Peters, O Kennard, WDS Motherwell, JR Rodgers, and DG Watson, The Cambridge Crystallographic Data Centre: computer-based search, retrieval, analysis, and display of information, *Acta Crystallogr.*, B35, 2331–2339, 1979.

16. FH Allen, The Cambridge Structural Database: a quarter of a million crystal structures and rising, *Acta Crystallogr.*, B 58, 380–388, 2002.

17. GJ Kleywegt, K Henrick, EJ Dodson, and DMF van Aalten, Pound-wise but penny-foolish: How well do micromolecules fare in macromolecular refinement? Structure 11, 1051–1059, 2003.

18. EF Meyer Jr., Storage and retrieval of macromolecular structural data. *Biopolymers*, 13, 419–422, 1974.

19. WR Pearson and DJ Lipman, Improved tools for biological sequence comparison, *Proc. Natl. Acad. Sci. U.S.*, 85, 2444–2448, 1988.

20. RA Laskowski, PDBsum: summaries and analyses of PDB structures, *Nucleic Acids Res.*, 29, 221–222, 2001.

21. Y Wang, KJ Addess, L Geer, T Madej, A Marchler-Bauer, D Zimmerman, and SH Bryant, MMDB: 3D structure data in Entrez, *Nucleic Acids Res.*, 28, 243–245, 2000.

22. Y Wang, JB Anderson, J Chen, LY Geer, S He, DI Hurwitz, CA Liebert, T Madej, GH Marchler, A Marchler-Bauer, AR Panchenko, BA Shoemaker, JS Song, PA Thiessen, RA Yamashita, and SH Bryant, MMDB: Entrez's 3D-structure database, *Nucleic Acids Res.*, 30, 249–252, 2002.

23. J Reichert, A Jabs, P Slickers, and J Suhnel, The IMB Jena Image Library of biological macromolecules, *Nucleic Acids Res.*, 28, 246–249, 2000.

24. RW Hooft, C Sander, M Scharf, and G Vriend, The PDBFINDER database: a summary of PDB, DSSP and HSSP information with added value, *Comput. Appl. Biosci.*, 12, 525–529, 1996.

25. T Etzold, A Ulyanov, and P Argos, SRS: information retrieval system for molecular biology data banks, *Methods Enzymol.*, 266, 114–128, 1996.

26. L Holm and C Sander, The FSSP database of structurally aligned protein fold families, *Nucleic Acids Res.*, 22, 3600–3609, 1994.

27. AG Murzin, SE Brenner, T Hubbard, and C Chothia, SCOP: a structural classification of proteins database for the investigation of sequences and structures, *J. Mol. Biol.*, 247, 536–540, 1995.

28. CA Orengo, AD Michie, S Jones, DT Jones, MB Swindells, and JM Thornton, CATH: a hierarchic classification of protein domain structures, *Structure*, 5, 1093–1108, 1997.

29. C Hadley and DT Jones, A systematic comparison of protein structure classifications: SCOP, CATH and FSSP, *Structure*, 7, 1099–1112, 1999.

30. M Novotny, D Madsen, and GJ Kleywegt, An evaluation of protein-fold-comparison servers, *Proteins Struct. Funct. Genet.*, 54, 260–270, 2004.

31. AC Wallace, RA Laskowski, and JM Thornton, Derivation of 3D coordinate templates for searching structural databases: Application to Ser-His-Asp catalytic triads in the serine proteases and lipases, *Prot. Sci.*, 5, 1001–1013, 1996.

32. GJ Kleywegt, Recognition of spatial motifs in protein structures, *J. Mol. Biol.*, 285, 1887–1897, 1999.

33. D Madsen and GJ Kleywegt, Interactive motif and fold recognition in protein structures, *J. Appl. Crystallogr.*, 35, 137–139, 2002.

34. A Stark, S Sunyaev, and RB Russell, A model for statistical significance of local similarities in structure, *J. Mol. Biol.*, 326, 1307–1316, 2003.

35. M Hendlich, Databases for protein-ligand complexes, *Acta Crystallogr.*, D54, 1178–1182, 1998.

36. GJ Kleywegt and TA Jones, Databases in protein crystallography, *Acta Crystallogr.*, D54, 1119–1131, 1998.

37. AS Siddiqui, U Dengler, and GJ Barton, 3Dee: a database of protein structural domains, *Bioinformatics*, 17, 200–201, 2001.

38. N Echols, D Milburn, and M Gerstein, MolMovDB: analysis and visualization of conformational change and structural flexibility, *Nucleic Acids Res.*, 31, 478–482, 2003.

39. KN Degtyarenko, AC North, and JB Findlay, PROMISE: a database of bioinorganic motifs, *Nucleic Acids Res.*, 27, 233–236, 1999.

40. C Sander and R Schneider, Database of homology-derived protein structures and the structural meaning of sequence alignment, *Proteins Struct. Funct. Genet.*, 9, 56–68, 1991.

41. K Henrick and JM Thornton, PQS: a protein quaternary structure file server, *Trends Biochem. Sci.*, 23, 358–361, 1998.

42. DF Burke, CM Deane, and TL Blundell, Browsing the SLoop database of structurally classified loops connecting elements of protein secondary structure, *Bioinformatics*, 16, 513–519, 2000.

43. LC Allcorn and AC Martin, SACS — self-maintaining database of antibody crystal structure information, *Bioinformatics*, 18, 175–181, 2002.

44. J Vondrasek and A Wlodawer, HIVdb: a database of the structures of human immunodeficiency virus protease, *Proteins Struct. Funct. Genet.*, 49, 429–431, 2002.

45. VL Murthy and GD Rose, RNABase: an annotated database of RNA structures, *Nucleic Acids Res.*, 31, 502–504, 2003.

46. O Ivanciuc, CH Schein, and W Braun, SDAP: database and computational tools for allergenic proteins, *Nucleic Acids Res.*, 31, 359–362, 2003.

47. PM Coutinho and B Henrissat, Carbohydrate-active enzymes: an integrated database approach, in HJ Gilbert, G Davies, B Henrissat, and B Svensson, Eds., *Recent Advances in Carbohydrate Bioengineering*, The Royal Society of Chemistry, Cambridge, 1999, 3–12.

48. X Cousin, T Hotelier, K Giles, P Lievin, JP Toutant, and A Chatonnet, The α/β fold family of proteins database and the cholinesterase gene server ESTHER, *Nucleic Acids Res.*, 25, 143–146, 1997.

49. F Horn, G Vriend, and FE Cohen, Collecting and harvesting biological data: the GPCRDB and NucleaRDB information systems, *Nucleic Acids Res.*, 29, 346–349, 2001.

50. F Horn, E Bettler, L Oliveira, F Campagne, FE Cohen, and G Vriend, GPCRDB information system for G protein-coupled receptors, *Nucleic Acids Res.*, 31, 294–297, 2003.

51. SA Islam, D Carvin, MJ Sternberg, and TL Blundell, HAD, a data bank of heavy-atom binding sites in protein crystals: a resource for use in multiple isomorphous replacement and anomalous scattering, *Acta Crystallogr.*, D 54, 1199–1206, 1998.

52. GL Gilliland, M Tung, DM Blakeslee, and J Ladner, The Biological Macromolecule Crystallization Database, Version 3.0: new features, data, and the NASA Archive for Protein Crystal Growth Data, *Acta Crystallogr.*, D50, 408–413, 1994.

53. JS Garavelli, Z Hou, N Pattabiraman, and RM Stephens, The RESID Database of protein structure modifications and the NRL-3D Sequence-Structure Database, *Nucleic Acids Res.*, 29, 199–201, 2001.

54. G Wang and RL Dunbrack, PISCES: a protein sequence culling server, *Bioinformatics*, 12, 1589–1591, 2003.

55. U Pieper, N Eswar, AC Stuart, VA Ilyin, and A Sali, MODBASE, a database of annotated comparative protein structure models, *Nucleic Acids Res.*, 30, 255–259, 2002.

56. AC Stuart, VA Ilyin, and A Sali, LigBase: a database of families of aligned ligand binding sites in known protein sequences and structures, *Bioinformatics*, 18, 200–201, 2002.

57. A Yamaguchi, M Iwadate, E Suzuki, K Yura, S Kawakita, H Umeyama, and M Go, Enlarged FAMSBASE: protein 3D structure models of genome sequences for 41 species, *Nucleic Acids Res.*, 31, 463–468, 2003.

58. DW Buchan, SC Rison, JE Bray, D Lee, F Pearl, JM Thornton, and CA Orengo, Gene3D: structural assignments for the biologist and bioinformaticist alike, *Nucleic Acids Res.*, 31, 469–473, 2003.

59. J Gough, The SUPERFAMILY database in structural genomics, *Acta Crystallogr.*, D 58, 1897–1900, 2002.

60. J Qian, B Stenger, CA Wilson, J Lin, R Jansen, SA Teichmann, J Park, WG Krebs, H Yu, V Alexandrov, N Echols, and M Gerstein, PartsList: a web-based system for dynamically ranking protein folds based on disparate attributes, including whole-genome expression and interaction information, *Nucleic Acids Res.*, 29, 1750–1764, 2001.

61. J Westbrook, Z Feng, L Chen, H Yang, and HM Berman, The Protein Data Bank and structural genomics, *Nucleic Acids Res.*, 31, 489–491, 2003.

62. CI Brändén and TA Jones, Between objectivity and subjectivity, *Nature*, 343, 687–689, 1990.

63. GJ Kleywegt, Validation of protein crystal structures, *Acta Crystallogr.*, D56, 249–265, 2000.

64. GJ Kleywegt and TA Jones, Where freedom is given, liberties are taken, *Structure*, 3, 535–540, 1995.

65. GJ Kleywegt, Use of non-crystallographic symmetry in protein structure refinement, *Acta Crystallogr.*, D52, 842–857, 1996.

66. TA Jones, JY Zou, SW Cowan, and M Kjeldgaard, Improved methods for building protein models in electron density maps and the location of errors in these models, *Acta Crystallogr.*, A47, 110–119, 1991.

67. GJ Kleywegt and AT Brünger, Checking your imagination: applications of the free R value, *Structure*, 4, 897–904, 1996.

68. E Dodson, GJ Kleywegt, and KS Wilson, Report of a workshop on the use of statistical validators in protein X-ray crystallography, *Acta Crystallogr.*, D52, 228–234, 1996.

69. GJ Kleywegt and TA Jones, Model-building and refinement practice, *Methods Enzymol.*, 277, 208–230, 1997.

70. TA Jones and M Kjeldgaard, Electron density map interpretation, *Methods Enzymol.*, 277, 173–208, 1997.

71. P Schultze and J Feigon, Chirality errors in nucleic acid structures, *Nature*, 387, 668–668, 1997.

72. U Das, S Chen, M Fuxreiter, AA Vaguine, J Richelle, HM Berman, and SJ Wodak, Checking nucleic acid crystal structures, *Acta Crystallogr.*, D57, 813–828, 2001.

73. DMF van Aalten, R Bywater, JBC Findlay, M Hendlich, RWW Hooft, and G Vriend, PRODRG, a program for generating molecular topologies and unique molecular descriptors from coordinates of small molecules, *J. Comput. Aided Mol. Design,* 10, 255–262, 1996.

74. AM Davis, SJ Teague, and GJ Kleywegt, Applications and limitations of X-ray crystallographic data in structure-based ligand and drug design, *Angew Chem. Int. Ed.,* 42, 2718–2736, 2003.

75. GJ Kleywegt, H Hoier, and TA Jones, A re-evaluation of the crystal structure of chloromuconate cycloisomerase, *Acta Crystallogr.,* D52, 858–863, 1996.

76. O Herzberg and J Moult, Analysis of steric strain in the polypeptide backbone of protein models, *Proteins Struct. Funct. Genet.,* 11, 223–229, 1991.

77. AT Brünger, Free *R* value: a novel statistical quantity for assessing the accuracy of crystal structures, *Nature,* 355, 472–475, 1992.

78. R Lüthy, JU Bowie, and D Eisenberg, Assessment of protein models with three-dimensional profiles, *Nature,* 356, 83–85, 1992.

79. GJ Kleywegt and TA Jones, Phi/Psi-chology: Ramachandran revisited, *Structure,* 4, 1395–1400, 1996.

80. RA Laskowski, MW MacArthur, and JM Thornton, Evaluation of protein coordinate data sets, in S Bailey, R Hubbard, D Waller, Eds., *From First Map to Final Model,* SERC Daresbury Laboratory, Warrington, 1994, 149–159.

81. C Ramakrishnan and GN Ramachandran, Stereochemical criteria for polypeptide and protein chain conformations. II. Allowed conformations for a pair of peptide units, *Biophys. J.,* 5, 909–933, 1965.

82. J Janin, S Wodak, M Levitt, and B Maigret, Conformation of amino acid side-chains in proteins, *J. Mol. Biol.,* 125, 357–386, 1978.

83. JW Ponder and FM Richards, Tertiary templates for proteins. Use of packing criteria in the enumeration of allowed sequences for different structural classes, *J. Mol. Biol.,* 193, 775–791, 1987.

84. G Vriend and C Sander, Quality control of protein models: directional atomic contact analysis, *J. Appl. Crystallogr.,* 26, 47–60, 1993.

85. J Kuszewski, AM Gronenborn, and GM Clore, Improving the quality of NMR and crystallographic protein structures by means of a conformational database potential derived from structure databases, *Protein Sci.,* 5, 1067–1080, 1996.

86. JF Doreleijers, G Vriend, ML Raves, and R Kaptein, Validation of nuclear magnetic resonance structures of proteins and nucleic acids: hydrogen geometry and nomenclature, *Proteins Struct. Funct. Genet.,* 37, 404–416, 1999.

87. A Bax, Weak alignment offers new NMR opportunities to study protein structure and dynamics, *Protein Sci.,* 12, 1–16, 2003.

88. GM Clore, MA Robien, and AM Gronenborn, Exploring the limits of precision and accuracy of protein structures determined by nuclear magnetic resonance spectroscopy, *J. Mol. Biol.,* 231, 82–102, 1993.

89. D Zhao and O Jardetzky, An assessment of the precision and accuracy of protein structures determined by NMR. Dependence on distance errors, *J. Mol. Biol.,* 239, 601–607, 1994.

90. K Wüthrich, *NMR of Proteins and Nucleic Acids,* John Wiley & Sons, New York, 1988.

91. GM Clore and AM Gronenborn, Determination of three-dimensional structures of proteins and nucleic acids in solution by nuclear magnetic resonance spectroscopy. *Crit. Rev. Biochem. Mol. Biol.,* 24, 479–564, 1989.

92. GM Clore and AM Gronenborn, NMR structure determination of proteins and protein complexes larger than 20 kDa, *Curr. Opinions Chem. Biol.*, 2, 564–570, 1998.

93. GM Clore and AM Gronenborn, New methods of structure refinement for macromolecular structure determination by NMR, *Proc. Natl. Acad. Sci. U.S.*, 95, 5891–5898, 1998.

94. P Güntert, Structure calculation of biological macromolecules from NMR data, *Q. Rev. Biophys.*, 31, 145–237, 1998.

95. MW MacArthur and JM Thornton, Conformational analysis of protein structures derived from NMR data, *Proteins Struct. Funct. Genet.*, 17, 232–251, 1993.

96. JF Doreleijers, JAC Rullman, and R Kaptein, Quality assessment of NMR structures: a statistical survey, *J. Mol. Biol.*, 281, 149–164, 1998.

97. RA Laskowski, JAC Rullmann, MW MacArthur, R Kaptein, and JM Thornton, AQUA and PROCHECK-NMR: programs for checking the quality of protein structures solved by NMR, *J. Biomol. NMR*, 8, 477–486, 1996.

98. JF Doreleijers, ML Raves, T Rullmann, and R Kaptein, Completeness of NOEs in protein structure: a statistical analysis of NMR, *J. Biomol. NMR*, 14, 123–132, 1999.

99. GM Clore, JG Omichinski, K Sakaguchi, N Zambrano, H Sakamoto, E Appella, and AM Gronenborn, Interhelical angles in the solution structure of the oligomerization domain of p53: correction, *Science*, 267, 1515–1516, 1995.

100. G Cornilescu, J Marquardt, M Ottiger, and A Bax, Validation of protein structure from anisotropic carbonyl chemical shifts in a dilute liquid crystalline phase, *J. Am. Chem. Soc.*, 120, 6836–6837, 1998.

101. M Ottiger and A Bax, Characterization of magnetically oriented phospholipid micelles for measurement of dipolar couplings in macromolecules, *J. Biomol. NMR*, 12, 361–372, 1998.

102. DS Garrett and GM Clore, R-factor, free R, and complete cross-validation for dipolar coupling refinement of NMR structures, *J. Am. Chem. Soc.*, 121, 9008–9012, 1999.

103. AC Drohat, N Tjandra, DM Baldisseri, and DJ Weber, The use of dipolar couplings for determining the solution structure of rat apo-S100B($\beta\beta$), *Protein Sci.*, 8, 800–809, 1999.

104. AC Wallace, RA Laskowski, and JM Thornton, LIGPLOT: a program to generate schematic diagrams of protein-ligand interactions, *Prot. Eng.*, 8, 127–134, 1995.

105. RA Laskowski, MW MacArthur, DS Moss, and JM Thornton, PROCHECK: a program to check the stereochemical quality of protein structures, *J. Appl. Crystallogr.*, 26, 283–291, 1993.

106. G Vriend, WHAT IF: a molecular modeling and drug design program, *J. Mol. Graph.*, 8, 52–56, 1990.

107. N Guex and MC Peitsch, SWISS-MODEL and the Swiss-PdbViewer: an environment for comparative protein modeling, *Electrophoresis*, 18, 2714–2723, 1997.

108. AA Vaguine, J Richelle, and SJ Wodak, *SFCHECK*: a unified set of procedures for evaluating the quality of macromolecular structure-factor data and their agreement with the atomic model, *Acta Crystallogr.*, D55, 191–205, 1999.

109. EU 3-D Validation Network. Who checks the checkers? Four validation tools applied to eight atomic resolution structures, *J. Mol. Biol.*, 276, 417–436, 1998.

110. GJ Kleywegt, Validation of protein models from Cα coordinates alone, *J. Mol. Biol.*, 273, 371–376, 1997.

111. CM Duarte and AM Pyle, Stepping through an RNA structure: a novel approach to conformational analysis, *J. Mol. Biol.*, 284, 1465–1478, 1998.

112. M Nayal and E Di Cera, Valence screening of water in protein crystals reveals potential Na+ binding sites, *J. Mol. Biol.*, 256, 228–234, 1996.

113. MS Weiss and R Hilgenfeld, A method to detect nonproline cis peptide bonds in proteins, *Biopolymers*, 50, 536–544, 1999.

114. SS Sheik, P Sundararajan, AS Hussain, and K Sekar, Ramachandran plot on the web, *Bioinformatics,* 18, 1548–1549, 2002.

115. C Colovos and TO Yeates, Verification of protein structures: patterns of nonbonded atomic interactions, *Prot. Sci.,* 2, 1511–1519, 1993.

116. O Dym, D Eisenberg, and TO Yeates, Detection of errors in protein models, in MG Rossmann, E Arnold, Eds., *International tables for Crystallography. Volume F. Crystallography of Biological Macromolecules*, Kluwer Academic Publishers, Dordrecht, 2001, 520–525.

117. D Eisenberg, R Lüthy, and JU Bowie, VERIFY3D: Assessment of protein models with three-dimensional profiles, *Methods Enzymol.,* 277, 396–404, 1997.

118. SC Lovell, IW Davis, WB Arendall, 3rd, PI De Bakker, JM Word, MG Prisant, JS Richardson, and DC Richardson, Structure validation by Cα geometry: ϕ, ψ and Cβ deviation, *Proteins Struct. Funct. Genet.*, 50, 437–450, 2003.

119. F Melo and E Feytmans, Assessing protein structures with a non-local atomic interaction energy, *J. Mol. Biol.*, 277, 1141–1152, 1998.

120. O Carugo and S Pongor, The evolution of structural databases, *Trends Biotechnol.*, 20, 498–501, 2002.

121. GJ Kleywegt, T Bergfors, H Senn, P Le Motte, B Gsell, K Shudo, and TA Jones, Crystal structures of cellular retinoic acid binding proteins I and II in complex with all-*trans*-retinoic acid and a synthetic retinoid, *Structure*, 2, 1241–1258, 1994.

122. R Koradi, M Billeter, and K Wüthrich, MOLMOL: a program for display and analysis of macromolecular structures, *J. Mol. Graph.*, 14, 51–55, 1996.

11 Problems in Computational Structural Genomics

Ruben Abagyan

CONTENTS

11.1 INTRODUCTION AND MOTIVATION

Protein X-ray crystallography and NMR continue to double the number of biological structures at atomic resolution approximately every three years and that pace is not showing any signs of slowing. The January 6, 2004 release of the Protein Data Bank [1] (PDB) contained 23,813 structures and over 5000 structures were added in 2004. Most of these structures were determined with X-ray crystallography (85%) and most of the remaining 15% were determined with NMR. An increasing number of structures came from structural genomics centers [2].

The main motivation for *computational* structural proteomics stems from the fact that these structures represent only a handful of snapshots of the Great Biological 3D Movie. The PDB contains about 4100 human protein entries, covering *parts of about 2000 genes* (conservatively less than 10% of human genes, and typically only partial coverage inside a gene). Many protein families are severely underrepresented: just one G-protein coupled receptor (bovine rhodopsin [3]), and only about 10% of protein kinases. For eukaryotic proteins or membrane proteins, the current situation is not going to be changed immediately by the structural genomics initiative since the expression-purification-crystallization cycle has much higher yield for simple and soluble bacterial proteins.

More importantly, if and when every single amino acid is covered by a structure due to *experimental* structural proteomics, some questions will remain and some of them will have to be answered computationally:

- The structures need to be better refined, especially at low resolutions, built into complete full atom models, and properly annotated
- Macromolecules associate, sometimes in multiple and combinatorial ways, and these complexes need to be and can be predicted
- Structures undergo normal variations and variations upon binding (induced fit);
- Most proteins are molecular machines and have multiple functional states, it is difficult to capture some of those states in a crystal
- Bio-macromolecules can be rationally, i.e., computationally redesigned, and small molecules can be designed to interact with bio-macromolecules

11.1.1 FUNDING

While computational structural genomics can expand the amount and the quality of useful structural information by several orders of magnitude, virtually *all* of the funding to structural proteomics goes exclusively to the experimental effort. The computational effort is trivialized and reduced to "could you please compile a target database and post it on the web," or "help us to track the reagent purchasing" kind of assignments. Better refinement technology, protein docking, expanding models with functional predictions, homology modeling, etc. are simply not funded even at a 5% level of the total structural genomics spending. Here is my plea to you: tell your representative to vote "No" on the "All the money goes to the experimental structural proteomics sweatshops" proposal, give 10% to computational charities and they may return a 1000% interest. Of course, it is only an opinion, probably somebody up there knows better.

Structure and function prediction methods hold a unique place in structural proteomics. It is the computational approaches that can overcome the incompleteness and paucity of structural information, thus extending this information to the scale of the whole proteome. Several good reviews of the area were published recently [4–8]. Furthermore, these methods can guide molecular design to create new therapeutics [9], and molecular research tools, and parts for nanotechnology of the future.

In this chapter we will discuss the following tasks of computational proteomics:

- Improving structural models and extending structural information
- Building and evaluating homology models
- Predicting functional features from a set of coordinates, detecting binding sites
- Automation of structure based drug screening and design
- Predicting protein association and peptide-protein association
- Improving underlying physical models

The progress in those areas in conjunction with structural proteomics efforts holds the promise of revolutionizing our understanding of biological function and boosting rational molecular design.

11.2 COMPLETING, IMPROVING, AND CORRECTING EXPERIMENTAL STRUCTURES

Atomic models in the PDB are incomplete and are frequently peppered with errors and omissions [10,11]. Most of them can be partially or fully fixed with computational methods. The errors and omissions are method specific.

Crystallographic structures comprising 85% of PDB suffer from two sorts of errors: (i) uncertainties due to derivation of an atomic model from faceless electron density, which at typical resolutions does not show hydrogens and cannot tell apart two different atoms with roughly the same electron density; and (2) gradually growing man-made "fantasy spots" in deposited models with poor quality of phases and/or resolution worse than 2.1 to 2.2A [12]. Note that about 50% of all crystallographic structures in PDB (10,000 entries in 2003), have resolution worse than 2.1A. By 2.5–2.7A resolution, as many as one-third of the backbone peptide groups can be flipped [13].

In practice, for detailed predictions, such as molecular docking or surface calculations, one needs to build a new full atomic model from a partial set of heavy atom coordinates with some guesses. What further complicates this model building is the fact that some protein regions, from individual side-chains to large regions, can be genuinely moveable, but may become ordered upon association [14]. However, even in those cases, we need to fill the void created by omission of the disordered region in a model.

To create a workable full atom model from a set of coordinates deposited in PDB, we need to consider the following aspects of those sets.

11.2.1 Symmetry and Biological Unit

A biologically meaningful structure may differ from what we find in a PDB entry. The deposited coordinates represent a so-called "asymmetric unit," a set of coordinates that can be multiplied according to crystallographic symmetry to occupy 3-D space. This asymmetric unit can be both smaller than the actual biological molecule (e.g., entry 1c8e containing one copy of the viral coat protein that needs to be multiplied according to the viral symmetry), and greater than the actual biological molecule (e.g., entry 1j7n containing two independent copies of the monomeric toxin). In the late 90's, PDB started to add the biological unit annotation to the entries (see http://www.rcsb.org/pdb/biounit_tutorial.html). See Figure 11.1.

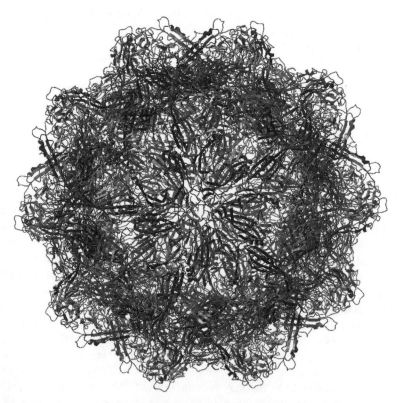

FIGURE 11.1 Biological structure built with the Molsoft-ICM-make-biomolecule function by applying 60 symmetry transformations to a single 534-residue coat protein in 1c8e entry.

11.2.2 ADDING MISSING OR ZERO-OCCUPANCY ATOMS, RESOLVING UNCERTAINTIES

If the electron density of a side-chain or a segment of protein chain is not strong enough, an honest crystallographer has a choice: either to create a *fantasy spot* and place atoms somehow with occupancy parameters set to zero, or skip these atoms/residues altogether. The omissions can occur because of genuine dynamics or experimental artifacts (e.g., loss of significant low-angle reflections). In both cases the atoms, the missing side-chains or loops need to be re-built. Prediction of side-chain conformations of the missing atoms can be done on the basis of energy optimization and is a relatively simple task [15–19].

We also need to deal with (i) 180° uncertainty in χ_2 angle of histidine and asparagine sidechains, and χ_3 angle of glutamine side chains (N and C are hard to distinguish by density at average resolutions); (ii) protonation state of histidines; (iii) charged state of aspartic and glutamic acids, lysine, and arginine (in some cases, e.g., in HIV protease, entry 1ida, Asp 25 and 25' are protonated); (iv) predict positions of rotatable polar hydrogens; (v) decide which water molecules are real; and (vi) decide what to do with deposited alternative conformations.

In most cases the above uncertainties can be solved by a short local energy optimization. This simulation may produce several possible local conformations. Multiple conformations can then be used for predictions around this protein model [20,21].

11.2.2.1 Detecting Errors: Energy Strain

The set of coordinates produced from the deposited coordinates and the above procedures can be further evaluated for errors. As a first approximation, one can use programs detecting atom clashes, strongly distorted covalent geometry or unusual values of torsion angles, as implemented in programs [22,23]. However, many errors are subtle: the covalent geometry may be perfect and all torsion angles acceptable, yet the total energy makes the conformation highly unlikely. The energy strain concept attempts to calculate a more sensitive error-indicator by comparing individual residue energies with the distribution of energies of the same residue type calculated in diverse refined high-resolution structures [24]. The normalized relative energies are sensitive to clashes and are best applied to energy-minimized structures. See Figure 11.2.

11.2.3 NMR STRUCTURES

Nuclear Magnetic Resonance (NMR) technology is the second main source of structural information (15% of the PDB), but the models derived from these data have more uncertainties due to the nature of the experiment. The main source of positional information in NMR is somewhat loose distance ranges between groups of protons, a long shot compared with precise locations of atomic density blobs in crystallography. Comparison of variation of residue contact geometry shows a threefold higher variation than normally expected in NMR structures [25]. Gert Vriend and colleagues have convincingly demonstrated that the NMR restraints are too loose

A   B   C

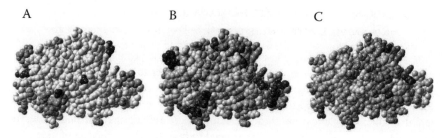

FIGURE 11.2 Thyroid hormone receptor beta 1 (1nq0) Resolution 2.4A. A. Residues with zero occupancies are shown in black, and residues with B-factors above 80 are shown in gray. B. Relative energy strain calculated with the calcEnergyStrain (From Maiorov, V. and R. Abagyan, Energy strain in three-dimensional protein structures, *Fold Des.*, 3, 4, 1998. With permission.) procedure: residues are shaded depending on relative strain. Red residues have energies more than nine standard deviations above the expected in good quality models. C. Restrained ICM side-chain energy optimization generates a better model.

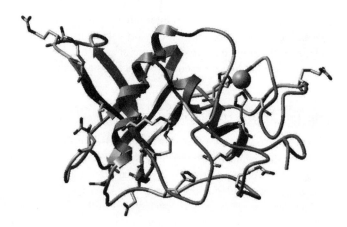

FIGURE 11.3 Problematic NMR structure of peptide deformylase (1def). The side chains with strongly distorted covalent geometries (e.g., xi5 angles of arginines deviate significantly from 0. or 180.) are shown.

to restrain the structure to a necessary level [26]. The majority of protein modelers who practice structure-based drug design prefer not to use NMR structures for docking, and crystallographers are hesitant about using NMR coordinates for molecular replacement (e.g., see the paper entitled "Does NMR mean 'Not for Molecular Replacement'?" [27]). The "structure in solution" argument should be separated from methodological uncertainties. It is true some parts are genuinely disordered, but this is also true for X-ray structures in which disordered loops are not visible. (See Figure 11.3.)

However, the situation can be radically improved with better quality experimental data and better energy optimization. It was recently demonstrated that a fully automated assignment of hetero-nuclear NOESY peaks followed by automated structure calculation using torsion dynamics produced a decent model stable to omission of up to 50% of cross peaks [28]. The NMR structures have much more to gain from a more rigorous restrained global energy optimization method that would optimize both the energy and NMR restraints. If such a procedure is validated, it can be applied to re-determine all NMR entries in PDB for which the experimental restraint files were deposited along with the coordinates. This may be particularly important for massive depositions originating from structural proteomics centers with a strong NMR component (e.g., RIKEN).

11.3 MODELING BY HOMOLOGY

Homology modeling plays a major role in structural biology. About 15,500 fragments of genomic sequences (65% of the total number of PDB entries) were present in PDB by the end of 2003, with the average length of a crystallized domain being around 230 residues. However, today over 1000 complete genomes have been sequenced with at least 500,000 unique genes with known or inferred function, and eukaryotic genes are frequently much larger than 230 residues. Comparative (or homology) modeling can bridge this 100 fold gap in sequence coverage and generate some useful structural models.

Given the current techniques available for homology modeling, the modeling tasks in most cases are either trivial or impossible. Precomputing a database of models doesn't make sense because generating a model on-the-fly is trivial and only improves in accuracy with an increased number of PDB structures. The current methodology consists of the following steps [29]:

Step 1. *Finding the template*, which is usually a PDB entry with the higher sequence similarity and better resolution. If A is a 2.5A-resolution template that is 90% identical to your query Q, and B is a 1.8A-resolution template of 80% sequence identity, you have a touch choice to make (I may choose B). In some cases you may prepare a hybrid template in which some domains or loops are grafted to the main template from fragments of other structures that happen to have better *local* similarity. This grafting can be done through a series of loop/domain end superpositions and adjustments. Once the template, single or chimeric, is prepared, you are ready to go to step 2.

Step 2. *Building an alignment* between your genomic sequence and the template sequence. Despite sophisticated and impressive efforts to improve the alignments (e.g., [30–32]), we have to confess that even though these methods may improve the *recognition* of distance homologs, most of the *alignments* will be either *trivial* or will be highly *problematic* for any algorithm [33]. We really need to zoom in on the problematic areas to see the difference between a good and a bad algorithm. Recently Brian Marsden and I tried to improve the alignments using an optimized mixture of sequence and structural information, and the gain we reached was probably not worth the effort. You may still be unlucky and the misaligned area will coincide with the area of interest. Usually, if the alignment is as highly ambiguous, as described in [30], the model is unusable for anything serious (e.g., pocket prediction, protein-, peptide- or ligand-docking). Furthermore, for some structurally dissimilar parts of two homologous structures, the very *notion of alignment breaks down*. How do you structurally align two helices hanging off conserved cores, if one helix is rotated by 20 degrees, tilted and shifted by 2A? So, let us not worry too much about distant homologues and do not build these models.

Step 3. Marking the loop boundaries (deformation regions), copying the backbone for non-loops and predicting the side-chain conformations. Out of these three sub-steps only the first one is nontrivial: the deformation regions are not merely sequence insertions; they may occur in the middle of a perfectly aligned fragment [29]. However, this difficulty is usually ignored and postulated to be just that. The side-chain placement step attracts a lot of undeserved attention (e.g., [18]). The problem has essentially been solved six to ten years ago [15,16,34–36]. This is the most pleasant part of the whole comparative modeling procedure, because, yes, it works fine for most of the side-chains.

Step 4. *Predicting loops.* There are loops and looo..ooops. Short loops (shorter than 7–8 residues) can be found in the PDB database or predicted in a relatively short simulation using internal coordinates with loop closure for sampling [37-39]. However, if the template entry points are distorted compared to the unknown correct backbone geometry, even short loops can go wrong. As loops get longer (say, larger than 12 residues), the complexity of the problem grows exponentially [40] and becomes essentially a *de novo* structure prediction problem: a still unsolved problem of computational biophysics. For multiple reasons it is still better to have a complete model with an incorrect loop occupying some space, as long as this loop is annotated as having "zero occupancy," rather than having a void.

Step 5. *Predicting how the inherited backbone deviates from the template.* This fails with rare exceptions. Usually the backbone moves away from the right answer. Generally it is better to keep the backbone exactly as in the template. An extreme illustration of this point would be refining the backbone of the template itself. In this case the desired outcome would be that the backbone does not move at all, but it generally moves by about 1.5A.

Step 6. *Predicting local quality of the homology model* is a critical step that dramatically increases the value of the model. In contrast to the alignments and loop predictions where it is either trivial or too difficult, the reliability prediction is actually doable. There is a big difference between knowing that half of the spots on the surface are wrong and knowing which particular spots are wrong. Let us consider the reliability prediction in more detail.

11.3.1 Predicting Local Model Reliability

Errors in models by homology stem from two main sources: (i) errors in the template itself; and (ii) the alignment mediated homology modeling process. If the template is a low resolution model, and/or has a wrong side chain or "fantasy spot," the model is not going to be any better in this area even if the alignment is 100% correct. For that reason choosing the template and between quality of the template and sequence distance from the template is a delicate exercise. Our goal is to predict where and by what extent the backbone conformation, and, consequently, local residue contacts deviate from the correct structure.

The main local error determinants can be classified as follows [41]:

Template errors: window and space averaged *local B-factor* in the template.

Sequence similarity in a window in a query sequence-template alignment.

Contact sequence similarity of a sequence fragment. If we find all residues in the alignment that are *in 3-D contact* with the fragment of interest, they can form a pseudo-sequence. The level of conservation of this contact pseudo-sequence is a strong predictor of a possible backbone deviation.

3-D-distance-to-gap: the nearest distance, as measured on the 3-D template, between the trial fragment and projections of alignment gaps on the structure. This measure implies a similar idea as the previous determinant: if you are next to a bad region, the prediction is going to be bad.

Local spatial density measured as the number of atoms in a sphere layer around the trial fragment. Low density means that the fragment is exposed and may deviate more easily.

See Figure 11.4.

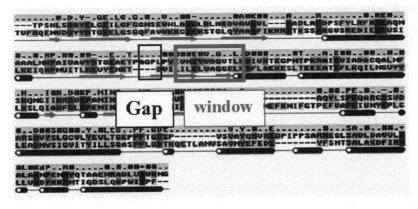

FIGURE 11.4 Window alignment, through-space proximity to gaps, and sequence conservation of residues in 3-D-contact with the window fragment are key determinants of the model errors.

In order to derive a combined function for predicting local deviations of the model backbone from its template, we studied the distribution of the deviation determinants between homologous pairs of protein structures and analyzed Log–Log correlations between each parameter and the observed deviation. A window of 10 residues was used. Meaningful ranges for the parameters were identified, and weights for the corresponding statistical pseudo-energy terms were derived. The final deviation score has more than twice the predictive power than local sequence identity or B-factor alone.

The backbone error score predicted on the basis of the template structure and the local quality of the sequence-structure alignment might further be combined with the energy strain, which evaluates a set of coordinates on their own merit regardless of their source. Now the model is built and locally annotated. If the area of immediate interest has favorable scores, this area of the model can be for further predictions, e.g., small molecule docking or protein docking.

11.3.2 HOW GOOD ARE HOMOLOGY MODELS?

The answer to this frequently asked question is simple: some homology models are better than experimental structures and some are worse. It is important to recognize that the so-called experimental coordinates are also models built via interpretation, in many case ambiguous, of experimental data such as density maps or resonance peaks. The process of experimental model building includes assignments, sampling or positional optimization, and in some cases arbitrary decisions.

Chotia and Lesk in 1986 [42] established a linear dependence between the Logarithm of the root-mean-square deviation (RMSD) and the fraction of mutated residues. Let us define Quality as a value proportional to -Log(RMSD) and shifted to the [0,100] range for simplicity, where RMSD is calculated with the correct structure.

The distributions of the Quality values of models built on experimental structures and models built on top of other models overlap. It really depends on two main factors: the quality of the template coordinates (the model is unlikely be any better than the template coordinates) and the level of sequence identity of the sequence-template alignment. The analysis of preservation of residue contacts shows that the overall quality of the homology models between 50% and 80% identity is similar to that of the NMR structures [25]. Let me suggest a simple linear quality "guesstimate" on a scale from 0 to 100 for a model:

$$Model\_Quality = SeqId - 50\ (Resolution-1.7),$$

where the Resolution parameter is truncated into the [1.7–3.7] range, and an NMR structure is assumed to be equivalent to a 3.0A resolution crystallographic structure. The experimental structure itself will be described by the same formula with SeqId = 100. While the functional dependence and the weight could be improved by fitting to a massive benchmark, who would read such a paper? Use any weight you like, it will not change the idea. Here is our answer. Globally, a structure itself (SeqID = 100%) has the quality of 100 if Resolution is below 1.7 and starts deteriorating to 0 for Resolution equal to 3.7. This is not entirely fair because at 3.7A the overall features can still be visible. A 50% sequence identity model with 1.7A template is is similar to a 2.7A structure (except for the insertions). The low sequence identity region probably needs to be expanded.

Secondly, it is important to stress that the deterioration of model accuracy compared to the crystallographic template is not even throughout the structure but rather patchy (see the previous section). Some parts may be as good as a crystallographic template structure and others may be totally wrong.

11.4 ANNOTATING PROTEIN SURFACE

Once we have a set of coordinates, the model can be used to predict the functions of the protein. At the conceptual level the repertoire of possibilities is limited. Proteins are going to bind other molecules, undergo conformational or chemical transitions and bind to molecules again. Basically, predicting function is mostly about predicting where molecules or ions may bind to a given protein structure, how they bind, what rearrangements they initiate, and what the ligands are. The binding sites must at least be specific to a certain degree. The binders can be classified into the following general categories:

- Metals
- Small organic molecules: metabolites and drugs
- Peptides and unfolded loops
- DNA or RNA
- Proteins

Each type of binder would require a special binding site. Predicting these binding sites is one of the main postmodel-building tasks.

There are several signals that help to identify biologically significant binding sites. Some of them are general and do not depend on a specific binder or even the binder type. Others are specific to a particular ligand.

11.4.1 SEQUENCE CONSERVATION

The first signal does not really depend on the binder type and is *evolutionary conservation*. Projecting alignment *conservation* strength to the surface may highlight interesting sights. In a simplest case, the *strength* can be defined as the fraction of identical residues in an alignment of family members with a similar function.

In a case of a larger class of functionally divergent families having, however, conserved backbone framework (e.g., GPCRs), the previously described evolutionary conservation would be watered down. The *evolutionary trace method* gives a much better prediction in that case [43,44]. Olivier Lichtarge and his colleagues kept refining the method and recently proposed with a spatial clustering of the evolutionary trace signal [45] that further enhanced its sensitivity.

Janet Thornton and her colleagues characterized many specific classes of protein binding sites over the last 20 years (see a good recent review in [46]). The residue signal characterizing a particular type of site can further be enhanced with neural networks and spatial clustering [47].

11.4.2 Structural Residue Pattern: Catalytic Triads

Specific residue-surface patterns is the second signal. Predicting catalytic binding sites consists of two main components: (i) finding a specific sequence-structure pattern such as catalytic triad, and (ii) finding a binding cleft [48]. Several groups developed algorithms to search for a structural residue pattern. Porter et al. [49] compiled a database of catalytic sites and created a Catalytic Site Atlas (http://www.ebi.ac.uk/thornton-srv/databases/CSA) in which one can find both known and predicted catalytic sites for PDB entries.

Unusual residue pK values or strained conformation is the third signal. Identification of catalytic residues can be approached from a different point of view, i.e., by calculating theoretical microscopic titration curves for each residue and detecting unusual local conditions (e.g., Asp residues of the HIV protease binding cleft are next to each other and are uncharged) [50].

11.4.3 Electrostatic Potential Indicates Binding Site for a Charged Ligand

If ligands are charged, e.g., DNA, RNA, nucleotides and many cell metabolites, the surface electrostatic potential becomes a strong determinant of binding. It is further enhanced by the long-distance nature of electrostatic forces that can steer a charged molecule to the right place. Barry Honig and Anthony Nicholls pioneered the electrostatic surface concept implemented in the GRASP program [51] and demonstrated how this potential can be used to predict binding sites.

The electrostatic potential can be calculated by solving the Poisson equation:

$$-\nabla(\varepsilon(\vec{r})\nabla\phi(\vec{r})) = \rho(\vec{r})$$

where ε is the dielectric constant, ϕ is the electric potential and ρ is the charge density.

The charge distribution outside the protein surface for low ionic strength solutions can be approximated as a linear function of the potential ϕ and the equation can be rewritten in the linear Poisson–Boltzmann form. This equation can be solved by either 3-D grid-based finite difference algorithms like DELPHI, ZAP, MEAD, or APBS (see a rigorous comparison of these algorithm in [52]), or calculated via a boundary element algorithm, such as REBEL [53].

REBEL uses a fast analytical calculation of the Connolly surface ([54]) and further accelerates the boundary element calculation by grouping small surface elements into atomic patches. In the boundary element formulation of the equation, each surface element has induced surface charge density σ_s on the molecular boundary. The electrostatic potential therefore can be written as shown below, and to find the resulting potential, we need to find these surface charge density values.

$$\phi(\vec{r}) = \sum_i \frac{q_i}{\varepsilon_{in} \mid \vec{r} - \vec{r}_i \mid} + \oiint \frac{\sigma(\vec{r}_s)}{\mid \vec{r} - \vec{r}_s \mid} ds$$

To find the surface charge densities σ_s for each atomic patch, one needs to solve a linear system of equations $R(\vec{r}) \bullet \sigma = e(\vec{r}, q)$, where matrix R are functions of molecular geometry and vector e is a function of atomic charges and geometrical parameters. The full form of the equation is shown below.

$$\sigma_k A_k - \left(\frac{\varepsilon_{in} - \varepsilon_{out}}{2\pi(\varepsilon_{in} - \varepsilon_{out})} \right) \sum_j \sigma_j \oiiint \frac{(\vec{r}_{sk} - \vec{r}_{sj}) \bullet \vec{n}_{sk}}{\mid \vec{r}_{sk} - \vec{r}_{sj} \mid^3} ds_j ds_k =$$

$$\left(\frac{\varepsilon_{in} - \varepsilon_{out}}{2\pi(\varepsilon_{in} - \varepsilon_{out})} \right) \frac{1}{\varepsilon_{in}} \sum_i q_i \oiint \frac{(\vec{r}_{sk} - \vec{r}_i) \bullet n_{sk}}{\mid \vec{r}_{sk} - \vec{r}_i \mid^3} ds_k$$

Preserving the exact shape of the molecular surface in this calculation but approximating the density values belonging to the same atom by the same value reduces the dimensionality of the problem and speeds up the algorithm. The REBEL algorithm does not need a grid; it uses the molecular surface and the exact positions of the charges.

For a simple task of surface mapping, techniques of this sophistication are probably overkill, but in structure prediction, binding energy prediction or calculation of transfer energies, the accuracy and speed of electrostatic calculation can become a bottleneck.

11.4.4 BINDING CLEFT IDENTIFICATION

The structure itself gives us another strong signal independent of alignments and conservation. Proteins surfaces cavities and clefts and some of they are indeed deep and attractive enough to bind natural ligands or drugs. Here we will try to identify features that do not really depend on the nature of the ligand. The small molecule mapping algorithms were reviewed recently by Sotriffer and Klebe [55].

Identification of peptide binding sites without any clear evolutionary signal still is accompanied by a large level of false positives. There are several classes of methods designed to predict general binding sites (regardless of their specificity): geometrical methods (e.g., SITE[48], LIGSITE [56], APROPOS [57], and PASS [58]) and mapping preferences for binding of small molecular probes/fragments (e.g., SUPERSTAR [59]). Recently Vajda and colleagues demonstrated that binding sites can also be identified by solvent mapping [60].

We recently developed a method that detects ligand binding sites using an averaged van der Waals potential. This method was used in analysis of rhodopsin and bacteriorhodopsin in which the ligand binding pocket is unambiguously detected [61]. The predicted sites overlap with 96.8% of 11535 pockets, in 82% the predicted pockets cover greater than 80% of the ligand atoms while the average fraction of protein surface covered by predicted and actual pockets is about the same.

11.4.5 POCKETOME

Here is an interesting idea. Let us introduce a new "-ome", namely a "pocketome." Let us define as pocketome a set of all small molecule binding pockets in a cell in a given organism. It is the pocketome to which drugs and metabolites are exposed when they enter the cell. In contrast to the protein–protein binding sites that are usually rather specific to a particular partner, pocketome is a useful concept since a pocket can be targeted by a virtually infinite number of chemical compounds. Pocketome and drugs or metabolites do not have a one-to-one correspondence, it is rather a many-to-many relationship: a pocket can be targeted by different molecules, and a molecule can go into different pockets. Pocketome may help us to understand crossreactivity and side effects.

As opposed to, say, the catalytic site classification, pocketome will not depend on substrate and will include allosteric and orphan binding pockets. Compiling the pocketome, clustering it into classes and categories, is within reach. We attempted classification of the binding pockets based on their shape and properties regardless on the substrate. See Figure 11.5.

11.4.6 PROTEIN–PROTEIN INTERACTION INTERFACE PREDICTION

Predicting protein–protein interfaces is difficult: the interfaces are large (averages at about 2000A^2 for homo-dimers) and, for transient interactions, look almost indistinguishable from a regular solvent exposed surface. Several groups analyzed protein interfaces, namely, transient protein–protein interfaces [62], 122 homodimer interfaces [63], and different classes [64] and structural basis of interactions [65]. Even though

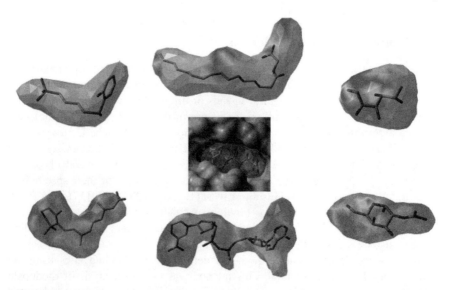

FIGURE 11.5 Highly represented pockets in a human pocketome. The volumes range from 160A^3 to 570A^3.

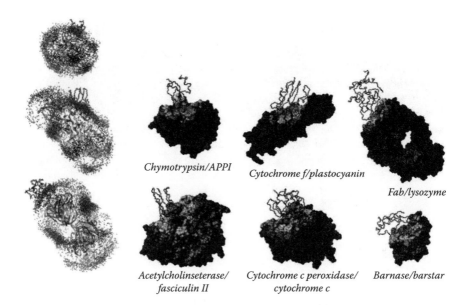

Chymotrypsin/APPI

Cytochrome f/plastocyanin

Fab/lysozyme

Acetylcholinseterase/ fasciculin II

Cytochrome c peroxidase/ cytochrome c

Barnase/barstar

FIGURE 11.6 Predicting protein–protein interfaces via protein docking. Locations and energies of the rigid body docking solutions (left) are converted into binding propensity P by accumulation of Boltzmann weighted residue surface area differences. Six examples show the predicted propensity by shade. Native ligands positions are shown in grey for comparison.

some general preferences exist, it is difficult to turn these preferences into a 100% reliable predictive method, especially for transient interactions.

Transient protein–protein interfaces can also be predicted with the heavy artillery of full-scale protein–protein docking simulations [66]. In this case, instead of trying to find a single correct docking geometry, we analyze the whole ensemble of predicted low-energy trial docking solutions and derive some sort of surface patch propensity from these ensembles. On a test set of 21 known protein–protein complexes not used in the training set we achieved 81% overlap of predicted high-propensity patch residues and the correct binding sites. See Figure 11.6.

11.4.7 SURFACE FLEXIBILITY VIEWS AND DOCKING

When we look at protein surfaces and decide how to design an inhibitor, it is useful to know which parts of the surface are flexible and which fixed. At the first approximation, we can analyze only the side chain flexibility. Simple coloring of Ca, C, N, and Cb, as well as all atoms of proline residues white and the rest by some other color is going to give a rough picture of which parts are more movable (see Figure 11.7a and Figure 11.7b).

For stronger binders it is better to rely on solid backbone surface because there will be no entropy loss associated with binding to such a surface and it will be easier to develop strong binders. On the other hand, using the side-chains atoms could be important to design inhibitors specific to a particular protein family member. Avoid-

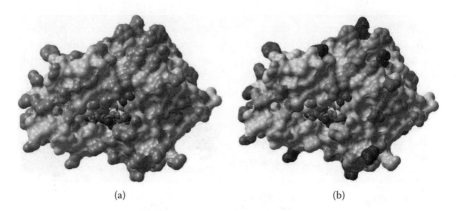

(a)   (b)

FIGURE 11.7 Surface flexibility view of enoyl reductase (PDB entry 1p45) bound to triclosan. (a) All potentially movable surface side chains shaded. (b) Predicted mean displacement shaded by the amplitude (dark for over 5A devinations).

ing flexible side chains is partially possible for small molecules, but is virtually unavoidable for much larger protein–protein interfaces.

A more quantitative view of side-chain mediated surface flexibility is given by Boltzmann averaging of atom displacements in an ensemble of low energy conformations after a short run of side-chain optimization (Totrov, personal communication). These considerations are critical for formulating receptor flexibility model for predicting induced fit upon ligand binding [21].

11.5 PROTEIN–PROTEIN DOCKING

Predicting protein–protein association remains one of the biggest challenges in computational biology. Over the last 5 to 10 years there has been significant progress, but we're still quite far from solving that problem. We are not going to discuss the problem preceding protein docking, which is whether two proteins interact or not, or whether a protein forms a multimeric complex. These are very important initial questions, and there are a variety of ideas about properties that can be combined into such a prediction. Let us assume that there is an experimental proof that proteins A and B form a complex. These are the main classes of macromolecular complexes:

- Two proteins with relatively small backbone changes upon association (our favorite class)
- Protein and a flexible peptide (e.g., peptide binding to SH2/PTB domains)
- Protein interactions leading to large conformational changes involving domain rearrangements (e.g., helix swapping)
- Protein association with nucleic acids
- Association of multiple domains

The major problem of predicting macromolecular association is the induced conformational changes upon association. At present we learned how to deal with the side chain rearrangements, but as the induced changes involve the backbone and increase in scale, the problem rapidly becomes unapproachable with the current methods.

To simplify the problem we need to use as much prior information about the association as we can. That may include mutation data, evolutionary conservation considerations (see above), patch predictions, antibody mapping data, hydrogen exchange data, etc. These restraints will narrow down the sampling task and simplify the problem.

11.5.1 DECIDING HOW TO DOCK

There are two molecular representations that can be used in macromolecular docking. In one, we represent the molecular model by its atoms with their types and positions, while in the other, we convert that information into 3-dimensional maps or grid potentials. Using maps of grid potentials accelerates the calculation, but it has two problems: 1) not every pairwise physical interaction can be accurately presented as a function of two grid maps; 2) the map is, by definition, static, and does not allow easy incorporation of conformational rearrangements. Consequently, the docking techniques can be divided into three classes:

- Atoms to Atoms
- Atoms to Grids
- Grids to Grids

Techniques of all three types are currently in use. In 1994, we attempted *ab initio* prediction of association geometry between an uncomplexed structure of lysozyme and an antibody with flexible side chains and a full atom approximation [67]. The lowest energy conformation was only 1.6A away from the correct structure, and what was even more surprising was that the global optimization of the interface side chains and the lysozyme position improved the RMSD from 5.5A to 1.6A. Later on, the same technique successfully predicted the association of beta-lactamase with its protein inhibitor in the first Docking Challenge of 1995 [68]. See Figure 11.8.

The global optimization procedure consisted of concerted moves of groups of torsion angles according to a continuous statistical distribution and so-called pseudo Brownian moves in which both the rotation and translation of the molecule would be changed. Gradient local minimization was applied after each random move and the solvation terms were added to the minimized energies. Despite the initial success, the method was not widely used since these calculations were too computationally expensive. The first simulation lasted for several months and some friendships were irreparably damaged.

11.5.2 ATOMS TO GRIDS

The idea to precompute the force field from the receptor and replace N by M pairwise calculations by M grid lookup operations emerged in 1985 [69,70]. Initially it was meant for use in small molecule receptor mapping or manual docking, but soon it was applied to protein–protein docking too. Let us describe the atom grid docking

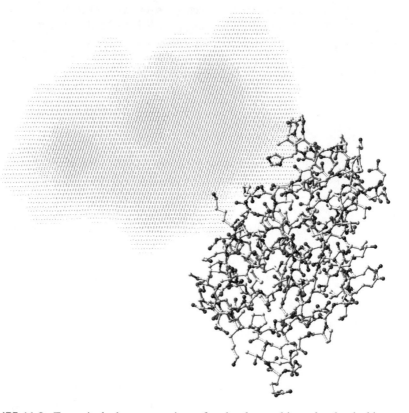

FIGURE 11.8 Two principal representations of molecules used in molecular docking.

algorithm called DISCO (Docking and Interface Side-Chain Optimization) as implemented in the ICM program. First, we need to develop a sampling strategy.

"Sticking pins" at a desired resolution to the two models creates an initial sampling grid. Then, based on the prior knowledge of the binding patches, some of the pins can be removed to prevent over-sampling. Each pair of pins will then be superimposed during the simulation and used to generate starting positions for the ligand for subsequent positional optimization. See Figure 11.9.

In the second step we will calculate a set of grid potentials for the receptor, which is defined as the larger molecule of the two. The interaction energy is represented by five types of precalculated grid potentials. The van der Waals potentials for a hydrogen atom probe and for a heavy atom probe, consist of an electrostatic potential where the partial atomic charges of the receptor are corrected by the induced surface solvent charge density to account for the solvent screening effect on the intermolecular pairwise electrostatic interactions, the hydrogen-bonding potential, and a simple hydrophobic potential roughly proportional to the number of hydrophobic atoms of the ligand in contact with the hydrophobic surface of the receptor. Energy balance between the different terms was optimized previously on a training set of the known protein–protein complexes. The solvation energy was calculated using an atomic solvent-accessible surface (ASA)-based model with per-atomic

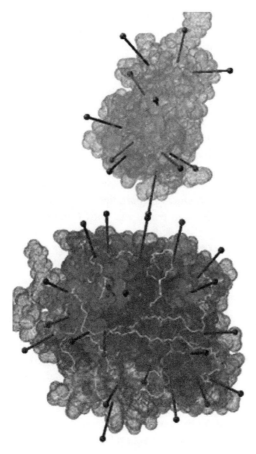

FIGURE 11.9 Overall docking sampling strategy: from each starting conformation a simulation is performed.

parameters derived from experimental vapor–water transfer energies for side-chain analogues and additional experimental data for charges solutes. This solvation term was added to the total energy to re-evaluate the docking solutions [66]. The grid potential penalizes the displacements of the ligand beyond the 10A margin in addition to the size of the ligand.

The docking itself consists of the following steps:

- Rigid body atoms-to-grids docking from each of the starting positions of the ligand
- Merging and compressing the best solutions from each simulation into a single set of conformations
- Rescoring the conformations in the set to improve the ranks of near native solutions

- Selecting a few hundred best candidates for global side-chain optimization and performing optimization
- Merging and rescoring the conformational sets again and selecting a few final models

11.5.2.1 How Successful is Protein Docking?

Somewhere between 20 and 50% of qualifying tasks. If one artificially separates the receptor and the ligand from the complex, it is not difficult to unambiguously predict their association. The real problem lies in the ability to make this prediction in a real case when both surfaces need to be deformed to become sufficiently complementary. In 2002, we collected the benchmark of 24 protein–protein complexes in which the 3-dimensional structures of participating proteins were available in both the free and the bound state and tested the DISCO procedure on that benchmark [71]. Later Zhiping Weng extended the benchmark with the antibody complexes [72]. In 3 of 24 complexes, a large backbone change upon association made it impossible to find the correct solution. In one-third of the remaining cases, the new native solutions had the lowest energy after refinement and consequently were ranked number one. The other two-thirds had ranks varying between 2 and 212. This 30% success rate was an achievement but still left large room for improvements.

Interestingly, about the same success rate was achieved in the CAPRI docking competition organized by Joel Janin, Shoshanna Wodak, and their colleagues. CAPRI, Critical Assessment of PRedicted Interactions [73], assembled 19 groups involved in the development of protein docking technologies. The participants during the first two rounds submitted predictions for seven protein–protein complexes. The structure of those complexes was known only to the organizations. The best models correctly predicted three out of seven target complexes, see Table 4 in Reference [74], but bear in mind that for each target up five models were submitted. See Figure 11.10a and Figure 11.10b.

11.5.3 GRID-TO-GRID DOCKING

The initial rigid body docking can also be performed using a combination of explicit rotational sampling and the Fast Fourier Transform (FFT) to determine the optimal translation [75]. Variations of this approach were implemented in a number of programs for optimal density matching, including DOT, ZDOCK [76–78].

The application of the FFT approach to protein docking is a derivative of the original optimal density superposition application. Recently, the FFT method was extended to fast rotational matching, an approach to finding the three rotational degrees of freedom for optimal density matching [79]. By recasting this problem into five angular parameters and one translational parameter, it was possible to apply the Fast Fourier Transform to five out of six degrees of freedom of the problem, thus reducing the sampling component of the matching to just one dimension. This promising method can now be adjusted and applied to protein docking.

One of the main problems of grid-to-grid algorithms is expressing the molecular mechanical interaction and solvation energy as a function of two maps. This trans-

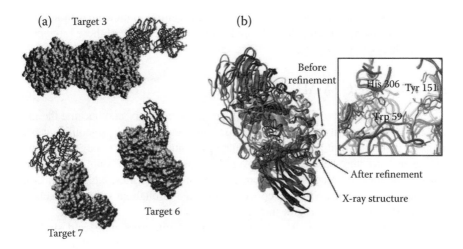

FIGURE 11.10 (a) Three best models predicted by ICM-DISCO for targets 3, 6, and 7 of the CAPRI 1, 2 round. (b) The global side-chain/positional refinement of a distant 7.4A –RMSD near-native solution improved as far as 1.5A, thus correctly predicting the induced fit.

formation is far from being straightforward, and inaccuracies due to this approximation reduce the significance of the generated scores. In any case, the FFT-related methods can be used as the first pass to generate hundreds to thousands of docking solutions.

Finally, both local and large-scale conformational changes are not adequately modeled by any of the existing protein docking algorithms and that is the area where the biggest improvements are needed.

11.6 PEPTIDE–PROTEIN DOCKING

Specific interactions between flexible peptides, frequently chemically modified, and their receptors is a large and important class of biological interactions. The problem is complicated by the large number of degrees of freedom and consequent combinatorial explosion of the conformational space of the peptide. Only a few attempts at peptide docking were made [80]. From the method point of view predicting the docked conformation of a peptide is a free peptide simulation performed in the vicinity of the receptor force fields. We collected several complexes between peptides and PTB or SH2 domains with known structures and performed global optimization simulations in the internal coordinate space for both phosphorylated and unphosphorylated peptides. The short peptides (4–6 residues) found the correct binding mode even though they were not restrained to any surface patch and sampled a large part of the SH2/PTB domain surface. Longer peptides (11 residues) needed at least one restraint to find the correct binding mode.

The main requirement for these kinds of simulations is convergence. It means that the same correct binding mode needs to be found in at least several simulations from completely different starting points. Achieving convergence is a necessary requirement for any prediction algorithm. As computer power increases, we might be able to take on larger peptides. This problem is central to molecular biology and will not go away because the diversity and combinatorial nature of those interactions leaves little hope that all of them will be characterized experimentally.

11.7 DOCKING SMALL MOLECULES

Molecular docking revolutionized the modern drug discovery and became the main link between structural proteomics and real life. As structural proteomics centers ramp up their structure production, an increasing number of new targets for quick and rational structure-based drug discovery become available. It is not a coincidence that the commercial structural proteomics/genomics companies, e.g., Syrrx, Structural GenomiX, Astex, and Plexxicon, combine high throughput structural proteomics with structure-based drug design as a business model.

The technology for small molecule docking, virtual screening, and structure-based design is quite mature but continues to expand into more challenging tasks. Improving the docking and scoring prediction accuracy, dealing with the induced conformational changes, understanding and designing for specificity rather than binding affinity only, and deorphanization are some of them. Several excellent reviews on the subject have been published recently [81–83]. See Figure 11.11.

The traditional issues in small molecule docking and scoring remain unchanged: improving flexible ligand pose prediction accuracy, improving ligand binding scores [84,85], learning how to search and design not only for potent binders but for specific binders as well, improving receptor pocket models. A number of success stories in

FIGURE 11.11 Results of docking and scoring of the 150,000 available compounds to the FGF receptor tyrosine kinase.

which novel (in some cases sub-micromolar) binders were found by virtual ligand screening have been reported [86–91].

The results of ligand docking depend strongly on the quality of the receptor structure. Shoichet and McCovern publish a rigorous study of deterioration of docking results in apo-structrures and homology models. In studying the same pocket in three different forms (holo-, apo- and modeled) for 10 different enzymes, 7 out of 10 times the holo-structure (bound to a ligand) was the best in selecting correct binders out of 95,000 small molecules.

The second evolving theme is docking of compounds to *multiple* receptors. This is important in evaluating potential cross-reactivity, or design for binding specificity [84,92,93]. Finally, the major challenges of receptor flexibility upon ligand binding was addressed.

The simplest way of dealing with possible receptor rearrangements is as follows:

- Generate or find in PDB multiple receptor conformations. They could be different receptor-ligand complexes, or even identical subunits A and B related by a noncrystallographic symmetry.
- Dock your library to each of these alternative pockets.
- Determine the ranks for each model and merge by rank (merging by energy is dangerous because there could be systematic score shifts).

If multiple bound structures are not available, the alternative conformations can be generated by docking flexible ligands to a flexible receptor pocket [21]).

11.7.1 DEORPHANIZATION

In silico docking and scoring of known biological small molecules may be attempted to discover the natural ligand, i.e., to de-orphanize the receptor. Cavasotto et al. [61] docked the KEGG-LIGAND library of about seven thousands of natural substrates [94] to bacteriorhodopsin and rhodopsin structures. Docking and scoring the LIGAND database to the identified binding pocket of a high resolution structure of bacteriorhodopsin (PDB entry: 1c3w, resolution 1.55A) gave the best binding score to retinal, the natural substrate of bacteriorhodopsin. Screening against bovine rhodopsin structure [3] of 2.88A resolution ranked retinal within 1.5% of the best scores. See Figure 11.12.

This method can become increasingly important in structural proteomics as completely uncharacterized proteins will be cloned, purified, and crystallized in a high throughput manner.

ACKNOWLEDGMENTS

I thank my long-term colleague Maxim Totrov, with whom we have been developing most of the ICM algorithms for the last 13 years, for stimulating discussions. I also thank Claudio Cavasotto, Andrew Orry, Jianghong An, Andrew Bordner, and Juan Fernandez-Recio for their contributions, and Colin Smith for helping with the manuscript preparation. Finally, I thank Michael Sundstrom for his patience and insight.

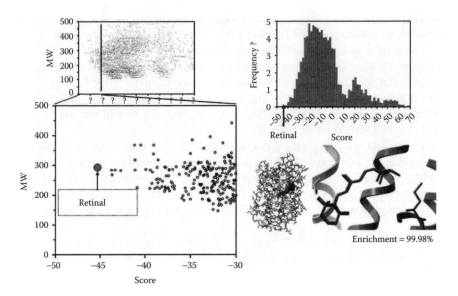

FIGURE 11.12 Illustration of how a natural substrate of bacteriorhodopsin could have been identified via virtual ligand screening of the LIGAND dataset from KEGG. Retinal has the best score out of ~7700 substrates.

REFERENCES

1. Berman, H.M., et al., The Protein Data Bank, *Nucleic Acids Res.,* 28, 1, 235–42, 2000.
2. Zhang, C. and S.H. Kim, Overview of structural genomics: from structure to function, *Curr. Opinions Chem. Biol.,* 7, 1, 28–32, 2003.
3. Palczewski, K., et al., Crystal structure of rhodopsin: A G protein-coupled receptor, *Science*, 289, 5480, 739–745, 2000.
4. Goldsmith-Fischman, S. and B. Honig, Structural genomics: computational methods for structure analysis, *Protein Sci.,* 12, 9, 1813–1821, 2003.
5. Sali, A., et al., From words to literature in structural proteomics, *Nature*, 422, 6928, 216–225, 2003.
6. Norin, M. and M. Sundstrom, Structural proteomics: lessons learnt from the early case studies, *Farmaco*, 57, 11, 947–951, 2002.
7. Frishman, D., What we have learned about prokaryotes from structural genomics, *Omics*, 7, 2, 211–224, 2003.
8. Claverie, J.M., et al., Recent advances in computational genomics, *Pharmacogenomics*, 2, 4, 361–372, 2001.
9. Anderson, S. and J. Chiplin, Structural genomics: shaping the future of drug design?, *Drug Discov. Today*, 7, 2, 105–107, 2002.
10. Kleywegt, G.J. and T.A. Jones, Homo crystallographicus — quo vadis? *Structure*, Camb, 10, 4, 465–472, 2002.
11. Marsden, B. and R. Abagyan, Identifying errors in three-dimensional models, in *Protein Structure Determination Analysis & Applications*, D. Chasman, Ed., Marcel Dekker, New York, 2003.

12. Davis, A.M., S.J. Teague, and G.J. Kleywegt, Application and Limitations of X-ray Crystallographic Data in Structure-Based Ligand and Drug Design, *Angew Chem. Int. Ed. Engl.*, 42, 24, 2718–2736, 2003.

13. Kleywegt, G.J. and T.A. Jones, Phi/psi-chology: Ramachandran revisited, *Structure*, 4, 12, 1395–1400, 1996.

14. Vucetic, S., et al., Flavors of protein disorder, *Proteins*, 52, 4, 573–584, 2003.

15. Eisenmenger, F., P. Argos, and R. Abagyan, A method to configure protein side-chains from the main-chain trace in homology modelling, *J. Mol. Biol.*, 231, 3, 849–860, 1993.

16. Dunbrack, R.L., Jr. and M. Karplus, Backbone-dependent rotamer library for proteins. Application to side-chain prediction, *J. Mol. Biol.*, 230, 2, 543–574, 1993.

17. Bower, M.J., F.E. Cohen, and R.L. Dunbrack, Jr., Prediction of protein side-chain rotamers from a backbone-dependent rotamer library: a new homology modeling tool, *J. Mol. Biol.*, 267, 5, 1268–1282, 1997.

18. Canutescu, A.A., A.A. Shelenkov, and R.L. Dunbrack, Jr., A graph-theory algorithm for rapid protein side-chain prediction, *Protein Sci.*, 12, 9, 2001–2014, 2003.

19. Fan, H. and A.E. Mark, Refinement of homology-based protein structures by molecular dynamics simulation techniques, *Protein Sci.*, 13, 1, 211–220, 2004.

20. Frimurer, T.M., et al., Ligand-induced conformational changes: improved predictions of ligand binding conformations and affinities, *Biophys. J.*, 84, 4, 2273–2281, 2003.

21. Cavasotto, C.N. and R.A. Abagyan, Protein flexibility in ligand docking and virtual screening to protein kinases, *J. Mol. Biol.*, 337, 1, 209–225, 2004.

22. Morris, A.L., et al., Stereochemical quality of protein structure coordinates, *Proteins*, 12, 4, 345–364, 1992.

23. Laskowski, R.A., Structural quality assurance, *Methods Biochem. Anal.*, 44, 273–303, 2003.

24. Maiorov, V. and R. Abagyan, Energy strain in three-dimensional protein structures, *Fold Des.*, 3, 4, 259–269, 1998.

25. Abagyan, R.A. and M.M. Totrov, Contact area difference, CAD, a robust measure to evaluate accuracy of protein models, *J. Mol. Biol.*, 268, 3, 678–685, 1997.

26. Spronk, C.A., et al., The precision of NMR structure ensembles revisited, *J. Biomol. NMR*, 25, 3, 225–234, 2003.

27. Chen, Y.W., E.J. Dodson, and G.J. Kleywegt, Does NMR mean "not for molecular replacement"? Using NMR-based search models to solve protein crystal structures, *Structure Fold Des.*, 8, 11, R213–220, 2000.

28. Jee, J. and P. Guntert, Influence of the completeness of chemical shift assignments on NMR structures obtained with automated NOE assignment, *J. Struct. Funct. Genomics*, 4, 2–3, 179–189, 2003.

29. Abagyan, R., et al., Homology modeling with internal coordinate mechanics: deformation zone mapping and improvements of models via conformational search, *Proteins*, Suppl. 1: 29–37, 1997.

30. John, B. and A. Sali, Comparative protein structure modeling by iterative alignment, model building and model assessment, *Nucleic Acids Res.*, 31, 14, 3982–3992, 2003.

31. Heger, A. and L. Holm, More for less in structural genomics, *J. Struct. Funct. Genomics*, 4, 2–3, 57–66, 2003.

32. Teodorescu, O., et al., Enriching the sequence substitution matrix by structural information, *Proteins*, 54, 1, 41–48, 2004.

33. Venclovas, C., Comparative modeling in CASP5: progress is evident, but alignment errors remain a significant hindrance, *Proteins*, 53 Suppl. 6: 380–388, 2003.

34. Lasters, I. and J. Desmet, The fuzzy-end elimination theorem: correctly implementing the side chain placement algorithm based on the dead-end elimination theorem, *Protein Eng.*, 6, 7, 717–722, 1993.
35. Desmet, J., M. De Maeyer, and I. Lasters, Theoretical and algorithmical optimization of the dead-end elimination theorem. *Pac. Symp. Biocomput.*, 122–133, 1997.
36. Dahiyat, B.I., C.A. Sarisky, and S.L. Mayo, De novo protein design: towards fully automated sequence selection, *J. Mol. Biol.*, 273, 4, 789–796, 1997.
37. Borchert, T.V., et al., The crystal structure of an engineered monomeric triosephosphate isomerase, monoTIM: the correct modeling of an eight-residue loop, *Structure*, 1, 205–213, 1993.
38. Rapp, C.S. and R.A. Friesner, Prediction of loop geometries using a generalized born model of solvation effects, *Proteins*, 35, 2, 173–183, 1999.
39. Canutescu, A.A. and R.L. Dunbrack, Jr., Cyclic coordinate descent: A robotics algorithm for protein loop closure, *Protein Sci.*, 12, 5, 963–972, 2003.
40. Kolodny, R. and M. Levitt, Protein decoy assembly using short fragments under geometric constraints, *Biopolymers*, 68, 3, 278–285, 2003.
41. Cardozo, T., S. Batalov, and R. Abagyan, Estimating local backbone structural deviation in homology models, *Comput. Chem.*, 24, 1, 13–31, 2000.
42. Chotia, C. and A.M. Lesk, The relation between the divergence of sequence and structure in proteins, EMBO J., 5, 4, 823–826, 1986.
43. Lichtarge, O., H.R. Bourne, and F.E. Cohen, An evolutionary trace method defines binding surfaces common to protein families, *J. Mol. Biol.*, 257, 2, 342–358, 1996.
44. Lichtarge, O., et al., Accurate and scalable identification of functional sites by evolutionary tracing, *J. Struct. Funct. Genomics*, 4, 2–3, 159–166, 2003.
45. Mihalek, I., et al., Combining inference from evolution and geometric probability in protein structure evaluation, *J. Mol. Biol.*, 331, 1, 263–279, 2003.
46. Bartlett, G.J., A.E. Todd, and J.M. Thornton, Inferring protein function from structure, *Methods Biochem. Anal.*, 44, 387–407, 2003.
47. Gutteridge, A., G.J. Bartlett, and J.M. Thornton, Using a neural network and spatial clustering to predict the location of active sites in enzymes, *J. Mol. Biol.*, 330, 4, 719–734, 2003.
48. Laskowski, R.A., et al., Protein clefts in molecular recognition and function, *Protein Sci.*, 5, 12, 2438–2452, 1996.
49. Porter, C.T., G.J. Bartlett, and J.M. Thornton, The Catalytic Site Atlas: a resource of catalytic sites and residues identified in enzymes using structural data, *Nucleic Acids Res.*, 32, 1, D129–133, 2004.
50. Shehadi, I.A., H. Yang, and M.J. Ondrechen, Future directions in protein function prediction, Mol. Biol. Rep., 29, 4, 329–335, 2002.
51. Honig, B. and A. Nicholls, Classical electrostatics in biology and chemistry, *Science*, 268, 5214, 1144–1149, 1995.
52. Feig, M., et al., Performance comparison of generalized born and Poisson methods in the calculation of electrostatic solvation energies for protein structures, *J. Comput. Chem.*, 25, 2, 265–284, 2004.
53. Totrov, M. and R. Abagyan, Rapid boundary element solvation electrostatics calculations in folding simulations: successful folding of a 23-residue peptide, *Biopolymers*, 60, 2, 124–133, 2001.
54. Totrov, M. and R. Abagyan, The contour-buildup algorithm to calculate the analytical molecular surface, *J. Struct. Biol.*, 116, 1, 138–143, 1996.

55. Sotriffer, C. and G. Klebe, Identification and mapping of small-molecule binding sites in proteins: computational tools for structure-based drug design. *Farmaco*, 57, 3, 243–251, 2002.

56. Hendlich, M., F. Rippmann, and G. Barnickel, LIGSITE: automatic and efficient detection of potential small molecule-binding sites in proteins, *J. Mol. Graph. Model*, 15, 6, 359–363, 389, 1997.

57. Peters, K.P., J. Fauck, and C. Frommel, The automatic search for ligand binding sites in proteins of known three-dimensional structure using only geometric criteria. *J. Mol. Biol.*, 256, 1, 201–213, 1996.

58. Brady, G.P., Jr. and P.F. Stouten, Fast prediction and visualization of protein binding pockets with PASS, *J. Comput. Aided Mol. Des.*, 14, 4, 383–401, 2000.

59. Verdonk, M.L., et al., SuperStar: improved knowledge-based interaction fields for protein binding sites, *J. Mol. Biol.*, 307, 3, 841–859, 2001.

60. Silberstein, M., et al., Identification of substrate binding sites in enzymes by computational solvent mapping, *J. Mol. Biol.*, 332, 5, 1095–113, 2003.

61. Cavasotto, C.N., A.J. Orry, and R.A. Abagyan, Structure-based identification of binding sites, native ligands and potential inhibitors for G-protein coupled receptors, *Proteins*, 51, 3, 423–433, 2003.

62. Nooren, I.M. and J.M. Thornton, Structural characterisation and functional significance of transient protein-protein interactions, *J. Mol. Biol.*, 325, 5, 991–1018, 2003.

63. Bahadur, R.P., et al., Dissecting subunit interfaces in homodimeric proteins, *Proteins*, 53, 3, 708–719, 2003.

64. Ofran, Y. and B. Rost, Predicted protein-protein interaction sites from local sequence information, FEBS Lett., 544, 1–3, 236–239, 2003.

65. Wodak, S.J. and J. Janin, Structural basis of macromolecular recognition, *Adv. Protein Chem.*, 61, 9–73, 2002.

66. Fernandez-Recio, J., M. Totrov, and R. Abagyan, Identification of protein-protein interaction sites from docking energy landscapes, *J. Mol. Biol.*, 335, 3, 843–865, 2004.

67. Totrov, M. and R. Abagyan, Detailed ab initio prediction of lysozyme-antibody complex with 1.6 A accuracy, *Nat. Struct. Biol.*, 1, 4, 259–263, 1994.

68. Strynadka, N.C., et al., Molecular docking programs successfully predict the binding of a beta-lactamase inhibitory protein to TEM-1 beta-lactamase, *Nat. Struct. Biol.*, 3, 3, 233–239, 1996.

69. Goodford, P.J., A computational procedure for determining energetically favorable binding sites on biologically important macromolecules, *J. Med. Chem.*, 28, 7, 849–857, 1985.

70. Pattabiramen, N., et al., Computer graphics in real-time docking with energy calculations and minimization, *J. Comput. Chem.*, 6, 5, 432–436, 1985.

71. Fernandez-Recio, J., M. Totrov, and R. Abagyan, Soft protein-protein docking in internal coordinates, *Protein Sci.*, 11, 2, 280–291, 2002.

72. Chen, R., et al., A protein-protein docking benchmark, *Proteins*, 52, 1, 88–91, 2003.

73. Janin, J., et al., CAPRI: a Critical Assessment of PRedicted Interactions, *Proteins*, 52, 1, 2–9, 2003.

74. Mendez, R., et al., Assessment of blind predictions of protein-protein interactions: current status of docking methods, *Proteins*, 52, 1, 51–67, 2003.

75. Katchalski-Katzir, E., et al., Molecular surface recognition: determination of geometric fit between proteins and their ligands by correlation techniques, *Proc. Natl. Acad. Sci. U.S.*, 89, 6, 2195–2199, 1992.

76. Mandell, J.G., et al., Protein docking using continuum electrostatics and geometric fit, *Protein Eng.*, 14, 2, 105–113, 2001.

77. Chen, R., L. Li, and Z. Weng, ZDOCK: an initial-stage protein-docking algorithm, *Proteins*, 52, 1, 80–87, 2003.

78. Vakser, I.A., O.G. Matar, and C.F. Lam, A systematic study of low-resolution recognition in protein — protein complexes, *Proc. Natl. Acad. Sci. U.S.*, 96, 15, 8477–8482, 1999.

79. Kovacs, J.A., et al., Fast rotational matching of rigid bodies by fast Fourier transform acceleration of five degrees of freedom, *Acta Crystallogr. D. Biol. Crystallogr.*, 59, Pt 8, 1371–1376, 2003.

80. Zhou, Y. and R. Abagyan, How and why phosphotyrosine-containing peptides bind to the SH2 and PTB domains, *Fold Des.*, 3, 6, 513–522, 1998.

81. Wong, C.F. and A.J. McCammon, Protein simulation and drug design, Adv. *Protein Chem.*, 66, 87–121, 2003.

82. Brooijmans, N. and I.D. Kuntz, Molecular recognition and docking algorithms, *Annu. Rev. Biophys. Biomol. Struct.*, 32, 335–373, 2003.

83. Schneider, G. and H.J. Bohm, Virtual screening and fast automated docking methods, *Drug Discov. Today*, 7, 1, 64–70, 2002.

84. Charifson, P.S., et al., Consensus scoring: A method for obtaining improved hit rates from docking databases of three-dimensional structures into proteins, *J. Med. Chem.*, 42, 25, 5100–5109, 1999.

85. Stahl, M. and M. Rarey, Detailed analysis of scoring functions for virtual screening, *J. Med. Chem.*, 44, 7, 1035–1042, 2001.

86. Perola, E., et al., Successful virtual screening of a chemical database for farnesyl-transferase inhibitor leads, *J. Med. Chem.*, 43, 3, 401–408, 2000.

87. Schapira, M., et al., In silico discovery of novel retinoic acid receptor agonist structures, *BMC Struct. Biol.*, 1, 1, 1, 2001.

88. Filikov, A.V., et al., Identification of ligands for RNA targets via structure-based virtual screening: HIV-1 TAR, *J. Comput. Aided Mol. Des.*, 14, 6, 593–610, 2000.

89. Schapira, M., et al., Rational discovery of novel nuclear hormone receptor antagonists, *Proc. Natl. Acad. Sci. U.S.*, 97, 3, 1008–1013, 2000.

90. Doman, T.N., et al., Molecular docking and high-throughput screening for novel inhibitors of protein tyrosine phosphatase-1B, *J. Med. Chem.*, 45, 11, 2213–2221, 2002.

91. Schapira, M., et al., Discovery of diverse thyroid hormone receptor antagonists by high-throughput docking, *Proc. Natl. Acad. Sci. U.S.*, 100, 12, 7354–7359, 2003.

92. Schapira, M., R. Abagyan, and M. Totrov, Nuclear hormone receptor targeted virtual screening, *J. Med. Chem.*, 46, 14, 3045–3059, 2003.

93. Bissantz, C., et al., Protein-based virtual screening of chemical databases. II. Are homology models of G-Protein Coupled Receptors suitable targets? *Proteins*, 50, 1, 5–25, 2003.

94. Goto, S., T. Nishioka, and M. Kanehisa, LIGAND database for enzymes, compounds and reactions, *Nucleic Acids Res.*, 27, 1, 377–379, 1999.

12 Applied Structural Genomics

Michael Sundström

CONTENTS

12.1 INTRODUCTION

Proteins are one of the major building blocks of all living organisms. They utilise and dissipate energy sources, and mould and form most structural elements as well as initiate and control signals. Even though DNA is the "blueprint of life," it is the proteins that carry out the messages encoded in the DNA. Without an understanding at the molecular level of protein function, we will never truly understand the biochemical and physiological processes in cells and organisms. We need to know this level of detail if we are to alter and control protein activity through therapeutic intervention.

The analysis of protein structural properties and features enables the generation of detailed models of the protein(s) tertiary structure(s). This knowledge describes the molecular aspects of the surface of proteins and their cavity properties such as size, shape, and charge. These features are important for the understanding of the overall functional properties of specific proteins that are key targets in drug discovery and other biotechnological applications. At a more basic level, structural information can be used to generate hypotheses for protein mechanism of action as well as biological function, to be verified in *in vitro* or *in vivo,* in biochemical and/or cell based assays.

In brief, the benefits from Structural Genomics efforts applied in the development of new therapeutic approaches are:

- Structural information for previously unannotated proteins, providing novel intervention points and tools for drug discovery.
- Structural information for established key targets in drug discovery and other rational ligand design efforts.
- Availability of structural descriptors to understand candidate drug selectivity towards antitargets and associated toxicity issues by generating protein family structural "maps" to guide selectivity design in medicinal chemistry.
- Access to experimental determined high quality protein structures to guide engineering of therapeutic proteins. Factors of particular importance are optimization of physical and drug properties such as stability, solubility, immunogenicity, and antigenicity.
- A significant increase in the production rate and availability of DNA and protein reagents for the research community.
- Significant impact on the technology and methodology development in high throughput/output approaches in molecular biology, protein biochemistry, structural biology, and associated research areas.

This chapter attempts to summarize and describe a subset of potential practical applications from Structural Genomics efforts. An overview of key factors contributing to the initiation of the research area including expected contributions is detailed in Figure 12.1.

12.2 COMPLETENESS OF STRUCTURAL SPACE

3-D protein structures are currently being generated at an unprecedented rate; however, the lack of completeness in the coverage of representative structures has

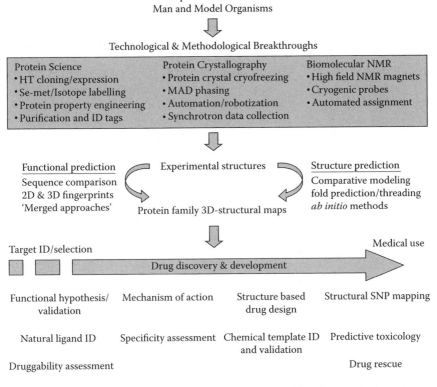

FIGURE 12.1 A schematic representation of the factors enabling Structural Genomics programs to be initiated, its current major deliverables, and potential applications in drug discovery.

prompted major initiatives in Structural Genomics. An initial goal of Structural Genomics was to provide family representative structural templates to allow computational approaches to predict structural characteristics of entire proteomes. The initial target was to generate enough 3-D structures to model all proteins at the 30% sequence identity threshold. Most selected targets were globular and soluble unannotated proteins, preferably predicted to contain a new fold. Thus, one hoped to reach a rapid coverage of protein sequence and structural space (reviewed by Norin and Sundstrom [1]) as exemplified by the initial efforts within the Protein Structure Initiative (PSI) (www.nigms.nih.gov/psi).

Various assessments have been made on how many proteins should be targeted to provide a basic 3-D structural space map of proteins, allowing reasonable comparative models to be generated of the full proteome(s) [2]. A study by the North East Structural Genomics Consortium (NESG) (www.nesg.org) assessed that 6 to 38% of all proteins currently have a structural template. When the remaining proteins were analyzed using various thresholds set for accuracy of comparative models, excluding nonglobular protein regions, it was found that as many as 48% of the remaining proteins would be excellent targets for structural genomics [3].

The Protein Data Bank (www.rcsb.org/pdb) currently (May 2005) consists of approximately 28,000 protein, virus, and peptide structures, although the number of unique structures is significantly lower, approximately 10,000 (http://www.rcsb.org/pdb/holdings.html). Of the deposited structures, approximately 85% are derived from protein crystallography and 15% from biomolecular NMR. The typical coverage of representative structures of soluble protein families is in the range of 5 to 15%, if one allows data from multiple species to be combined such as yeast, *E. coli*, thermophiles, mouse, and man. The proportion is often significantly smaller when considering any proteome from a single species.

An increased coverage of protein structural space is expected to have a significant impact on basic and applied research in analogy with the genome sequencing projects.

12.3 PUBLIC STRUCTURAL GENOMIC INITIATIVES

In the first phase of structural genomics, the major efforts involved approaches towards model organism proteomes, especially thermophilic organisms such as *Pyrococcus furiosus* and *Methanobacterium thermoautotrophicum* as well as prokaryotic pathogens (e.g., *Mycobacterium tuberculosis*) to ensure that representative structural templates would be determined by X-ray crystallography or NMR, and rapidly increase the protein fold space (see also Chapter 1 and Chapter 9 in this volume).

The current public large scale projects are:

- Protein Structure Initiative, PSI (http://www.nigms.nih.gov/psi)
- RIKEN Structural Genomics Initiative, RSGI (http://www.rsgi.riken.go.jp)
- Structural Proteomics in Europe, SPINE (http://www.spineurope.org)
- Structural Genomics Consortium, SGC (http://www.thesgc.com)

They are gaining momentum towards determination of protein structures of relevance to applied research. However, a large fraction is still devoted to simpler model organisms such as prokaryotes and fungi.

12.3.1 RSGI (JAPAN)

RIKEN Structural Genomics Initiative is funded by the Japanese Science and Technology Agency. This research program was initiated in 1998 and is focused on the high-throughput determination of three-dimensional structures of proteins from mouse, *Arabidopsis thaliana*, and *Thermus thermophilus HB8*. Major infrastructure investments have been made to allow high throughput including synchrotron facilities, high field NMR spectroscopy, and cell free protein production. Recently, the RSGI was moving resources toward eukaryotic genomes as well.

12.3.2 PSI (UNITED STATES)

The Protein Structure Initiative (PSI) is a joint industrial-academic effort aimed at reducing the costs and improving timeframes for structure determination of proteins. The goal of the PSI is to make the three-dimensional atomic-level structures of most proteins easily obtainable from knowledge of their corresponding DNA sequences. The program was initiated in 2000 and will be funded until 2005. PSI is a multi-centre initiative, consisting of:

- Berkeley Structural Genomics Center
- Center for Eukaryotic Structural Genomics
- The Joint Center for Structural Genomics
- The Midwest Center for Structural Genomics
- New York Structural Genomics Research Consortium
- Northeast Structural Genomics Consortium
- The Southeast Collaboratory for Structural Genomics
- Structural Genomics of Pathogenic Protozoa Consortium
- TB Structural Genomics Consortium

For a brief outline of each centre see: http://www.nigms.nih.gov/psi/centers.html. The NIH recently funded the continuation of PSI for an additional five years.

12.3.3 SPINE (EUROPEAN UNION)

Structural Proteomics in Europe (SPINE), funded by the European Commission, is an integrated research project adopting the decentralized approach often used by the European Union framework programs. SPINE was initiated in 2003 and is funded for three years of operation. The goals of the SPINE program are to develop new methods and technologies for high-throughput structural biology. The project aims to speed up the structure determination process by optimizing approaches in protein production and purification, crystallization, and structure analysis. SPINE efforts are directed towards proteins implicated in disease states with particular focus on cancer and neurodegenerative diseases as well as certain pathogens.

12.3.4 SGC (CANADA AND UNITED KINGDOM)

To date, few of the structural genomics efforts have focused on human proteins; an area where significant impact is expected regards using 3-D structures as templates for drug discovery and development. A major exception is the Structural Genomics Consortium (SGC), which is an international collaboration of British and Canadian scientists, funded by the Wellcome Trust, a consortium of Canadian funding agencies and the private sector. The primary objective of the SGC is to determine the 3-D structures of human proteins of known therapeutic or medical relevance and release them into the public domain. The SGC, initiated in 2003, operates from the Universities of Toronto and Oxford, and has currently (May 2005) deposited ~60 human and malarial protein structures.

12.4 COMMERCIAL STRUCTURAL GENOMICS OPERATIONS

Structural biology as a business model was initially adopted by companies like Agouron (now part of Pfizer, www.pfizer.com) and Vertex Pharmaceuticals (www.vpharm.com). These companies fully integrated the use of target protein structural information in drug discovery programs.

Other companies, with a more conventional approach to drug discovery, utilized structural information more or less consciously, either through protein crystallographic iterative design (where a large number of co-crystal structures are solved in the quest to convert a promising chemical hit into a drug candidate) or indirect computational methods using active site descriptors derived from the target or its structural homologues.

When Structural Genomics efforts were gaining initial momentum, several companies were formed and attempted to provide novel protein structures of drug targets to pharmaceutical companies, to facilitate informed selection and rational design of hit and lead compounds. Examples of such companies are SYRRX (www.syrrx.com), Structural GenomiX (www.stromix.com), Integrative Proteomics (now Affinium Pharmaceuticals, www.afnm.com), and Astex Technology (www.astex-technology.com).

12.4.1 STRUCTURAL GENOMIX (UNITED STATES)

SGX was the pioneer in the creation of a Structural Genomics business operation. The initial concept was focused on the structure determination of hitherto not published drug targets, aiming to create a database of structures for which potential users could subscribe. However, SGX has re-focused to a more traditional structure-based drug design (SBDD) approach aimed at the use of protein structures for lead discovery and optimization.

SGX aims to solve protein structures at high throughput to enhance internal or collaboration drug-discovery projects (for an example see Peat et al. [4]). Announced collaborations include Urogene S.A. (kinase inhibitor drug discovery program for urological cancer), Boehringer Ingelheim Pharmaceuticals (drug discovery), OSI Pharmaceuticals (anticancer drug discovery), Eli Lilly (drug discovery), and others.

12.4.2 SYRRX (UNITED STATES)

Started as a technology platform company in which high-throughput crystallization robotics was one of the primary goals. The business model adopted by SYRRX aimed to capture the value of methodologies that quickly assess the feasibility of protein targets to form well-diffracting crystals. SYRRX aims to couple improved protein biotechnology approaches with nanovolume crystallization as a platform to support its small-molecule drug-discovery program (exemplified in Hosfield et al. and Aparacio et al. [5,6]). Recently announced collaborations include PPD Inc. (Dipeptidyl Peptidase IV inhibitors for diabetes drug discovery) and RTS Life Science International (marketing of high throughput structural biology automation). Recently, Takeda Pharmaceutical Company Ltd. acquired Syrrx.

12.4.3 AFFINIUM PHARMACEUTICALS (CANADA)

Integrative Proteomics (now Affinium) was founded as a Structural Genomics-based company. Over the last few years, Affinium's business model changed to become a

SBDD company focused on infectious disease drug discovery and development. The company announced drug discovery collaboration with Pfizer Inc. in 2002 and recently acquired the rights to an antiinfective research program (staphylococcal infections) from Glaxo Smith Kline.

12.4.4 ASTEX TECHNOLOGY (UNITED KINGDOM)

Astex was not founded as a pure Structural Genomics company but rather has attempted to be profiled in high throughput structural chemistry with strong resources towards structure determination of novel and difficult targets. Astex aims to use protein 3-D structures for drug discovery, especially by using protein crystallography as a screening tool to identify chemical starting points (fragment-based) for drug discovery [7]. Another area of particular focus is their published success in determining the structures of cytochrome P450 structures [7–9]. Astex recently merged with Metagen Pharmaceuticals and has also initiated collaborations with AstraZeneca (Alzheimers disease drug discovery), Fujisawa (P450, drug rescue), and Schering AG (anticancer drug discovery).

12.5 STRUCTURE GUIDED FUNCTIONAL VALIDATION

Considerable efforts have been directed towards theoretical approaches to assign protein function from bioinformatics/sequence analyses as well as functional assignments from homology modelled and/or threaded protein models. Here the goal is to understand the role of gene products in cell signalling and disease and to utilize information on their biological function for basic research. Subsequently, structural annotation of novel targets allows the initiation of protein and chemistry-based therapeutics discovery programs [10].

Other approaches have relied on the (rather cumbersome) analysis of protein–protein interactions/networks exemplified by yeast two-hybrid analyses [11]. Although modelled structures provide valuable information, a high degree of uncertainty is bound to be present especially in cases of low homology to well characterized template genes/proteins in modelling studies. In addition, it is a well-established fact that protein–protein interaction studies by yeast two-hybrid analysis often provide a large fraction of false positive interactions (as well as false negatives) [12].

An obvious use of protein structures of unannotated proteins is to generate functional hypotheses based on the 3-D structural features and subsequently attempt to verify such hypotheses using experimental approaches, e.g., by biochemical and cell-based assays. Structural Genomics could thus provide an opportunity to assign functions (class, mechanism of action, natural ligands, active site features, etc.) to the many unannotated sequences that to date have been generated within the genome sequencing programs, and which often have a poor degree of functional validation. To date, the indirect methods have dominated over large scale direct (experimental) methods.

An early and good example of direct functional annotation from structure came from Zarembinski and coworkers [13]. In this study, the crystal structure of the unannotated protein, MJ0577, from *Methanococcus jannaschii* revealed a bound ATP in the high resolution electron density maps, suggesting that the protein was an ATP-ase or an ATP-mediated molecular switch. The structure-based hypothesis could subsequently be confirmed by biochemical experiments and the novel ATP binding motif could be used to search for other putative ATP binding sequences. A relatively large amount of similar studies have thereafter, sometimes successfully, used structure-assisted assignment of function for unannotated proteins (for a selection of recent results see [14–21]).

Structural Genomics efforts will function as a source of purified proteins of which only a fraction will yield well-diffracting crystals or NMR spectra of high quality. Thus, a large number of samples (protein reagents) will be available for functional validation using biochemical and/or cell-based readouts.

12.6 SCREENING FOR NATURAL LIGANDS TO INFER FUNCTION

The establishment of protein functional hypotheses can often be facilitated by the identification of ligands and small molecules that co-purify with the protein and are subsequently identified when the protein structure has been determined.

Current biochemical approaches have not systematically aimed to develop sensitive and rapid methods to identify and purify ligands (metals, co-factors, substrate analogues, or other substances) derived from the samples. This is also true for the analysis and identification of posttranslational modifications, such as glycosylation, lipid modifications, and phosphorylation. Thus, the direct structural identification of natural ligands and post translational modifications from 3-D structures should become a major source of experimental data when combined with biochemical and biophysical analyses. As the number of novel proteins with unknown function expressed and purified continuously increases, the arena is set for functional proteomics efforts to rapidly yield characterization data in pace with the developments from large-scale proteomics efforts.

12.7 SCREENING FOR ENZYMATIC ACTIVITY

One of the easy methodologies for functional validation of unannotated proteins is undoubtedly to test the samples of unknown function for enzymatic activity. However, very few systematic efforts have been set up in the academic environment. One exception is the structural genomics efforts ongoing at the University of Toronto (Canada) where a set of general screens for catalytic activity are being used in the validation and characterization of prokaryotic proteins of unknown function [22,23].

When performing enzymatic screens, one must be aware that it is highly likely to obtain a large proportion of false negatives due to enzymes often not being in active form when expressed recombinantly. This could be due to a number of reasons, including:

- The domain chosen is not biologically relevant or misfolded
- The protein is dependent on interaction partners for activity
- Posttranslational modifications are absent or erroneous
- The catalytic activity requires very specific reaction conditions
- That the substrate in the assay is not compatible with the enzyme

12.8 THE DRUGGABLE GENOME

A major limitation on the size of the pool of biologically relevant targets that are subjected to drug discovery efforts is due to (from a medicinal chemistry perspective) the fact that only a subset of the protein families in the proteome are easily approachable (druggable). In the near future, systematic experimental structure determination and modelling efforts of the structural proteome should provide a roadmap for the selection of novel and druggable targets.

Recent estimates suggest that the human genome contains approximately 3000 to 5000 genes/proteins that directly or indirectly are linked to disease conditions. A second pool of targets of about the same size are considered druggable (such as kinases, proteases, nuclear receptors, G-protein coupled receptors and ion channels) [24,25]. The disease causing and druggable protein pools coincide in relatively few cases, at present estimated at 600 to 1500 druggable and medically relevant targets [25], which thus should be the current estimate of the "druggable proteome."

The derived three-dimensional experimental (or predicted) structures of novel proteins will provide powerful knowledge for the selection of targets amenable for structure-based drug discovery. In the subsequent steps of the drug discovery process, the access to timely and highly accurate structural data enables prioritization between targets at distinct decision points and provides a framework for rational compound design. Experimental high-resolution protein structures as well as low-resolution protein models are valuable in the prioritization of protein targets by providing a rationale for assessing target druggability. Clusters of mutations/variations of amino-acid residues obtained from comparative genomics experiments, single nucleotide polymorphism analyses, or mutational scanning experiments can be combined with structural information to identify functional sites of novel proteins. The best-case scenario for the design of low molecular weight candidates is when the target profile includes an antagonistic/inhibitory activity directed at a well-defined cavity. In contrast, the worst-case scenario is an agonistic activity combined with a flat receptor surface.

12.9 STRUCTURE BASED DRUG DISCOVERY

Protein structures are established and well-proven tools for the detailed understanding of protein properties and function. SBDD has for selected targets been essential to increase speed and reduce costs in medicinal chemistry for drug discovery and development programs, e.g., HIV protease, thrombin, and farnesyl transferase inhibitors (reviewed in [26,27]) as well as for protein engineering of macromolecular therapies (reviewed in Weng and DeLisi [28]).

The classical SBDD approach in the pharmaceutical and biotechnology industry has been proven to significantly facilitate selected compound design programs. However, it is probably fair to state that few companies integrated SBDD as a key discipline in the quest to turn biological hypotheses into useful pharmaceutical agents. In traditional pharmaceutical companies, a major bottleneck often relates to insufficient resources and investments in protein production expertise and facilities, causing delays in the generation of structural data by NMR and protein crystallography.

Structural Genomics efforts aim to reduce time frames for structure determination and to cut costs per protein structure determined (five to tenfold compared to classical structural biology, the often mentioned goal is 100 kUSD/protein structure). As structural space becomes populated, primarily through the results of the ongoing public efforts, a significant impact both directly (iterative design) and indirectly (improved computational approaches towards ligand design), will undoubtedly be observed due to a greater opportunities for systematic structural analyses.

In addition, it is not unlikely that Structural Genomics efforts will significantly affect the drug discovery industry by depositing industry-confidential structures into the public domain as well as making reagents and protocols available to the research community, which were previously only accessed by scientists in the pharmaceutical sector. Providing structural information in the public domain on next generation targets will function as a precompetitive research effort, allowing freedom to operate and promote rapid advancement of the discovery and optimization of drugs targeting human disease conditions with unmet medical need.

12.10 IDENTIFICATION OF CHEMICAL STARTING POINTS

As Structural Genomics is generating protein structures at unprecedented speed, one of the obvious applications is to utilize this wealth of new data to systematically screen potential and validated targets as well as antitargets for low molecular weight chemical inhibitors/antagonists and/or agonists. *Antitargets* are defined as critical proteins involved in signalling or other metabolic functions that should be undisturbed by a chemical compound, allowing a normal physiological process, i.e., not contributing to unwanted side effects.

The use of virtual screening could be integrated in the Structural Genomics discovery process to generate chemical maps of chemical scaffolds with the potential to interact favourably with the target protein. Discovery of novel chemical starting points and lead compounds through virtual screening/docking of one or more conformer of chemical compounds in a virtual library against protein structures is well established (reviews in [29,30]). Currently, a large number of methods is available (reviewed e.g., in [29,31,32]) exemplified by ICM-Dock (www.molsoft.com); for recent case studies see Schapira et al. [33,34].

Primary criteria in virtual screening are site identification/description, docking of a virtual chemical library, and interaction scoring, as well as screening utility and speed. Virtual screening methods are, however, dependent on the quality of the protein structures used and the ability to identify and accurately describe suitable binding pockets (cavities). Thus, Structural Genomics projects should carefully ensure that the quality of deposited structures are of sufficiently high quality, with particular emphasis on potential suitable intervention points/active sites/hot-spots [35] (see also Chapter 10 Kleywegt and Hansson in this volume).

Traditionally, chemical libraries tend to be relatively large in size. However, an alternative approach could be to generate a small diverse, drug-like and LMW weight (Mr 200–300) compound collection that is likely to be useful for the generation of directed chemical libraries. Such drug-like compounds could be prioritized using virtual screening and then used in direct ligand binding assays to probe specific but most likely low affinity interactions. Thus, one could generate virtual chemical maps for each protein family and chemically annotate protein structures already at the point when the structural data has been generated. This approach is becoming increasingly popular among biotech companies and is referred to as Fragment Based Drug Discovery (for a review see [36]).

12.11 VALIDATING CHEMICAL STARTING POINTS

Screening by NMR has developed into one of the more promising methods for the identification of chemical binders to proteins. One major limitation with the methodologies when first tested was the low sensitivity, which led to requirements of protein sample in the high end and limitations regarding compound solubility. Despite this limitation, the results obtained with screening by NMR were highly reliable and led to less screening artefacts compared with other methods (for a recent review see [37]).

More recently developed methods such as site specific screening [38,39], Water-Logsy, and 3-FABS protocols [40,41] have enabled NMR to be used routinely in the biotechnology and pharmaceutical sectors. One particularly interesting case of applying nontraditional screening is the combination of virtual screen to identify chemical scaffolds that bind specifically to a single and predefined site on the target protein, followed by, for example, WaterLogsy experiments to verify if indeed the potential binders prioritized from the virtual screen could also bind experimentally [40]. When applying such workflow, it could be a significant albeit quite feasible task to perform for smaller companies as well as laboratory scale activities in the academic sector, especially if the compound library used is designed to cover well-behaving (drug like, diverse, and soluble) compounds that are chemically tractable for further modifications/expansion in few-step synthesis protocols. Thus one could expect that this development could enable Structural Genomics research groups or units to perform structure guided screens with relatively small efforts, allowing identification of chemotypes that could interact with individual targets and target families. Other biophysical direct ligand binding methods, such as surface plasmon resonance and thermoflour, could potentially be used to reach similar results if combined with virtual screening, as already has been reported for traditional screening [42,43].

12.12 PROTEIN–PROTEIN INTERACTIONS

Protein–protein interactions constitute a largely unexploited target pool for drug discovery. The interacting surfaces are often flat and solvent exposed without the classical features of sites suitable for inhibition by low molecular weight chemicals. With the exception of integrin antagonists (see e.g., Feuston et al. [41]) very few success stories have been reported for inhibition of protein–protein interactions in which structural data have been exploited. An increased number of structures of hitherto nontractable targets (read phosphatases, protein–protein and protein–nucleic acid complexes) will hopefully allow the research community to systematically analyze structural features, including protein–protein interaction hot-spots [42]. Subsequently, one would use such knowledge to build adapted chemical libraries based on common structural features of a large set of representative structures.

12.13 STRUCTURE BASED SELECTIVITY DESIGN

The structural analysis of homologous proteins is a powerful tool to determine potential selectivity liabilities of targeted interaction sites and to identify related proteins that could be included in secondary and tertiary assays to monitor selectivity of lead compounds and candidate drugs. For example, comparative computer modelling was successfully applied in the design and modification of a nonselective compound, flurbiprofen, into inhibitors highly selective for the cyclooxygenase isoform COX-2 over COX-1 [46], leading to the identification of a potentially more safe and efficacious compound. However, recently safety concerns on these compounds led to them being taken off the market or only available with restrictions.

More complete structural maps and an increased understanding of which targets are important to avoid inhibiting or stimulating (antitargets) could lead to a case where a partially complete structural map of a target class and its nearest neighbours are available and used to model the interaction of hit and lead chemical compounds with near and remote neighbours. As an example, if an adenine mimetic motif would be used to generate chemical inhibitors (e.g., for protein kinases), one could computationally assess how a particular inhibitor would interact with other identified critical kinases involved in target specific adverse events, as well as including other adenine binding enzyme families (e.g., ATP, CoA, NAD, NADP, and FAD dependent enzymes [47,48]) in the analysis. With more complete structural maps of the proteome, it would be very valuable to perform full structural analysis of all experimentally derived 3-D structures within a proteome, complement the information with model built structures, structurally annotate the proteome and computationally screen all binding motifs against a representative subset of chemical compounds.

12.14 DRUG METABOLISM SCREENS

Cytochrome P450 (CYP450) proteins are membrane-associated proteins that metabolize drug (and other) compounds. Human CYP450s such as CYP1A2, CYP2C9, CYP2C19, CYP2D6, and CYP3A4 are the major drug-metabolizing proteins and contribute to the metabolism of >90% of the drugs in current clinical trials. Astex Technology (discussed above) have solved the structures of several P450 isoforms [8], providing rational design possibilities aimed at rescue of new derivatives of drug candidates prone to be P450-metabolized and thus increase their *in vivo* efficacy.

Structural Genomics and classical structural biology efforts should, in time, generate additional P450 structures that will further enhance such efforts and provide a public resource that will enable the full scientific community to investigate the reasons for such causes of limitation *in vivo*.

12.15 STRUCTURAL MAPPING OF DISEASE CAUSING MUTATIONS

A number of proteins are either aberrant in their expression profiles or carry specific mutations that inactivate or constitutively activate their intracellular signalling properties (or other function). In such cases, it seems obvious that Structural Genomics, Single Nucleotide Polymorphism (SNP) -mapping and genetic diseases research groups/initiatives should join forces to annotate the structural proteome primarily regarding known disease-causing mutations and potential, but not yet validated, functional SNPs to create structural-functional SNP-maps.

Prototype studies have been published directed towards smaller protein families without the direct use of structural information, such as for PKC isoforms [49], but to my knowledge, no systematic large-scale effort is ongoing, although the potential for the methodology has been discussed in literature [50,51]. In the longer term, such maps could be used to suggest/identify protein epitopes containing functional SNPs, e.g., for the production of specific antibodies used as reagents in diagnostic tests.

12.16 AREAS FOR IMPROVEMENT

12.16.1 MEMBRANE PROTEINS

Structural data delivered from Structural Genomics are, in the next few years, unlikely to include data on larger numbers of fully membrane integrated receptors, channels and transporters. First, the protein expression and biochemistry approaches are most often not generating samples in adequate quantities (unless produced as inclusion bodies) and quality to facilitate a serious large scale approach. However, with increased efforts and improvements in methodologies for these notoriously difficult target classes, we should see an increase in total numbers structures of transmembrane targets. However, the fraction of such targets in the PDB is likely to (on a relative scale) become a smaller and smaller fraction of the total pool of solved structures, unless new technologies/methodologies provide a true breakthrough. What this breakthrough constitutes is at present not easily predicted.

However, the research community has reached a point where we start to know how to handle membrane proteins biochemically. The real difficulty is that besides *E. coli*, there is no obvious alternative expression system that will produce a properly folded membrane protein at the required quantities for 3-D crystallisation. The *E. coli* system has, nevertheless, allowed the number of membrane protein structures solved to increase on a yearly basis, especially for prokaryotic targets [52,53].

In May 2005, 91 experimentally determined and unique membrane protein structures were available in the PDB (~170 entries). For an updated and detailed overview, see http://blanco.biomol.uci.edu/Membrane_Proteins_xtal.html.

12.16.2 "Not Hot" Targets

An obvious risk is that we will see a disproportional enrichment of human (or mammal) soluble enzymes/soluble receptor domains belonging to the most attractive targets from the pharmaceutical perspective (kinases, proteases, and nuclear receptors). However, the community needs to ensure that work on the next generation of potential medically relevant targets is carried out in an appropriate proportion such as key intervention points in signalling pathways devoid of suitable chemistry at present (e.g. phosphatases) as well as the hitherto unannotated or poorly annotated new protein families, which have been clustered but not adequately characterized *in vitro* and *in vivo*.

12.16.3 Biased Sampling

Current initiatives in Structural Genomics will provide (and are providing) new structures at a rate that was not realized just a few years ago. However, a majority of the initiatives is focusing on targets from prokaryotic organisms, primarily because they sometimes constitute important human pathogens and are easy to process from clone to structure.

12.17 PROMISES OF STRUCTURAL GENOMICS

Large-scale and high-throughput structural biology approaches were facilitated by technological and methodological advances in biological, structural, and computational sciences. As the Structural Genomics programs have progressed, they now strongly contribute to rapid advances in regards to technology development and systematic information acquisition and analysis.

The overarching goal is to determine the functions of all proteins encoded by the genomes. To succeed, it is of the utmost importance to merge structural information with reliable and high quality data in regards to temporal, spatial, and physiological regulation of proteins, their interaction partners, natural ligands and substrates, biochemical mechanism of action, cellular function, and posttranslational modifications, as well as animal model knock-out and over-expression phenotype (to mention a few).

The major challenge of understanding the sequence genomics information led to the emergence of a new discipline, functional genomics, to determine the functions of genes and protein from entire genomes/proteomes. Structural Genomics will reveal three-dimensional structures and the structure-based inferences of functions for representative members of (hopefully) all protein families. The combination of structural, sequence and other biological information of a protein offers the most reliable method to deduce the biological function(s) of any unannotated protein, which subsequently also will facilitate the annotation of sequences and structures of related proteins. Furthermore, Structural Genomics can provide remote evolutionary relationships among sequence families that are unrecognizable by sequence comparison alone.

Thus, Structural Genomics is a critical component of a systems biology view of the cellular components created during evolution to regulate the machinery of life.

ACKNOWLEDGEMENTS

I would like to thank Victoria Barnsley, Declan Doyle, Aled Edwards, and Martin Norin for suggestions and critical reading of the manuscript.

REFERENCES

1. Norin, M. and M. Sundstrom, Structural proteomics: developments in structure-to-function predictions, *Trends Biotechnol.*, 20, 2, 79-84, 2000.
2. Vitkup, D., et al., Completeness in structural genomics, *Nat. Structural Biol.*, 8, 6, 559–566, 2001.
3. Liu, J. and B. Rost, Target space for structural genomics revisited, *Bioinformatics*, 18, 7, 922–933, 2002.
4. Peat, T., et al., From information management to protein annotation: preparing protein structures for drug discovery, *Acta Crystallogr. D. Biol. Crystallogr.*, 58, Pt 11, 1968–1970, 2002.
5. Hosfield, D., et al., A fully integrated protein crystallization platform for small-molecule drug discovery, *J. Struct. Biol.*, 142, 1, 207–217, 2003.
6. Aparicio, A., et al., Structural genomics of the human protein kinase family, *Cell. Mol. Biol. Lett.*, 8, 2A, 600, 2003.
7. Carr, R. and H. Jhoti, Structure-based screening of low-affinity compounds, *Drug Discov. Today*, 7, 9, 522–527, 2002.
8. Williams, P.A., et al., Crystal structure of human cytochrome P450 2C9 with bound warfarin, *Nature*, 424, 6947, 464–468, 2003.
9. Sharff, A. and H. Jhoti, High-throughput crystallography to enhance drug discovery, *Curr. Opinions Chem. Biol.*, 7, 3, 340–345, 2003.
10. Rost, B., B. Honig, and A. Valencia, Bioinformatics in structural genomics, *Bioinformatics*, 18, 7, 897–898, 2002.
11. Schwikowski, B., P. Uetz, and S. Fields, A network of protein-protein interactions in yeast., *Nature Biotechnol.*, 18, 1257–1261, 2000.
12. Edwards, A.M., et al., Bridging structural biology and genomics: assessing protein interaction data with known complexes, *Trends Genet.*, 18, 10, 529–536, 2002.
13. Zarembinski, T., et al., Structure-based assignment of the biochemical function of a hypothetical protein: A test case of structural genomics, *PNAS*, USA, 95, 151189–15193, 1998.
14. Christendat, D., et al., The crystal structure of hypothetical protein MTH1491 from *Methanobacterium thermoautotrophicum*, *Protein Sci.*, 11, 6, 1409–1414, 2002.
15. Schulze-Gahmen, U., et al., Crystal structure of a hypothetical protein, TM841 of Thermotoga maritima, reveals its function as a fatty acid-binding protein, *Proteins*, 50, 4, 526–530, 2003.
16. Teplyakov, A., et al., Crystal structure of the YjeE protein from *Haemophilus influenzae*: a putative Atpase involved in cell wall synthesis, *Proteins*, 48, 2, 220–226, 2002.
17. Martinez-Cruz, L.A., et al., Crystal structure of MJ1247 protein from *M. jannaschii* at 2.0 A resolution infers a molecular function of 3-hexulose-6-phosphate isomerase, *Structure*, 10, 2, 195–204, 2002.
18. Parsons, J.F., et al., From structure to function: YrbI from *Haemophilus influenzae* (HI1679) is a phosphatase, *Proteins*, 46, 4, 393–404, 2002.

19. Burman, J.D., et al., Crystallization and preliminary X-ray analysis of the *E. coli* hypothetical protein TdcF, *Acta Crystallogr. D. Biol. Crystallogr.*, 59, Pt 6, 1076–1078, 2003.

20. Handa, N., et al., Crystal structure of the conserved protein TT1542 from *Thermus thermophilus* HB8, *Protein Sci.*, 12, 8, 1621–1632, 2003.

21. Teplyakov, A., et al., Crystal structure of the YajQ protein from Haemophilus influenzae reveals a tandem of RNP-like domains, *J. Struct. Funct. Genomics*, 4, 1, 1–9, 2003.

22. Kuznetsova, E., M. Proudfoot, S.A. Sanders, J. Reinking, A. Savchenko, Enzyme genomics: application of general enzymatic screens to discover new enzymes, *FEMS Microbiol. Rev.*, 29, 2, 263–279, 2005.

23. Yakunin, A.F., A.A. Yee, A. Savchenko, A.M. Edwards, C.H. Arrowsmith, Structural proteomics: a tool for genome annotation, *Curr. Opin. Chem. Biol.*, 8, 1, 42–48, 2004.

24. Zambrowicz, B.P. and A.T. Sands, Knockouts model the 100 best-selling drugs — will they model the next 100?, *Nat. Rev. Drug Discov.*, 2, 1, 38–51, 2003.

25. Hopkins, A.L. and C.R. Groom, The druggable genome, *Nat. Rev. Drug Discov.*, 1, 9, 727–730, 2002.

26. Klebe, G., Recent developments in structure-based drug design, *J. Mol. Med.*, 78, 5, 269–281, 2000.

27. Kan, C.C., Impact of recombinant DNA technology and protein engineering on structure-based drug design: case studies of HIV-1 and HCMV proteases, *Curr. Top. Med. Chem.*, 2, 3, 247–269, 2002.

28. Weng, Z. and C. DeLisi, Protein therapeutics: promises and challenges for the 21st century, *Trends Biotechnol.*, 20, 1, 29–35, 2002.

29. Schneider, G. and H.J. Bohm, Virtual screening and fast automated docking methods, *Drug Discov. Today*, 7, 1, 64–70, 2002.

30. Bissantz, C., et al., Protein-based virtual screening of chemical databases. II. Are homology models of G-Protein Coupled Receptors suitable targets?, *Proteins*, 50, 1, 5–25, 2003.

31. Toledo-Sherman, L.M. and D. Chen, High-throughput virtual screening for drug discovery in parallel, *Curr. Opin. Drug Discov. Devel.*, 5, 3, 414–421, 2002.

32. Taylor, R.D., P.J. Jewsbury, and J.W. Essex, A review of protein-small molecule docking methods, *J. Comput. Aided Mol. Des.*, 16, 3, 151–166, 2002.

33. Schapira, M., R. Abagyan, and M. Totrov, Nuclear hormone receptor targeted virtual screening, *J. Med. Chem.*, 46, 14, 3045–3059, 2003.

34. Schapira, M., et al., Discovery of diverse thyroid hormone receptor antagonists by high-throughput docking, *Proc. Natl. Acad. Sci. U.S.*, 100, 12, 7354–7359, 2003.

35. Davis, A.M., S.J. Teague, and G.J. Kleywegt, Application and Limitations of X-ray Crystallographic Data in Structure-Based Ligand and Drug Design, *Angew Chem. Int. Ed. Engl.*, 42, 24, 2718–2736, 2003.

36. Hartshorn, M.J., C.W. Murray, A. Cleasby, M. Frederickson, I.J. Tickle, H. Jhoti, Fragment based lead discovery using X-ray crystallography, *J. Med. Chem.*, 27, 48, 2, 403–413, 2005.

37. Vogtherr, M. and K. Fiebig, NMR-based screening methods for lead discovery, *Exs*, 93, 183–202, 2003.

38. Weigelt, J., et al., Site-selective labeling strategies for screening by NMR, *Comb. Chem. High Throughput Screen*, 5, 8, 623–630, 2002.

39. van Dongen, M.J., et al., Structure-based screening as applied to human FABP4: a highly efficient alternative to HTS for hit generation, *J. Am. Chem. Soc.*, 124, 40, 11874–11880, 2002.

40. Dalvit, C., et al., WaterLOGSY as a method for primary NMR screening: practical aspects and range of applicability, *J. Biomol. NMR*, 21, 4, 349–359, 2001.

41. Dalvit, C., et al., A general NMR method for rapid, efficient, and reliable biochemical screening, *J. Am. Chem. Soc.*, 125, 47, 14620–14625, 2003.

42. Jenkins, J.L., R.Y. Kao, and R. Shapiro, Virtual screening to enrich hit lists from high-throughput screening: a case study on small-molecule inhibitors of angiogenin, *Proteins*, 50, 1, 81–93, 2003.

43. Keseru, G.M., A virtual high throughput screen for high affinity cytochrome P450cam substrates. Implications for *in silico* prediction of drug metabolism, *J. Comput. Aided Mol. Des.*, 15, 7, 649–657, 2001.

44. Feuston, B.P., et al., Binding model for nonpeptide antagonists of alpha(v)beta(3) integrin, *J. Med. Chem.*, 45, 26, 5640–5648, 2002.

45. Clackson, T. and J. Wells, A hot spot of binding energy in a hormone-receptor interface, *Science*, 267, 383–386, 1995.

46. Bayly, C., et al., Structure-based design of COX-2 selectivity into flurbiprofen, *Bioorganic Med. Chem. Lett.*, 1999. 9, 307–312, 1999.

47. Denessiouk, K.A. and M.S. Johnson, When fold is not important: a common structural framework for adenine and AMP binding in 12 unrelated protein families, *Proteins*, 38, 3, 310–326, 2000.

48. Denessiouk, K.A., V.V. Rantanen, and M.S. Johnson, Adenine recognition: a motif present in ATP-, CoA-, NAD-, NADP-, and FAD-dependent proteins, *Proteins*, 44, 3, 282–291, 2001.

49. Kofler, K., et al., Molecular genetics and structural genomics of the human protein kinase C gene module, *Genome Biol.*, 3, 3, research0014.1 – research0014.10, 2002.

50. Sunyaev, S., W. Lathe, 3rd, and P. Bork, Integration of genome data and protein structures: prediction of protein folds, protein interactions, and "molecular phenotypes" of single nucleotide polymorphisms, *Curr. Opinions Struct. Biol.*, 11, 1, 125–130, 2001.

51. Sunyaev, S., et al., Prediction of deleterious human alleles, *Hum. Mol. Genet.*, 10, 6, 591–597, 2001.

52. MacKinnon, R. and D.A. Doyle, Prokaryotes offer hope for potassium channel structural studies, *Nat. Struct. Biol.*, 4, 11, 877–879, 1997.

53. Kuo, A., et al., Crystal structure of the potassium channel KirBac1.1 in the closed state, *Science*, 300, 5627, 1922–1926, 2003.

INDEX